Treatise on Clean Surface Technology
Volume 1

Treatise on Clean Surface Technology
Volume 1

EDITED BY
K. L. MITTAL
IBM Corporate Technical Institutes
Thornwood, New York

PLENUM PRESS • NEW YORK AND LONDON

Library of Congress Cataloging in Publication Data

Treatise on clean surface technology.

Includes bibliographies.
1. Surfaces (Technology) 2. Surface contamination. 3. Surface preparation. I. Mittal, K. L., 1945–
TA407.T63 1987   620.4   87-2402
ISBN 0-306-42420-7 (v. 1)

© 1987 Plenum Press, New York
A Division of Plenum Publishing Corporation
233 Spring Street, New York, N.Y. 10013

All rights reserved

No part of this book may be reproduced, stored in a retrieval system, or transmitted in any form or by any means, electronic, mechanical, photocopying, microfilming, recording, or otherwise, without written permission from the Publisher

Printed in the United States of America

Contributors

Morton Antler, AT&T Bell Laboratories, Columbus, Ohio

Jack Brous, Alpha Metals, Inc., Newark, New Jersey

K. P. Homewood, Joint Laboratory of Physics and Electrical Engineering, University of Manchester Institute of Science and Technology, Manchester, United Kingdom. *Present address:* Department of Electronic and Electrical Engineering, University of Surrey, Guildford, Surrey, United Kingdom

J. L. Jellison, Process Metallurgy, Sandia National Laboratories, Albuquerque, New Mexico

David C. Jolly, Consultant, Brookline, Massachusetts

Pedro Lilienfeld, MIE, Inc., Bedford, Massachusetts

Joseph R. Monkowski, MRI, San Diego, California

Buddy D. Ratner, National ESCA and Surface Analysis Center for Biomedical Problems (NESAC/BIO), Center for Bioengineering and Department of Chemical Engineering, University of Washington, Seattle, Washington

Eric B. Sansone, Environmental Control and Research Program, NCI-Frederick Cancer Research Facility, Program Resources, Inc., Frederick, Maryland

John C. Scott, Fluid Mechanics Research Institute, University of Essex, Colchester, Essex, United Kingdom. *Present address:* Admiralty Research Establishment, Southwell, Portland, Dorset, United Kingdom

John R. Vig, U.S. Army Electronics Technology and Devices Laboratory, LABCOM, Fort Monmouth, New Jersey

Tuan Vo-Dinh, Advanced Monitoring Development Group, Health and Safety Research Division, Oak Ridge National Laboratory, Oak Ridge, Tennessee

Robert Walker, Department of Materials Science and Engineering, University of Surrey, Guildford, Surrey, United Kingdom

Preface to the Treatise

This multivolume treatise is intended to provide a comprehensive source of information on clean surface technology. The impetus and justification for the treatise were provided by the excellent reviews received by the two-volume set on *Surface Contamination: Genesis, Detection, and Control* (published by Plenum in 1979). That set chronicled the proceedings of a symposium, held in Washington, D.C., in 1978, which was quite comprehensive in its own right. However, after the symposium it was felt that there was a definite need for a series of volumes containing state-of-the-art chapters on various aspects of surface contamination and cleaning written by experts and active practitioners. So this treatise was conceived, and when potential authors were initially contacted, the response was most gratifying. The general consensus was that the proposed treatise was both timely and needed.

The ubiquitous nature of surface contamination makes the subject of clean surface technology of paramount importance in many and diverse technologies, and all signals indicate that its importance is going to increase. Surface contamination has always been a *bête noire* to people working in areas such as adhesion, composites, adsorption, tribology, soldering, device fabrication, and printed circuit boards, and the proper level of cleanliness has always been a desideratum. For example, one commonly hears the statements "cleanliness may be next to godliness, but it sure precedes adhesion" or "if it ain't clean, it won't stick." Also, in a world of shrinking dimensions, surface contamination and cleaning become of cardinal importance. A few years ago, a micrometer-size particle was considered innocuous (or at worst cosmetically unappealing), but in the era of submicrometer geometries in microelectronics, this same particle can be disastrous. So there is patently a need to understand why surfaces get contaminated and how to clean them and keep them clean.

In order to cover the subject of clean surface technology in a holistic manner, one must consider all of its essential aspects: sources, causes, and mechanisms of surface contamination; techniques for cleaning; techniques for characterizing the degree of cleanliness; kinetics of recontamination or storage of clean parts; and implications of surface contamination. All these ramifications are intended to be covered in this treatise. Both film and particulate contaminants will be addressed, and all sorts of surfaces (metal, alloy, glass, ceramic, plastic, and liquid) will be considered. The authors have been urged to cover the topic under consideration as completely as possible. In the references, the titles of papers have been included, so that one may be able to find something of interest merely by looking over the reference lists. For those chapters that discuss a technique, the recommended outline is: introduction (historical development), equipment needed (with all the requisite details to carry out experiments), results (showing the utility of the technique), potentialities and limitations, and directions for future development.

The intent and hope, over the course of this treatise, are to cover the topic of clean surface technology *in toto*. It may even be desirable, in a few years, to update selected volumes of the treatise.

The authors invited to contribute chapters are experts in their specialties and they hail from the groves of academia as well as from industrial and governmental research and development laboratories.

In closing, I feel very strongly that this treatise is both timely and needed, and that it should fill the lacuna in the existing body of literature on clean surface technology. These volumes should provide a rich source of information and should serve as a *vade mecum* for veterans as well as neophytes interested in the wonderful world of surfaces.

I now have the pleasant task of acknowledging the enthusiasm, efforts, and patience of the authors, without which this treatise would not have seen the light of day. I would also like to thank the appropriate management of IBM Corporation for allowing me to edit this treatise, and special thanks are due to S. B. Korin for his interest and support.

K. L. Mittal
IBM Corporate Technical Institutes
500 Columbus Avenue
Thornwood, New York 10594

Preface to Volume 1

This is the premier volume in the *Treatise on Clean Surface Technology*. The rationale for and scope of this treatise have been delineated in the Preface to the Treatise.

This particular volume contains thirteen chapters covering a number of topics, ranging from UV/ozone cleaning of surfaces to measurement of ionic contamination to implications of surface contamination to the application of pellicles in clean surface technology.

Since even a capsule description of each chapter would make this preface prohibitively long, only a few highlights of each chapter are noted.

The opening chapter by John R. Vig discusses the UV/ozone cleaning of surfaces. The author provides a comprehensive description of what this technique is, what it can do, and how to use it. In the last few years UV/ozone cleaning of surfaces has evoked a great deal of interest.

As pointed out in the Preface to the Treatise, the treatise will cover all kinds of surfaces, and so the next chapter by John C. Scott deals with techniques for cleaning liquid surfaces, which are very important in fluid mechanics and other areas. Robert Walker, in Chapter 3, discusses the Hydroson cleaning of surfaces and compares it with ultrasonic and megasonic techniques. Examples are cited to illustrate its commercial applications, and some recent developments are highlighted.

The measurement of ionic contamination on surfaces is discussed by Jack Brous in Chapter 4. Various techniques have been developed by a number of workers to monitor levels of ionic contamination, and the author discusses their potentialities and limitations. Chapter 5 by Tuan Vo-Dinh covers the characterization of surface contaminants by luminescence using UV excitation, and illustrates the application of this approach to the monitoring of a variety of contaminants on skin.

In a world of shrinking device dimensions, the importance of particle contamination is quite apparent, and Chapter 6 by Joseph R. Monkowski concentrates on particle surface contamination and device failures. The author cites examples of device failures attributed to particulates and emphasizes the need to keep these detrimental particles away from devices.

Chapters 7–12 deal with the effects or implications of surface contamination in various areas of human endeavor. Chapter 7 by David C. Jolly discusses the effect of surface contamination on the performance of HVDC insulators and test methods to evaluate insulator performance. Morton Antler in Chapter 8 covers the effects of surface contamination on electric contact performance. Contact resistance probes for the detection and characterization of contamination are also discussed. The role of surface contamination in solid-state welding of metals is the topic of Chapter 9 by J. L. Jellison, and mechanisms for the elimination of surface barriers during solid-state welding are also discussed. In Chapter 10, K. P. Homewood discusses how surface contamination can influence contact electrification, a topic of great importance in many technologies, for example, xerography.

Nowadays there is a great deal of interest in biomaterials (e.g., prosthetics in the human body) and the question of how their function and behavior are affected by surface contamination is quite important. Chapter 11 by Buddy D. Ratner reviews biomaterials contamination studies.

The redispersion of indoor surface contamination and its implications is the topic of Chapter 12 by Eric B. Sansone, and measurements of redispersion or resuspension factors are discussed. This chapter's emphasis is on radioactive contamination.

The volume concludes with a chapter by Pedro Lilienfeld on the application of pellicles in clean surface technology, and he cites examples to illustrate how pellicles can be effectively used to keep surfaces clean. If contamination can be prevented from depositing on a surface in the first place, then it will not be necessary to clean it. Preventive approaches and techniques are thus of great importance in clean surface technology.

As stated in the Preface to the Treatise, the treatise is intended to cover all aspects of clean surface technology, and the diverse topics covered in this premier volume set the tone for future volumes.

I sincerely hope that this first volume will receive a warm welcome by those involved in surfaces; their comments or suggestions would be most welcome.

K. L. Mittal
IBM Corporate Technical Institutes
500 Columbus Avenue
Thornwood, New York 10594

Contents

1. UV/Ozone Cleaning of Surfaces
John R. Vig

1. Introduction 1
2. The Variables of UV/Ozone Cleaning 2
   - 2.1. The Wavelengths Emitted by the UV Sources 2
   - 2.2. Distance between the Sample and the UV Source 7
   - 2.3. The Contaminants 8
   - 2.4. The Precleaning 9
   - 2.5. The Substrate 10
   - 2.6. Rate Enhancement Techniques 11
3. The Mechanism of UV/Ozone Cleaning 13
4. UV/Ozone Cleaning in Vacuum Systems 14
5. Safety Considerations 14
6. UV/Ozone Cleaning Facility Construction 16
7. Applications 17
8. Effects Other Than Cleaning 20
   - 8.1. Oxidation 20
   - 8.2. UV-Enhanced Outgassing 20
   - 8.3. Other Surface/Interface Effects 21
   - 8.4. Etching 21
9. Summary and Conclusions 21
   - References and Notes 22

2. Techniques for Cleaning Liquid Surfaces
John C. Scott

1. Introduction 27
2. The Origin of Dynamic Liquid Surface Phenomena 28
3. The Notion of Surface Cleanliness 32
4. The History of Clean Surfaces 32
5. The Nature of Surface Contamination 33
6. Cleaning Techniques 36
   - 6.1. Primary Distillation 37
   - 6.2. Further Distillation 39
   - 6.3. Bubble Cleaning 40

	6.4. Surface Skimming and Talc Cleaning	43
	6.5. Solid Adsorption Techniques	44
	6.6. Laser Burning	44
	6.7. Solution Preparation	45
	6.8. Surface Cleaning in Engineering Applications	46
7.	Materials for Clean-Surface Experiments	47
	7.1. Principles	47
	7.2. Construction of Apparatus	47
	7.3. Water Storage Materials	48
8.	Cleaning of Apparatus	49
	8.1. "Hard" Materials—Chromic Acid	49
	8.2. Perspex—Detergents	49
9.	General Design Considerations	50
10.	Summary	50
	References	51

3. Hydroson Cleaning of Surfaces
Robert Walker

1.	Introduction	53
2.	The Hydroson System	53
3.	Experimental Investigation of Mechanisms in the Tank	54
4.	Commercial Applications of Hydroson Cleaning	57
5.	Safety and Economy	59
6.	Recent Developments	60
	6.1. Wire Cleaning	60
	6.2. Coil Strip Cleaning	62
	6.3. Barrel Cleaning and Rinsing	62
	6.4. Cleaning Molds in the Glass Industry	63
	6.5. Nuclear Industry	63
	6.6. Phosphating	63
	6.7. Electrodeposition	64
7.	Size and Cost of Equipment	65
8.	Comparison with Ultrasonic and Megasonic Cleaning	65
9.	Conclusion	67
	References	68

4. Methods of Measurement of Ionic Surface Contamination
Jack Brous

1.	Introduction	71
2.	Ionic Measurement	72
3.	Static Extraction Methods	76
	3.1. Egan's Method	76
	3.2. Method of Hobson and DeNoon	81
	3.3. Omega Meter	83
	3.4. Ion Chaser	86
	3.5. Contaminometer	87
4.	Dynamic Extraction Method–Ionograph	90
5.	Applications of Ionic Contamination Measurements	97

6. Future Prospects 98
7. Summary 99
   References 99

5. Characterization of Surface Contaminants by Luminescence Using Ultraviolet Excitation
Tuan Vo-Dinh

1. Introduction 103
2. The Luminescence Technique for Surface Detection 104
3. Applications 107
   3.1. The Use of UV "Black Light" for Surface Detection 107
   3.2. Study of Workers' Skin Contamination by the "Skin-Wash" Method 108
   3.3. The Fiberoptic Luminoscope for Monitoring Occupational Skin Contamination 109
   3.4. Detection of Surface Contamination with the Spill Spotter 111
   3.5. Remote Sensing with Laser-Based Fluorosensors 111
   3.6. A Fluorescent Tracer Detection Technique 112
   3.7. Studies of Absorption of Carcinogenic Materials into Mouse Skin 113
   3.8. Chromogenic and Fluorogenic Spot Test Techniques 115
   3.9. Sensitized Fluorescence Spot Tests 116
   3.10. Surface Detection by Room Temperature Phosphorimetry 117
4. Conclusion 120
   References 121

6. Particulate Surface Contamination and Device Failures
Joseph R. Monkowski

1. Introduction 123
2. Sources of Particulate Contamination 124
   2.1. Air 124
   2.2. Chemicals 127
   2.3. Gases 128
   2.4. Wafer Handling 129
3. Effects on Device Performance 131
   3.1. Particulate Contamination on Photomasks 131
   3.2. Epitaxial Growth 135
   3.3. Failure Mechanism in MOS Gate Oxides 137
   3.4. Impurity Contamination in Silicon 144
4. Summary 145
   References 145

7. Effect of Surface Contamination on the Performance of HVDC Insulators
David C. Jolly

1. Introduction 149
2. General Overview of the Flashover Process 152

3. Deposition of Particles 154
   3.1. Introduction 154
   3.2. The Electrical Environment 155
   3.3. Charging Mechanisms 155
   3.4. Particle Deposition Rates 156
4. Deposition of Moisture 158
5. Thermal Processes 159
6. Localized Electrical Breakdown 159
7. Discharge Growth 160
   7.1. Introduction 160
   7.2. Extinction Theories 160
   7.3. Effect of Polarity 164
   7.4. Effect of Voltage Waveform 166
   7.5. Effect of the Composition of the Conducting Layer 166
   7.6. Effect of Ambient Pressure 167
   7.7. Effect of Ambient Temperature 167
   7.8. Effect of Ambient Gas 167
8. Test Methods to Evaluate Insulator Performance 168
9. Prevention of Flashover 168
   9.1. Deposition of Contamination 168
   9.2. Deposition of Moisture 169
   9.3. Dry-Band Formation and Electric Field Concentration 169
   9.4. Localized Electrical Breakdown across the Dry Band 170
   9.5. Discharge Growth 170
10. Conclusions and Future Prospects 170
    References 170

8. Effect of Surface Contamination on Electric Contact Performance
Morton Antler

1. Introduction 179
2. Sources of Contamination 182
   2.1. Oxidation and Corrosion 184
   2.2. Particulates 186
   2.3. Thermal Diffusion 189
   2.4. Fretting 190
   2.5. Manufacturing Processes 191
   2.6. Outgassing and Condensation on Contact Surfaces of Volatiles from Noncontact Materials 193
3. Effects of Contamination 194
   3.1. Direct 194
   3.2. Indirect 196
4. Contact Resistance Probes for the Detection and Characterization of Contamination .. 197
   4.1. Description of Probes 197
   4.2. Determination of Contact Resistance 198
   4.3. Modes of Operation 200
5. Summary 202
   References 202

Contents

9. The Role of Surface Contaminants in the Solid-State Welding of Metals
J. L. Jellison

1. Introduction 205
2. Role of Contaminants in Preventing Solid-State Welds 206
3. Classification of Surface Contaminants 209
   - 3.1. Inorganic Films 209
   - 3.2. Organic Contaminants 209
   - 3.3. Particulate Contaminants 210
4. Role of Contaminant Properties 210
   - 4.1. Inorganic Films 211
   - 4.2. Organic Films 213
   - 4.3. Particulate Contaminants 218
5. Mechanisms for Elimination of Surface Barriers during Solid-State Welding 219
   - 5.1. Thermal Mechanisms Occurring during Diffusion Welding ... 220
   - 5.2. Deformation Welding 222
     - 5.2.1. Mechanical Mechanisms 222
     - 5.2.2. Thermal Mechanisms 225
6. Surface Preparation 228
7. Concluding Remarks 231
   - References 231

10. Surface Contamination and Contact Electrification
K. P. Homewood

1. Introduction 235
2. Types of Contaminants 236
   - 2.1. Adsorbed Molecules 236
   - 2.2. Ionic Contamination 239
   - 2.3. Adsorbed Water 240
3. Identification of Extrinsic Traps 241
4. Effect of Contamination on the Metal 243
5. Conclusion 244
   - References 244

11. Surface Contamination and Biomaterials
Buddy D. Ratner

1. Introduction 247
2. General Principles of Surface Contamination 248
3. Review of Biomaterials Contamination Studies 250
   - 3.1. Cleaning Agent Residues 250
   - 3.2. Environmental Contaminants 253
   - 3.3. Biocompatible Contaminants 254
4. Conclusions 257
   - Acknowledgments 257
   - References 257

12. Redispersion of Indoor Surface Contamination and Its Implications
Eric B. Sansone

1. Introduction 261
2. Measurements of the Redispersion or Resuspension Factor (K) 262
3. Measurements of "Transferable" Surface Contamination 270
4. The Contribution of Resuspended Particulates to Exposure 283
5. Concluding Remarks 286
   Acknowledgment 286
   References 286

13. Application of Pellicles in Clean Surface Technology
Pedro Lilienfeld

1. Introduction 291
2. Pellicles in Integrated Circuit Fabrication 292
   2.1. Semiconductor Fabrication by Optical Microlithography 292
   2.2. Methods of Projection 293
   2.3. Feature Dimensions and Performance Limits of Optical Microlithography 297
   2.4. Masks and Reticles in Semiconductor Microlithography 299
   2.5. Particle Contamination in Integrated Circuit Fabrication 299
   2.6. Principles of Reticle/Mask Protection with Pellicles 302
   2.7. Advantages Resulting from the Use of Pellicles 307
   2.8. Optical Properties and Effects of Pellicles 308
      2.8.1. Transmission Loss Mechanisms 308
      2.8.2. Optical Lifetime 309
      2.8.3. Refraction 310
      2.8.4. Effects of Thickness Nonuniformity 311
   2.9. Pellicle Materials and Mechanical Properties 311
   2.10. Pellicle Mounting and Attachment 313
   2.11. Inspection of Reticles and Masks Protected by Pellicles 314
3. Protective Films on Optical Data Storage Media 318
   3.1. Principles and Methods of Optical Storage 318
   3.2. Optical Data Readout and Effect of Protective Layer 319
4. Summary and Future Prospects 321
   References 322

Index 327

1

UV/Ozone Cleaning of Surfaces

JOHN R. VIG

1. Introduction

The ability of ultraviolet (UV) light to decompose organic molecules has been known for a long time, but it is only during the past decade that UV cleaning of surfaces has been explored.

In 1972, Bolon and Kunz[1] reported that UV light had the capability to depolymerize a variety of photoresistant polymers. The polymer films were enclosed in a quartz tube that was evacuated and then backfilled with oxygen. The samples were irradiated with UV light from a medium-pressure mercury lamp that generated ozone. The several-thousand-angstroms-thick polymer films were successfully depolymerized in less than one hour. The major products of depolymerization were found to be water and carbon dioxide. Subsequent to depolymerization, the substrates were examined by Auger electron spectroscopy (AES) and were found to be free of carbonaceous residues. Only inorganic residues, such as tin and chlorine, were found. When a Pyrex filter was placed between the UV light and the films or when a nitrogen atmosphere was used instead of oxygen, the depolymerization was hindered. Thus, Bolon and Kunz recognized that oxygen and wavelengths shorter than 300 nm played a role in the depolymerization.

In 1974, Sowell et al.[2] described UV cleaning of adsorbed hydrocarbons from glass and gold surfaces, in air and in a vacuum system. A clean glass surface was obtained after 15 hours of exposure to the UV radiation in air. In a vacuum system at 10^{-4} torr of oxygen, clean gold surfaces were produced after about two hours of UV exposure. During cleaning, the partial pressure of O_2 decreased, while that of CO_2 and H_2O increased. The UV also desorbed gases from the vacuum chamber walls. In air, gold surfaces which had been

---

JOHN R. VIG • U.S. Army Electronics Technology and Devices Laboratory, LABCOM, Fort Monmouth, New Jersey 07703-5000.

contaminated by adsorbed hydrocarbons could be cleaned by "several hours of exposure to the UV radiation." Sowell et al. also noted that storing clean surfaces under UV radiation maintained the surface cleanliness indefinitely.

During the period 1974–1976, Vig et al.[3-5] described a series of experiments aimed at determining the optimum conditions for producing clean surfaces by UV irradiation. The variables of cleaning by UV light were defined, and it was shown that, under the proper conditions, UV/ozone cleaning has the capability of producing clean surfaces in less than one minute. Since 1976, use of the UV/ozone cleaning method has grown steadily. UV/ozone cleaners are now available commercially.

2. The Variables of UV/Ozone Cleaning

2.1. The Wavelengths Emitted by the UV Sources

To study the variables of the UV cleaning procedure, Vig and LeBus[5] constructed the two UV cleaning boxes shown in Figure 1. Both were made of aluminum, and both contained low-pressure mercury discharge lamps and an aluminum stand with Alzak[6] reflectors. The two lamps produced nearly equal intensities of short-wavelength UV light, about 1.6 mW/cm^2 for a sample 1 cm from the tube. Both boxes contained room air (in a clean room) throughout these experiments. The boxes were completely enclosed to reduce recontamination by air circulation.

Since only the light which is absorbed can be effective in producing photochemical changes, the wavelengths emitted by the UV sources are important variables. The low-pressure mercury discharge tubes generate two wavelengths

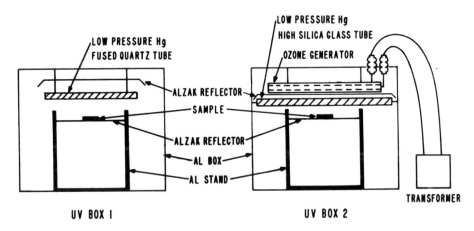

Figure 1. Apparatus for UV/ozone cleaning experiments.

Table 1. Principal Wavelengths of Low-Pressure Hg Discharge Lamps

| | Lamp envelope[a] | | |
Wavelength (nm)	Quartz	High-silica glass	Glass
184.9	T	O	O
253.7	T	T	O
>300.0	T	T	T

[a]T = transparent, O = opaque.

of interest: 184.9 nm and 253.7 nm. Whether or not these wavelengths are emitted depends upon the lamp envelopes. The emissions through the three main types of envelopes are summarized in Table 1.

The 184.9-nm wavelength is important because it is absorbed by oxygen, thus leading to the generation of ozone.[7] The 253.7-nm radiation is not absorbed by oxygen; therefore, it does not contribute to ozone generation, but it is absorbed by most organic molecules[8,9] and by ozone.[7] The absorption by ozone is principally responsible for the destruction of ozone in the UV box. Therefore, when both wavelengths are present, ozone is continually being formed and destroyed. An intermediate product, both of the formation and of the destruction processes, is atomic oxygen, which is a very strong oxidizing agent. The absorption spectra of oxygen and ozone are shown in Figures 2 and 3, respectively. The effects of the principal wavelengths generated by low-pressure mercury discharge lamps are summarized in Table 2.

The tube of the UV lamp[10] in box 1 consisted of 91 cm of "hairpin-bent" fused quartz tubing. The fused quartz transmits both the 253.7-nm and the 184.9-nm wavelengths. The lamp emitted about 0.1 mW/cm^2 of 184.9-nm radiation measured at 1 cm from the tube.

The lamp in box 2 had two straight and parallel 46-cm-long high-silica glass tubes made of Corning UV Glass No. 9823, which transmits at 253.7 nm but not at 184.9 nm. Since this lamp generated no measurable ozone, a separate Siemens-type ozone generator[11] was built into box 2. This ozone generator did not emit UV light. Ozone was produced by a "silent" discharge when high-voltage AC was applied across a discharge gap formed by two concentric glass tubes, each of which was wrapped in aluminum foil electrodes. The ozone-generating tubes were parallel to the UV tubes, and were spaced approximately 6 cm apart.

UV box 1 was used to expose samples, simultaneously, to the 253.7-nm and 184.9-nm wavelengths and to the ozone generated by the 184.9-nm wavelength. UV box 2 permitted the options of exposing samples to 253.7 nm plus ozone, 253.7 nm only, or ozone only.

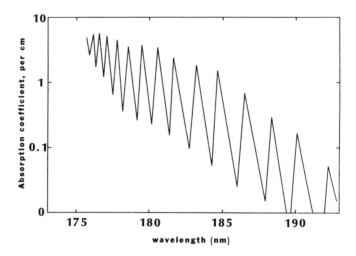

Figure 2. Absorption spectrum of oxygen.

Vig et al. used contact angle measurements, wettability tests, and Auger electron spectroscopy to evaluate the results of cleaning experiments. Most of the experiments were conducted on polished quartz wafers, the cleanliness of which could be evaluated by the "steam test," a highly sensitive wettability test.[5, 12, 15] Contact angle measurements and the steam test can detect fractional monolayers of hydrophobic surface contamination.

Also tested was a "black-light," long-wavelength UV source that emitted wavelengths above 300 nm only. This UV source produced no noticeable cleaning, even after 24 hours of irradiation.

In the studies of Vig et al., it was found that samples could be cleaned

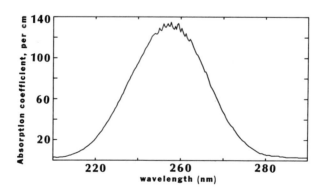

Figure 3. Absorption spectrum of ozone.

Table 2. Effects of the Principal Wavelengths Generated by Low-Pressure Hg Discharge Lamps

Wavelength (nm)	Effects
184.9	• Absorbed by O_2 and organic molecules • Creates atomic oxygen and ozone • Breaks contaminant molecule bonds
253.7	• Absorbed by organic molecules and O_3; not absorbed by O_2 • Destroys ozone • Breaks contaminant molecule bonds

consistently by UV irradiation only if gross contamination was first removed from the surfaces. Their precleaning procedure consisted of the following steps:

1. Scrubbing with a swab while the sample was immersed in ethyl alcohol.
2. Degreasing ultrasonically in a solvent such as trichlorotrifluoroethane.
3. Boiling in fresh ethyl alcohol, then agitating ultrasonically.
4. Rinsing in running ultrapure (18 MΩ-cm) water.
5. Spinning dry immediately after the running-water rinse.

Subsequent to this precleaning procedure, the steam test and contact angle measurements invariably indicated that the surfaces were contaminated. However, after exposure to UV/ozone in box 1, the same tests always indicated clean surfaces. The cleanliness of such UV/ozone-cleaned surfaces has been verified on numerous occasions, in the author's laboratory and elsewhere, by AES and electron spectroscopy for chemical analysis (ESCA).[1,3,4,13-16] Figure 4 shows Auger spectra before and after UV/ozone cleaning.[16] Ten minutes of UV/ozone cleaning reduced the surface contamination on an aluminum thin film to below the AES detectability level, about one percent of a monolayer. The effectiveness of UV/ozone cleaning has also been confirmed by ion scattering spectroscopy/secondary ion mass spectroscopy (ISS/SIMS).[17]

A number of quartz wafers were precleaned and exposed to the UV light in box 1 until clean surfaces were obtained. Each of the wafers was then thoroughly contaminated with human skin oil, which has been a difficult contaminant to remove. (The skin oil was applied by rubbing the wafer on the forehead of one of the researchers.) The wafers were precleaned again, groups of wafers were exposed to each of the four UV/ozone combinations mentioned earlier, and the time needed to attain a clean surface was measured, as indicated by the steam test. In each UV box, the samples were placed within 5 mm of the UV source (where the temperature was about 70°C).

The wafers exposed to 253.7 nm + 184.9 nm + ozone in UV box 1

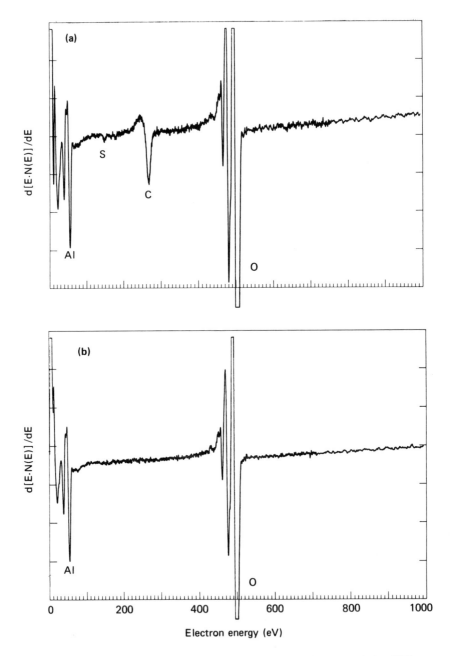

Figure 4. Auger spectra of evaporated aluminum film on silicon substrate: (a) before UV/ozone cleaning; (b) after UV/ozone cleaning.[16]

became clean in 20 s. The samples exposed to 253.7 nm + ozone in UV box 2 reached the clean condition in 90 s. Samples exposed to 253.7 nm without ozone and to ozone without UV light were cleaned within one hour and ten hours, respectively. The results are summarized in Table 3.

Therefore, one may conclude that, while both UV light without ozone and ozone without UV light can produce a slow cleaning effect in air, the combination of short-wavelength UV light and ozone, such as is obtained from a quartz UV lamp, produces a clean surface orders of magnitude faster. Although the 184.9-nm radiation is also absorbed by many organic molecules, it was not possible from these experiments to isolate the cleaning effect of the 184.9-nm radiation. The ozone concentrations had not been measured. As is discussed below, within each box the ozone concentrations vary with distance from the UV source.

2.2. Distance between the Sample and the UV Source

Another variable that can greatly affect the cleaning rate is the distance between the sample and the UV source. Because of the shapes of the UV tubes and of the Alzak reflectors above the tubes and below the samples, the lamps in both boxes were essentially plane sources. Therefore, one may conclude that the intensity of UV light reaching a sample would be nearly independent of distance. However, this is not so when ozone is present, because ozone has a broad absorption band[7,18,19] centered at about 260 nm, as is shown in Figure 3. At 253.7 nm, the absorption coefficient is 130 cm^{-1} atm^{-1}. The intensity I of the 253.7-nm radiation reaching a sample therefore decreases as

$$I = I_0 e^{-130pd} \tag{1}$$

where p is the average ozone pressure between the sample and the UV source in atmospheres at 0°C, and d is the distance to the sample in centimeters. When a quartz UV tube is used, both the ozone concentration and the UV radiation intensity decrease with distance from the UV source.

Table 3. Exposure Types vs. Cleaning Times

Exposure type	Cleaning time
"Black light" (>300 nm)	No cleaning
O_3, no UV	10 h
253.7 nm, no O_3	1 h
253.7 nm + O_3	90 s
253.7 nm + 184.9 nm + O_3	20 s

Two sets of identically precleaned samples were placed in UV box 2. One set was placed within 5 mm of the UV tube, the other was placed at the bottom of the box about 8 cm from the tube. With the ozone generator off, there was less than a 30% difference in the time it took for the two sets of samples to attain a minimal ($\sim 4°$) contact angle, about 60 min versus 75 min. When the experiment was repeated with the ozone generator on, the samples near the tube became clean nearly ten times faster (about 90 s versus 13 min). Similarly, in UV box 1, samples placed within 5 mm of the tube were cleaned in 20 s versus 20–30 min for samples placed near the bottom of the box at a distance of 13 cm. Therefore, to maximize the cleaning rate, the samples should be placed as close to the UV source as possible.

2.3. The Contaminants

Vig *et al.* tested the effectiveness of the UV/ozone cleaning procedure on a variety of contaminants. Among the contaminants were:

1. Human skin oils (wiped from the forehead of one of the researchers).
2. Contamination adsorbed during prolonged exposure to air.
3. Cutting oil.[20]
4. Beeswax and rosin mixture.
5. Lapping vehicle.[21]
6. Mechanical vacuum pump oil.[22]
7. DC 704 silicone diffusion pump oil.[23]
8. DC 705 silicone diffusion pump oil.[23]
9. Silicone vacuum grease.[23]
10. Acid (solder) flux.[24]
11. Rosin flux from a rosin-core lead–tin solder.
12. Cleaning solvent residues, including acetone, ethanol, methanol, isopropyl alcohol, trichloroethane, and trichlorotrifluoroethane.

The contaminants were applied with swabs to clean, polished quartz wafers. The amount of contaminants was not measured. However, each time a swab was used in the application, it was obvious to the unaided eye that the samples had been thoroughly contaminated. After contamination, the wafers were precleaned, then exposed to UV/ozone by placement within a few millimeters of the tube in UV box 1. After a 60-s exposure, the steam test and AES indicated that all traces of the contaminants had been removed.

Using AES, no differentiation could be made between the silicon peaks due to quartz and those due to the silicon-containing contaminants. The removal of silicone diffusion pump fluids was, therefore, also tested on Alzak, which normally has a silicon-free oxide surface, and on gold. Following UV/ozone cleaning, AES examination of both the Alzak and the gold surfaces showed no presence of silicon.

During the course of their studies, Vig *et al.* learned from colleagues working on ion implantation for integrated circuits that the usual wet-cleaning procedures (with hot acids) failed to remove the photoresist from silicon wafers that had been exposed to radiation in an ion-implantation accelerator, presumably because of cross-linking of the photoresist. Ion-implanted silicon wafers, each with approximately a 1-μm coating of exposed Kodak Micro Resist 747,[25] were placed within a few millimeters of the source in UV box 1. After an overnight (~ 10 h) exposure to UV/ozone, all traces of the photoresist were removed from the wafers, as confirmed by AES.

Films of carbon, vacuum-deposited onto quartz to make the quartz surfaces conductive for study in an electron microscope, were also successfully removed by exposure to UV/ozone. Inorganic contaminants, such as dust and salts, cannot be removed by UV/ozone and should be removed in the precleaning procedure.

UV/ozone has also been used for wastewater treatment and for destruction of highly toxic compounds.[26-29] Experimental work in connection with these applications has shown that UV/ozone can convert a wide variety of organic and some inorganic species to relatively harmless, mostly volatile products such as CO_2, CO, H_2O, and N_2. Compounds which have been destroyed successfully in water by UV/ozone include: ethanol, acetic acid, glycine, glycerol, palmitic acid; organic nitrogen, phosphorus and sulfur compounds; potassium cyanide; complexed Cd, Cu, Fe, and Ni cyanides; photographic wastes, medical wastes, secondary effluents; chlorinated organics and pesticides such as pentachlorophenol, dichlorobenzene, dichlorbutane, chloroform, malathion, Baygon, Vapam, and DDT. It has also been shown[30] that using the combination of UV and ozone is more effective than using either one alone in destroying microbial contaminants (*E. coli* and *Streptococcus faecalis*) in water.

2.4. The Precleaning

Contaminants such as thick photoresist coatings and pure carbon films can be removed with UV/ozone, without any precleaning, but, in general, *gross* contamination cannot be removed without precleaning. For example, when a clean quartz wafer was coated thoroughly with human skin oils and placed in UV box 1 (Figure 1) without any precleaning, even prolonged exposure to UV/ozone failed to produce a low-contact-angle surface, because human skin oils contain materials such as inorganic salts, which cannot be removed by photosensitized oxidation.

The UV/ozone removed silicones from surfaces which had been precleaned, as described earlier, and also from surfaces which had simply been wiped with a cloth to leave a thin film. However, when the removal of a thick film was attempted, the UV/ozone removed most of the film upon prolonged exposure, but it also left a hard, cracked residue on the surface, possibly because many chemicals respond to radiation in various ways, depending upon whether or not oxygen is present. For instance, in the presence of oxygen many polymers degrade when irradiated; whereas in the absence of oxygen (as would be the case for the bulk of a thick film) these same polymers cross-link. In the study of the radiation degradation of polymers in air, the "results obtained with thin films are often markedly different from those obtained using thick specimen. . . ."[31]

For the UV/ozone cleaning procedure to perform reliably, the surfaces must be precleaned: first, to remove contaminants such as dust and salts, which cannot be changed into volatile products by the oxidizing action of UV/ozone, and, second, to remove thick films, the bulk of which could be transformed into a UV-resistant film by the cross-linking action of the UV light that penetrates the surface.

2.5. The Substrate

The UV/ozone cleaning process has been used with success on a variety of surfaces, including glass, quartz, mica, sapphire, ceramics, metals, silicon, gallium arsenide, and a conductive polyimide cement.

Quartz and sapphire are especially easy to clean with UV/ozone, since these materials are transparent to short-wavelength UV. For example, when a pile of thin quartz plates, approximately two centimeters deep, was cleaned by UV/ozone, both sides of all the plates, even those at the bottom of the pile, were cleaned by the process. Since sapphire is even more transparent, it, too, could probably be cleaned the same way. When flat quartz plates were placed on top of each other so that there could have been little or no ozone circulation between the plates, it was possible to clean both sides of the plates by the UV/ozone cleaning method. (Reference 32 shows that photocatalytic oxidation of hydrocarbons, without the presence of gaseous oxygen, can occur on some oxide surfaces.)

When white alumina ceramic substrates were cleaned by UV/ozone, the surfaces were cleaned properly. However, the sides facing the UV turned yellow, probably due to the production of UV-induced color centers. After a few minutes at high temperatures ($> 160°C$), the white color returned.

Metal surfaces could be cleaned by UV/ozone without any problems, so long as the UV exposure was limited to the time required to produce a clean surface. (This time should be approximately one minute or less for surfaces which have been properly precleaned.) However, prolonged exposure of oxide-forming metals to UV light can produce rapid corrosion. Silver samples, for example, blackened within one hour in UV box 1. Experiments with sheets of Kovar, stainless steel (type 302), gold, silver, and copper showed that, upon extended UV irradiation, the Kovar, the stainless steel, and the gold appeared unchanged; the silver and copper oxidized on both sides, but the oxide layers were darker on the sides facing away from the UV source. When electroless gold-plated nickel parts were stored under UV/ozone for several days, a powdery black coating gradually appeared on the parts. Apparently, nickel diffused to the surface through pinholes in the gold plating, and the oxidized nickel eventually covered the gold nearly completely. The corrosion was also observed in UV box 2, even when no ozone was being generated. The rates of corrosion increased substantially when a beaker of water was placed in the UV boxes to increase the humidity. Even Kovar showed signs of corrosion under such conditions.

The corrosion may possibly be explained as follows: as is known in the science of air-pollution control, in the presence of short-wavelength UV light, impurities in air, such as oxides of nitrogen and sulfur, combine with water vapor to form a corrosive mist of nitric and sulfuric acids. Therefore, the use of controlled atmospheres in the UV box may minimize the corrosion problem.

Since UV/ozone dissociates organic molecules, it may be a useful means of cleaning some organic materials, just as etching and electropolishing are sometimes useful for cleaning metals. The process has been used successfully to clean quartz resonators which have been bonded with silver-filled polyimide cement.[33] Teflon (TFE) tape exposed to UV/ozone in UV box 1 for ten days experienced a weight loss of 2.5%.[34] Also, the contact angles measured on clean quartz plates increased after a piece of Teflon was placed next to the plates in a UV box.[35] Similarly, Viton shavings taken from an O-ring experienced a weight loss of 3.7% after 24 hours in UV box 1. At the end of the 24 hours, the Viton surfaces had become sticky. Semiconductor surfaces have been successfully UV/ozone cleaned without adversely affecting the functioning of the devices. For example, after a 4K static RAM integrated circuit was exposed to UV/ozone for 120 min in a commercial UV/ozone cleaner, the device continued to function without any change in performance. (This IC had been made using n-channel silicon gate technology, with 1- to 1.5-μm junction depths.[36])

2.6. Rate Enhancement Techniques

UV/ozone cleaning "rate enhancement" techniques have been investigated by Zafonte and Chiu.[37] Experiments on gas-phase enhancement techniques included a comparison of the cleaning rates in dry air, dry oxygen, moist air, and moist oxygen. The moist air and moist oxygen consisted of gases that had been bubbled through water. Oxygen that had been bubbled through hydrogen peroxide was also tried. Experiments on liquid enhancement techniques consisted of a dropwise addition either of distilled water or of hydrogen peroxide solutions of various concentrations to the sample surfaces. Most of the sample surfaces consisted of various types of photoresist on silicon wafers.

The gas-phase "enhancement" techniques resulted in negligible to slight increases in the rates of photoresist removal (3–20 Å/min without enhance-

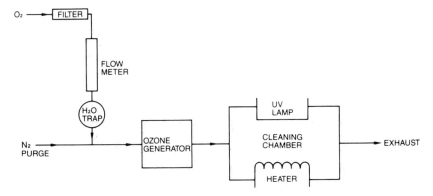

Figure 5. Schematic drawing of a UV/ozone cleaner that uses a silent-discharge ozone generator.[38]

ment vs. 3–30 Å/min with enhancement). The water and hydrogen peroxide liquid-phase enhancement techniques both resulted in significant rate enhancements (to 100–200 Å/min) for non-ion-implanted resists. The heavily ion-implanted resists (10^{15}–10^{16} atoms/cm^2) were not significantly affected by UV/ozone, whether "enhanced" or not.

Photoresist removal rates of 800–900 Å/min for positive photoresists and 1500–1600 Å/min for negative photoresists were reported by one manufacturer of UV/ozone cleaning equipment.[38] The fast removal rate was achieved at 300°C by using a 253.7-nm UV source, a silent-discharge ozone generator, and a heater built into the cleaning chamber and by using oxygen from a gas

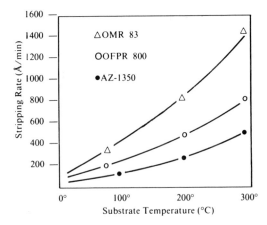

Figure 6. Photoresist stripping rate vs. substrate temperature for three types of photoresists.[38]

cylinder to generate the ozone. A schematic drawing of this UV/ozone cleaner is shown in Figure 5. The photoresist stripping rate versus substrate temperature for three different photoresists is shown in Figure 6.

3. The Mechanism of UV/Ozone Cleaning

The available evidence indicates that UV/ozone cleaning is primarily the result of photosensitized oxidation processes, as is represented schematically in Figure 7. The contaminant molecules are excited and/or dissociated by the absorption of short-wavelength UV light. Atomic oxygen and ozone[18,19] are produced simultaneously when O_2 is dissociated by the absorption of UV with wavelengths less than 245.4 nm. Atomic oxygen is also produced[18,19] when ozone is dissociated by the absorption of the UV and longer wavelengths of radiation. The excited contaminant molecules, and the free radicals produced by the dissociation of contaminant molecules, react with atomic oxygen to form simpler, volatile molecules, such as CO_2, H_2O, and N_2.

The energy required to dissociate an O_2 molecule into two ground-state O atoms corresponds to 245.4 nm. However, at and just below 245.4 nm the absorption of O_2 is very weak.[7,18,19] The absorption coefficient increases rapidly below 200 nm with decreasing wavelengths, as is shown in Figure 2. A convenient wavelength for producing O_3 is the 184.9 nm emitted by low-pressure Hg discharge lamps in fused quartz envelopes. Similarly, since most organic molecules have a strong absorption band between 200 and 300 nm, the 253.7-nm wavelength emitted by the same lamps is useful for exciting or dissociating contaminant molecules. The energy required to dissociate ozone corresponds to 1140 nm; however, the absorption by ozone is relatively weak above 300 nm. The absorption reaches a maximum near the 253.7-nm wavelength, as is shown in Figure 3. The actual photochemical processes occurring during UV/ozone cleaning are more complex than that shown in Figure 7. For example,

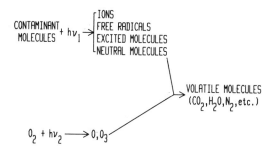

Figure 7. Simplified schematic representation of UV/ozone cleaning process.

the rate of production of ozone by 184.9-nm photons is promoted by the presence of other molecules, such as N_2 and CO_2.

As was described previously, the combination of short-wavelength UV light and ozone produced clean surfaces about 200 to 2,000 times faster than UV light alone or ozone alone. Similarly, Prengle et al.[26,29] had found in their studies of wastewater treatment that UV enhances the reaction with ozone 10^2- to 10^4-fold, and the products of the reactions are materials such as CO_2, H_2O, and N_2. Increasing the temperature increased the reaction rates. Mattox[39] also found that mild heat increases the UV/ozone cleaning rates. Bolon and Kunz,[1] on the other hand, had reported that the rate of UV/ozone depolymerization of photoresists did not change significantly between 100°C and 300°C. The rate of destruction of microorganisms was similarly insensitive to a temperature increase from room temperature to 40°C.[30] One manufacturer of UV/ozone cleaning equipment claims that the rate of photoresist stripping by UV/ozone increases severalfold as the temperature is raised from 20°C to 300°C.[38]

4. UV/Ozone Cleaning in Vacuum Systems

Sowell et al.[2] reported that when 10^{-4}-torr pressure of oxygen was present in a vacuum system, short-wavelength UV desorbed gases from the walls of the system. During UV irradiation, the partial pressure of oxygen decreased, while that of CO_2 and H_2O increased.

One must exercise caution in using a mercury UV source in a vacuum system because, should the lamp envelope break or leak, mercury can enter and ruin the usefulness of the system. Mercury has a high vapor pressure; its complete removal from a vacuum chamber is a difficult task. Other types of UV sources, such as xenon or deuterium lamps, may be safer to use in vacuum systems. The UV light can also be radiated into systems through sapphire or quartz windows, or through deep-UV fiber optic bundles. A small partial pressure of oxygen should be present during UV cleaning.

Caution must also be exercised when using UV/ozone in a cryopumped vacuum system, since cryopumped ozone is potentially explosive,[40] particularly during regeneration of the cryopump. A convenient method of dealing with this potential hazard is to use two kinds of UV sources, one an ozone-generating source, the other an "ozone killer" source.[41] (See next section.)

5. Safety Considerations

In constructing a UV/ozone cleaning facility, one should be aware of the safety hazards associated with exposure to short-wavelength UV light, which can cause serious skin and eye injury within a short time. In the UV boxes used by Vig et al., switches are attached to the doors in such a manner that when the doors are opened, the UV lamps are shut off automatically. If the application demands that the UV lamps be used without being completely enclosed (for example, as might be the case if a UV cleaning facility is incorporated into a wire bonder), then proper clothing and eye protection (e.g., UV safety glasses with side flaps) should be worn to prevent skin and eye damage.

Short-wavelength UV radiation is strongly absorbed by human cellular DNA. The absorption can lead to DNA–protein cross-links and can result in cancer, cell death, and cell mutation. It is now well known that solar UV radiation is the prime causative factor in human skin cancer[42,43] and is a significant risk factor in eye cancer.[44] The 290- to 320-nm portion of solar UV radiation has been found to be the most effective wavelength region for causing skin cancer. Because the atmosphere filters out the shorter wavelengths, humans are not normally exposed to wavelengths as short as 254 nm. However, in a study of the effects of UV radiation on skin cancer rates, it was found that the 254-nm wavelength was many times more effective in causing cell mutations than were the above-300-nm wavelengths. Therefore, it is essential that personnel not be exposed to the short wavelengths needed for UV/ozone cleaning because even low doses of these wavelengths can cause significant damage to human cells.

Another safety hazard is ozone, which is highly toxic. In setting up a UV cleaning facility, one must ensure that the ozone levels to which people are exposed do not exceed 0.1 ppm, the OSHA standard.[45] Ozone is a potential hazard in a cryopumped vacuum system because cryopumped ozone can become explosive under certain conditions.[40]

One method of minimizing the hazards associated with ozone is to use two types of short-wavelength ultraviolet sources for UV/ozone cleaning[41]: one, an ozone-generating UV lamp, e.g., a low-pressure mercury light in a fused quartz envelope; the other, a UV lamp that does not generate ozone but which emits one or more wavelengths that are strongly absorbed by ozone, e.g., a low-pressure mercury light in a high-silica glass tube which emits at 253.7 nm but not at 184.9 nm. Such a non-ozone-generating UV source can be used as an "ozone killer." For example, in one cryopumped vacuum system, UV/ozone cleaning was performed in up to 20 torr of oxygen. After the cleaning was completed and the ozone-generating UV lamp was turned off, ten minutes of "ozone killer" UV light reduced the concentration of ozone to less than 0.01 ppm, a level that is safe for cryopumping.[46] Therefore, with the "ozone killer" lamp, ozone concentrations were reduced by at least a factor of 100 within ten minutes. Without the "ozone killer" lamp, the half-life of ozone is three days at 20°C.[47]

The decomposition of ozone can also be greatly accelerated by using catalysts. For example, prior to 1980, ozone was found to be a causative factor for flight personnel and passengers in high-flying aircraft experiencing headaches; eye, nose, and throat irritations; and chest pains. This problem has been alleviated by passing the aircraft cabin air through a precious-metal catalytic converter, which reduces the ozone concentration from the 1- to 2-ppm level present in the troposphere to the low levels required for passenger comfort and safety.[48]

6. UV/Ozone Cleaning Facility Construction

The materials chosen for the construction of a UV/ozone cleaning facility should remain uncorroded by extended exposure to UV/ozone. Polished aluminum with a relatively thick anodized oxide layer, such as Alzak,[6] is one such material. It is resistant to corrosion, has a high thermal conductivity, which helps to prevent heat buildup, and is also a good reflector of short-wavelength UV. Most other metals, including silver, are poor reflectors in this range.

Initially, Vig *et al.* used an ordinary shop-variety aluminum sheet for UV box construction, which was found not to be a good material because, in time, a thin coating of white powder (probably aluminum oxide particles) appeared at the bottom of the boxes. Even in a UV box made of standard Alzak, after a couple of years' usage, white spots appeared on the Alzak, probably due to pinholes in the anodization. To avoid the possibility of particles being generated inside the UV/ozone cleaning facility, the facility should be inspected periodically for signs of corrosion. The use of "Class M" Alzak may also aid in avoiding particle generation, since this material has a much thicker oxide coating and is made for "exterior marine service," instead of the "mild interior service" specified for standard Alzak. Some commercially available UV/ozone cleaners are now constructed of stainless steel.[49, 50] To date, no corrosion problems have been reported with such cleaners.

Organic materials should not be present in the UV cleaning box. For example, the plastic insulation usually found on the leads of UV lamps should be replaced with inorganic insulation such as glass or ceramic. The box should be enclosed to minimize recontamination by circulating air, and to prevent accidental UV exposure and ozone escape.

The most widely available sources of short-wavelength UV light are the mercury arc lamps. Low-pressure mercury lamps in pure fused quartz envelopes operate near room temperature, emit approximately 90 percent at the 253.7-nm wavelength, and generate sufficient ozone for effective surface cleaning. Approximately five percent of the output of these lamps is at 184.9 nm. Medium- and high-pressure UV lamps[7] generally have a much higher output in the short-wavelength UV range. These lamps also emit a variety of additional wavelengths below 253.7 nm, which may enhance their cleaning action. However, they operate at high temperatures (the envelopes are near red-hot), have a shorter lifetime, higher cost, and present a greater safety hazard. The mercury tubes can be fabricated in a variety of shapes to fit different applications. In addition to mercury arc lamps, microwave-powered mercury vapor lamps are also available.[51]

Other available sources of short-wavelength UV include xenon lamps and deuterium lamps. These lamps must also be in an envelope transparent to short-wavelength UV, such as quartz or sapphire, if no separate ozone generator is to be used. In setting up a UV cleaning facility, one should choose a UV source which will generate enough UV/ozone to allow for rapid photosensitized oxidation of contaminants. However, too high an output at the ozone-generating wavelengths can be counterproductive because a high concentration of ozone can absorb most of the UV light before it reaches the samples. The samples should be placed as close to the UV source as possible to maximize the intensity reaching them. In the UV cleaning box 1 of Vig *et al.*, the parts to be cleaned are placed on an Alzak stand the height of which can be adjusted to bring the parts close to the UV lamp. The parts to be cleaned can also be placed directly onto the tube if the box is built so that the tube is on the bottom of the box.[52]

An alternative to using low-pressure mercury lamps in fused quartz envelopes is to use an arrangement similar to that of box 2, shown in Figure 1. Such a UV/ozone cleaner, now also available commercially,[38] uses silent-discharge-generated ozone and a UV source that generates the 253.7-nm wavelength, as is shown in Figure 5. The manufacturer of this cleaner claims a cleaning rate that is much faster than that which is achievable with UV/ozone cleaners that do not contain separate ozone generators. This cleaner also uses oxygen from a gas cylinder and a built-in sample heater that may further increase the cleaning rate.

7. Applications

The UV/ozone cleaning procedure is now used in numerous applications. A major use is substrate cleaning prior to thin film deposition, as is widely practiced in the quartz crystal industry during the manufacture of quartz crystal resonators for clocks and frequency control. There is probably no other device of which the performance is so critically dependent upon surface cleanliness. For example, the aging requirement for one 5-MHz resonator is that the frequency change no more then two parts in 10^{10} per week, whereas adsorption on or desorption from such a device of a monolayer of contamination changes the frequency by about one part in 10^6. The surface cleanliness must therefore be such that the rate of contamination transfer within the hermetically sealed resonator enclosure is less than 10^{-4} monolayers per week! In the author's quartz resonator fabrication laboratory, UV/ozone has been used at several points during the fabrication sequence, such as for cleaning and storing metal tools, masks, resonator parts, and storage containers.

The process is also being applied in a hermetic sealing method which relies on the adhesion between clean surfaces in an ultrahigh vacuum.[14,53-55] It has been shown that metal surfaces will weld together under near-zero forces if the surfaces are atomically clean. A gold gasket between gold-metallized (UV/ozone-cleaned) aluminum oxide sealing surfaces is currently providing excellent hermetic seals in the production of a ceramic flatpack enclosed quartz resonator. It has also been shown[53-55] that it is feasible to achieve hermetic seals by pressing a clean aluminum gasket between two clean, unmetallized aluminum oxide ceramic surfaces.

The same adhesion phenomenon between clean (UV/ozone-cleaned) gold surfaces has been applied to the construction of a novel surface contaminant detector.[56,57] The rate of decrease in the coefficient of adhesion between freshly cleaned gold contacts is used as a measure of the gaseous condensable contaminant level in the atmosphere.

The process has also been applied to improve the reliability of wire bonds, especially at reduced temperatures. For example, it has been shown[58,59] that the thermocompression bonding process is highly temperature dependent when organic contaminants are present on the bonding surfaces. The temperature dependence can be greatly reduced by UV/ozone cleaning of the surfaces just prior to bonding, as is shown in Figure 8. In a study of the effects of cleaning methods on gold ball bond shear strength, UV/ozone cleaning was found to be the most effective method of cleaning contaminants from gold surfaces.[60] UV/ozone is also being used for cleaning alumina substrate surfaces during the processing of thin film hybrid circuits.[61]

A number of cleaning methods were tested when the nonuniform appearance of thermal/flash protective electrooptic goggles was traced to organic contaminants on the electrooptic wafers. UV/ozone proved to be the most effective method for removing these contaminants, and thus it was chosen for use in the production of the goggles.[62]

Other applications which have been described are: photoresist removal[1,5,13,38]; the cleaning of vacuum chamber walls,[2] photomasks,[63] silicon wafers (for enhancing photoresist adhesion),[63] lenses,[63] mirrors,[63] solar panels,[63] sapphire (before the deposition of HgCdTe)[63] and other fine-linewidth devices,[63,64] inertial guidance subcomponents (glass, chromium-oxide surfaced-gas bearings, and beryllium),[63,65] and gallium-arsenide wafers[66]; the cleaning of stainless steel for studying a milk–stainless steel interface[67]; and the cleaning of adsorbed species originating from epoxy adhesives.[16] Since short-wavelength UV can generate radicals and ions, a side benefit of UV/ozone cleaning of insulator surfaces can be the neutralization of static charges.[68]

UV/ozone cleaning of silicon substrates in silicon molecular beam epitaxy (MBE) has been found to be effective in producing near defect-free MBE films.[69] By using UV/ozone cleaning, the above-1200°C temperatures required for removing surface carbon in the conventional method can be lowered

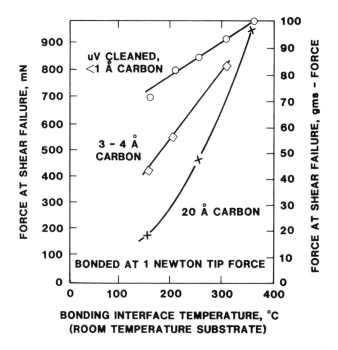

Figure 8. Effect of UV/ozone cleaning on gold-to-gold thermocompression bonding.[58] Data are for t/c ball bonds of 2 5 μm dia. gold wire bonded to gold-chromium metallization.

to below 1000°C. The slip lines resulting from thermal stresses and thermal pits that are often produced by the high-temperature treatment are minimized in the lower-temperature processing. Impurity redistribution in the substrate is also reduced.

In the processing of semiconductor wafers, a single UV/ozone exposure has been found to be capable both of "descumming" and of stabilizing.[70] After developing and rinsing the photoresist pattern, the UV/ozone removes the thin layers of organic photoresist residue (scum) from the "clear" regions. The photoresist stabilization is believed to be due to cross-linking produced by the short-wavelength (deep) UV radiation.[71] The stabilization rate is accelerated by increasing the temperature. For example, UV/ozone exposure times of 10 to 30 minutes from a 25 cm × 25 cm low-pressure mercury grid lamp at 100°C yields satisfactory results. The stabilized photoresist pattern exhibits: (1) improved adhesion to the substrate, (2) improved ability to maintain geometrical shape under thermal stress, and (3) improved ability to withstand exposure to the etchants and solvents used to create the desired patterns in the circuit coatings.[70]

8. Effects Other Than Cleaning

Short-wavelength UV, ozone, and the combination of the two can have effects other than surface cleaning. Some of the more significant of these effects are described in the following sections.

8.1. Oxidation

Ozone's oxidation power is second only to that of fluorine. Ozone can oxidize most inorganic compounds to their final oxidative state.[47] For most substrates, UV/ozone cleaning, for the minimum time necessary to obtain a clean surface, will not cause a significant amount of oxidation. However, extended storage under UV/ozone may be detrimental for some oxidizable surfaces. In some cases, the enhanced oxide formation may be beneficial. For example, whereas the "native" oxide on GaAs is only about 30 Å thick, UV/ozone produces an oxide layer that is 100 Å to 300 Å thick,[72] i.e., UV/ozone can produce a clean, enhanced "oxide passivated" surface. Ten minutes of UV/ozone cleaning increased the oxide thickness on silicon substrates from 0.9 nm to 1.2 nm.[69] Similarly, the native UV/ozone-produced oxide layer at the HgCdTe–SiO_2 interface has been found to enhance the interface properties.[73] Solar radiation and atmospheric ozone have been found to markedly enhance the sulfidation of copper.[74] Extended exposure to UV/ozone has been found to significantly increase the oxide layer thickness on aluminum surfaces.[75] Whereas the oxide thickness on air-exposed aluminum surfaces is normally limited to about 50 Å, UV/ozone exposure increased the oxide layer thickness significantly beyond the "normal" 50-Å limit, as is shown in Table 4.

8.2. UV-Enhanced Outgassing

Short-wavelength UV has been found to enhance the outgassing of glasses.[76] The UV light produced the evolution of significant quantities of hydrogen, and also water, carbon dioxide, and carbon monoxide. The hydrogen evolution was proportional to the amount of radiation incident to the samples.

Table 4. UV/Ozone Exposure vs. Oxide Thickness on Aluminum

Substrate treatment	Oxide thickness (Å)
Evaporate 1 μm of aluminum	47
10-minute UV/ozone cleaning	90
60-minute UV/ozone cleaning	200

For UV-opaque glasses, the evolution occurred from the side exposed to the UV; for high-transmission samples, the gas evolved from both sides.

8.3. Other Surface/Interface Effects

Energetic radiation such as UV and gamma radiation has been reported to produce dehydration and the formation of free radicals on silica surfaces.[77] However, dehydrated (or siloxinated) silica surfaces are hydrophobic,[78,79] whereas UV/ozone-cleaned silica (quartz) surfaces exhibit a very low (less than 4°) water contact angle, thus indicating that the UV/ozone does not dehydrate the surfaces, nor does it modify surface silanol groups the way high-temperature vacuum baking does.[80] Short-wavelength UV has also been found to produce a bleaching effect in $Si-Si_3$ interfaces with thin oxides,[81] and has also been found to produce yellowing (color centers) during the cleaning of aluminum oxide ceramics.[34] The yellowing can be readily bleached by heating the sample to above 160°C.

8.4. Etching

Short-wavelength (193 nm) UV laser irradiation of biological and polymeric materials has been shown to be capable of etching the materials with great precision, via "ablative photodecomposition," and without significant heating of the samples. Linewidths 5 μm wide have been etched onto a plastic film to demonstrate the capability of this technique.[82] Oxygen does not appear to have the same significance in this process as it does in UV/ozone cleaning. The etch depth vs. fluence was found to be the same in vacuum and in air.[83] UV/ozone has been found to etch Teflon,[34,35] and Viton,[34] and will likely etch other organic materials as well.[84,85] The susceptibility of polymers to degradation by ozone can be reduced by various additives and through the elimination of "the offending double bonds from the backbone structure of the polymers."[86]

9. Summary and Conclusions

The UV/ozone cleaning procedure has been shown to be a highly effective method of removing a variety of contaminants from surfaces. It is a simple-to-use dry process which is inexpensive to set up and operate. It can produce clean surfaces at room temperature, either in a room atmosphere or in a controlled atmosphere.

The variables of the UV cleaning procedure are: the contaminants initially present, the precleaning procedure, the wavelengths emitted by the UV source, the atmosphere between the source and sample, the distance between the source and sample, and the time of exposure. For surfaces which are properly precleaned and placed within a few millimeters of an ozone-producing UV source, the process can produce a clean surface in less than one minute. The combination of short-wavelength UV light plus ozone produces a clean surface substantially faster than either short-wavelength UV light without ozone or ozone without UV light. Clean surfaces will remain clean indefinitely during storage under UV/ozone, but prolonged exposure of oxide-forming metals to UV/ozone in room air can produce rapid corrosion.

The cleaning mechanism seems to be a photosensitized oxidation process in which the contaminant molecules are excited and/or dissociated by the absorption of short-wavelength UV light. Simultaneously, atomic oxygen is generated when molecular oxygen is dissociated and when ozone is dissociated by the absorption of short and long wavelengths of radiation. The products of the excitation of contaminant molecules react with atomic oxygen to form simpler molecules, such as CO_2 and H_2O, which desorb from the surfaces.

References and Notes

1. D. A. Bolon and C. O. Kunz, Ultraviolet depolymerization of photoresist polymers, *Polym. Eng. Sci.* 12, 109-111 (1972); also, D. A. Bolon, Method of Removing Photoresist from Substrate, U.S. Patent 3,890,176 (June 17, 1975).
2. R. R. Sowell, R. E. Cuthrell, D. M. Mattox, and R. D. Bland, Surface cleaning by ultraviolet radiation, *J. Vac. Sci. Technol.* 11, 474-475 (1974).
3. J. R. Vig, C. F. Cook, Jr., K. Schwidtal, J. W. LeBus, and E. Hafner, Surface studies for quartz resonators, in: Proceedings of the 28th Annual Symposium on Frequency Control, U.S. Army Electronics Command, Ft. Monmouth, N.J., AD 011113 (1974), pp. 96-108. Article reprinted as ECOM Tech. Rep. 4251, AD 785513 (1974).
4. J. R. Vig, J. W. LeBus, and R. L. Filler, Further results on UV cleaning and Ni electrobonding, in: Proceedings of the 29th Annual Symposium on Frequency Control (1975), pp. 220-229. Copies available from the National Technical Information Service, Springfield, Va., AD A017466.
5. J. R. Vig and J. W. LeBus, UV/ozone cleaning of surfaces, *IEEE Trans. Parts, Hybrids and Packag.* PHP-12, 365-370, (December, 1976).
6. Alzak is an aluminum reflector material with a corrosion-resistant oxide coating. The Alzak process is licensed to several manufacturers by the Aluminum Co. of America, Pittsburgh, Pa. 15219.
7. J. G. Clavert and J. N. Pitts, Jr., *Photochemistry*, pp. 205-209, 687-705, John Wiley and Sons, New York (1966).
8. V. S. Fikhtengolt's, R. V. Zolotareva, and Yu A. L'vov, *Ultraviolet Spectrum of Elastomers and Rubber Chemicals*, Plenum Press Data Div., New York (1966).
9. L. Lang, *Absorption Spectra in the Ultraviolet and Visible Region*, Academic Press, New York (1965).
10. Model No. R-52 Mineralight Lamp, UVP, Inc., San Gabriel, CA 91778.
11. See, e.g., *Encyclopaedic Dictionary of Physics*, Vol. 5, p. 275, Pergamon Press, New York (1962).
12. M. E. Schrader, Surface-contamination detection through wettability measurements, in: *Surface Contamination: Genesis, Detection and Control* (K. L. Mittal, ed.), Vol. 2, pp. 541-555, Plenum Press, New York (1979).

13. P. H. Holloway and D. W. Bushmire, Detection by Auger electron spectroscopy and removal by ozonization of photoresist residues, in: *Proceedings of the 12th Annual Reliability Physics Symposium*, pp. 180-186, Institute of Electrical and Electronic Engineers, Piscataway, N.J. (1974).
14. R. D. Peters, Ceramic flatpack enclosures for precision crystal units, in: Proceedings of the 30th Annual Symposium on Frequency Control, pp. 224-231 (1976). Copies available from the National Technical Information Service, Springfield, Va., AD A046089.
15. C. E. Bryson and L. J. Sharpen, An ESCA analysis of several surface cleaning techniques, in: *Surface Contamination: Genesis, Detection and Control*, (K. L. Mittal, ed.), Vol. 2, pp. 687-696, Plenum Press, New York (1979).
16. R. C. Benson, B. H. Nall, F. G. Satkiewitz, and H. K. Charles, Jr., Surface analysis of adsorbed species from epoxy adhesives used in microelectronics, *Applic. Surface Sci. 21*, 219-229 (1985).
17. W. L. Baun, ISS/SIMS characterization of UV/O_3 cleaned surfaces, *Appl. Surface Sci.* 6, 39-46, (1980).
18. J. R. McNesby and H. Okabe, Oxygen and ozone, in: *Advances in Photochemistry* (W. A. Noyes, Jr., G. S. Hammond, and J. N. Pitts, eds.), Vol. 3, pp. 166-174, Interscience Publishers, New York (1964).
19. D. H. Volman, Photochemical gas phase reactions in the hydrogen-oxygen system, in: *Advances in Photochemistry* (W. A. Noyes, Jr., G. S. Hammond, and J. N. Pitts, eds.), Vol. 1, pp. 43-82, Interscience Publishers, New York (1963).
20. P. R. Hoffman Co., Carlisle, Pa. 17013.
21. John Crane Lapping Vehicle 3M, Crane Packing Co., Morton Grove, IL 60053.
22. Welch Duo-Seal, Sargent-Welch Scientific Co., Skokie, IL 60076.
23. Dow Corning Corp., Midland, MI 48640.
24. Dutch Boy No. 205, National Lead Co., New York NY 10006.
25. Eastman Kodak Co., Rochester, NY 14650.
26. H. W. Prengle, C. E. Mauk, R. W. Legan, and C. G. Hewes, Ozone/UV process effective wastewater treatment, *Hydrocarbon Processing 54*, 82-87 (October, 1975).
27. H. W. Prengle, Jr., C. E. Mauk, and J. E. Payne, Ozone/UV oxidation of chlorinated compounds in water, Forum on Ozone Disinfection (1976), International Ozone Institute, Warren Bldg., Suite 206, 14805 Detroit Ave., Lakewood, OH 44107.
28. H. W. Prengle, Jr., and C. E. Mauk, Ozone/UV oxidation of pesticides in aqueous solution, Workshop on Ozone/Chlorine Dioxide Oxidation Products of Organic Materials, EPA/International Ozone Institute, Warren Bldg., Suite 206, 14805 Detroit Ave., Lakewood, OH 44107 (November, 1976).
29. H. W. Prengle, Jr., in: Proceedings of the International Ozone Institute Ozone Symposium, International Ozone Institute, Warren Bldg., Suite 206, 14805 Detroit Ave., Lakewood, OH 44107 (1978).
30. J. D. Zeff, R. R. Barton, B. Smiley, and E. Alhadeff, UV-Ozone Water Oxidation/Sterilization Process, U.S. Army Medical Research and Development Command, Final Report, Contract No. DADA 17073-C-3138 (September, 1974). Copies available from NTIS, AD A004205.
31. H. V. Boenig, *Structure and Properties of Polymers*, p. 246, Wiley, New York (1973).
32. V. N. Filimonov, in *Elementary Photoprocesses in Molecules* (B. S. Neporent, ed.), pp. 248-259, Consultants Bureau, New York (1968).
33. R. L. Filler, J. M. Frank, R. D. Peters, and J. R. Vig, Polyimide bonded resonators, in: Proceedings of the 32nd Annual Symposium on Frequency Control (1978), pp. 290-298. Copies available from Electronics Industries Assoc., 2001 Eye St., NW Washington, DC 20006.
34. J. W. LeBus and J. R. Vig, U.S. Army Electronics Technology and Devices Lab., Fort Monmouth, NJ 07703, unpublished information (1976).
35. J. Kusters, Hewlett Packard Co., Santa Clara, CA 95050, personal communication (1977).

36. E. Lasky, Aerofeed Inc., Chalfont, PA, personal communication (1978).
37. L. Zafonte and R. Chiu, Technical Report on UV-Ozone Resist Strip Feasibility Study, UVP, Inc., 5100 Walnut Grove Avenue, San Gabriel, CA 91778 (September, 1983); presented at the SPIE Santa Clara Conference on Microlithography in March, 1984.
38. Application Note, Photoresist Stripping with the UV-1 Dry Stripper, March Instruments Inc. Concord, CA 94520.
39. D. M. Mattox, Surface cleaning in thin film technology, *Thin Solid Films 53*, 81–96 (1978).
40. C. W. Chen and R. G. Struss, On the cause of explosions in reactor cryostats for liquid nitrogen, *Cryogenics 9*, 131–132 (April 1969).
41. J. R. Vig and J. W. LeBus, Method of Cleaning Surfaces by Irradiation with Ultraviolet Light, U.S. Patent 4,028,135 (June 7, 1977).
42. M. J. Peak, J. G. Peak, and C. A. Jones, Different (direct and indirect) mechanisms for the induction of DNA–protein crosslinks in human cells by far- and near-ultraviolet radiations (290 and 405 nm), *Photochem. Photobiol. 42*, 141–146 (1985).
43. H. E. Kubitschak, K. S. Baker, and M. J. Peak, Enhancement of mutagenesis and human skin cancer rates resulting from increased fluences of solar ultraviolet radiation, *Photochem. Photobiol. 43*, 443–447 (1986).
44. M. A. Tucker, J. A. Shields, P. Hartge, J. Augsburger, R. N. Hoover, and J. F. Fraumeni, Jr., Sunlight exposure as risk factor for intraocular malignant melanoma, *New Eng. J. Med. 313*, 789–792 (1985).
45. *Occupational Safety and Health Standards*, Vol. 1, General Industry Standards and Interpretations (October, 1972), Pt. 1910, 1000, Table Z-1, Air Contaminants, p. 642,4 as per change 10, June 26, 1975.
46. D. A. Ehlers, Ozone Generation and Decomposition by UV in the ERADCOM QXFF, Report No. PT81-004, General Electric Neutron Devices Dept., P.O. Box 2908, Largo, FL 34924 (Jan. 26, 1981).
47. Matheson Gas Data Book, published by Matheson Gas Products Co., East Rutherford, NJ, 6th Ed., pp. 574–577 (1980).
48. J. C. Bonacci, W. Egbert, M. F. Collins, and R. M. Heck, New catalytic abatement product decomposes ozone in jet aircraft passenger cabins, *Int'l Precious Metals Inst. Proceedings* (1982); reprint and additional literature on DEOXO Catalytic Ozone Converters is available from Englehard Corp., Specialty Chemicals Div., 2655 U.S. Rt. 22, Union, NJ 07083.
49. UVOCS Div., Aerofeed Inc., P.O. Box 303, Chalfont, PA 18914.
50. UVP, Inc., 5100 Walnut Grove Ave., San Gabriel, CA 91778.
51. A. N. Petelin and M. G. Ury, Plasma sources for deep-UV lithography, in: *VLSI Electronics: Microstructure Science, Vol. 8, Plasma Processing for VLSI* (N. G. Einspruch and D. M. Brown, eds.), Academic Press, Orlando, FL (1984).
52. R. D. Peters, General Electric Neutron Devices Dept., P.O. Box 2908, Largo, FL 34924, personal communication (1976).
53. J. R. Vig and E. Hafner, Packaging Precision Quartz Crystal Resonators, Technical Report ECOM-4134, U.S. Army Electronics R&D Command, Fort Monmouth, NJ (July, 1973). Copies available from NTIS, AD 763215.
54. E. Hafner and J. R. Vig, Method of Processing Quartz Crystal Resonators, U.S. Patent 3,914,836 (Oct. 28, 1975).
55. P. D. Wilcox, G. S. Snow, E. Hafner, and J. R. Vig, A new ceramic flatpack for quartz resonators, in: Proceedings of the 29th Annual Symposium on Frequency Control (1975), pp. 202–210. See Reference 4 above for availability information.
56. R. E. Cuthrell and D. W. Tipping, Surface contaminant detector, *Rev. Sci. Instrum. 47*, 595–599 (1976).
57. R. E. Cuthrell, Description and operation of two instruments for continuously detecting air-

borne contaminant vapors, in: *Surface Contamination: Genesis, Detection and Control* (K. L. Mittal, ed.), Vol. 2, pp. 831–841, Plenum Press, New York (1979).
58. J. L. Jellison, Effect of surface contamination on the thermocompression bondability of gold, *IEEE Trans. Parts, Hybrids and Packaging PHP-11*, 206–211 (1975).
59. J. L. Jellison, Effect of surface contamination on solid phase welding—An overview, in: *Surface Contamination: Genesis, Detection and Control* (K. L. Mittal, ed.), Vol. 2, pp. 899–923, Plenum Press, New York (1979).
60. J. A. Weiner, G. V. Clatterbaugh, H. K. Charles, Jr., and B. M. Romenesko, Gold ball bond shear strength, in: Proceedings of the 33rd Electronic Components Conference (1983), pp. 208–220.
61. R. Tramposch, Processing thin film hybrids, *Circuits Manufacturing 23*, 30–40 (March, 1983).
62. J. A. Wagner, Identification and elimination of organic contaminants on the surface of PLZT ceramic wafers, in: *Surface Contamination: Genesis, Detection and Control* (K. L. Mittal, ed.), Vol. 2, pp. 769–783, Plenum Press, New York (1979).
63. E. Lasky, UVOCS Div., Aerofeed Inc., Chalfont, PA 18914, personal communication (1983).
64. H. I. Smith, Massachusetts Institute of Technology, unpublished class notes on "Cleaning of Oxides," and personal communications (1982).
65. J. R. Stemniski and R. L. King, Jr., Ultraviolet cleaning: Alternative to solvent cleaning, in: *Adhesives for Industry*, pp. 212–228, Technology Conferences, El Segundo, Calif. (1980).
66. J. A. McClintock, R. A. Wilson, and N. E. Byer, UV-ozone cleaning of GaAs for MBE, *J. Vac. Sci. Technol. 20*, 241–242 (February, 1982).
67. K. A. Almas and B. Lund, Cleaning and characterization of stainless steel exposed to milk, *Surface Technol. 23*, 29–39 (1984).
68. D. H. Baird, Surface Charge Stability on Fused Silica, Final Technical Report, TR 76-807.1 (December, 1976). Copies available from NTIS, AD A037463.
69. M. Tabe, UV ozone cleaning of silicon substrates in silicon molecular beam epitaxy, *Appl. Phys. Lett. 45*, 1073–1075 (1984).
70. W. L. Gardner, Engelhard Millis Corp., Millis, MA 02054, personal communication (November, 1985).
71. J. C. Matthews and J. I. Wilmott, Jr., Stabilization of single layer and multilayer resist patterns to aluminum etching environments, presented at SPIE Conference, Optical Microlithography III, Santa Clara, CA (March 14–15, 1985); reprints available from Fusion Semiconductor Systems Corp., 7600 Standish Place, Rockville, MD 20855.
72. J. A. McClintock, Martin Marietta Laboratories, Baltimore, MD 21227, personal communication (1981).
73. B. K. Janousek and R. C. Carscallen, Photochemical oxidation of (HgCd)Te: Passivation process and characteristics, *J. Vac. Sci. Technol. A, 3*, 195–198 (January–February, 1985).
74. T. E. Graedel, J. P. Franey, and G. W. Kammlott, Ozone- and photon-enhanced atmospheric sulphidation of copper, *Science 224*, 599–601 (1984).
75. G. V. Clatterbaugh, J. A. Weiner, and H. K. Charles, Jr., Gold–aluminum intermetallics: Ball shear testing and thin film reaction couples, in: Proceedings of the 34th Electronic Components Conference (1984), pp. 21–30.
76. V. O. Altemose, Outgassing by ultraviolet radiation, in: *Vacuum Physics and Technology* (G. L. Weissler and R. W. Carlson, eds.), Vol. 14 of *Methods of Experimental Physics*, Chap. 7, pp. 329–333, Academic Press, New York (1979).
77. M. M. Tagieva and V. F. Kiseler, The action of radiation on the surface properties of silica, *Russian J. Phys. Chem. 35*, 680–681 (1961).
78. M. L. Hair, The molecular nature of adsorption on silica surfaces, in: Proceedings of the 27th Annual Symposium on Frequency Control, AD 771042 (1973), pp. 73–78.
79. M. L. White, Clean surface technology, in: Proceedings of the 27th Annual Symposium on

Frequency Control, AD 771042 (1973), pp. 79-88; also, The detection and control of organic contaminants on surfaces, in: *Clean Surfaces: Their Preparation and Characterization for Interfacial Studies* (G. Goldfinger, ed.), pp. 361-373, Marcel Dekker, Inc., New York (1970).
80. R. N. Lamb and D. N. Furlong, Controlled wettability of quartz surfaces, *J. Chem. Soc., Faraday Trans. 1*, *78*, 61-73 (1982).
81. P. J. Caplan, E. H. Poindexter, and S. R. Morrison, Ultraviolet bleaching and regeneration of $\cdot Si \equiv Si_3$ centers at the Si/SiO_2 interface of thinly oxidized silicon wafers, *J. Appl. Phys. 53*, 541-545 (1982).
82. R. Srinivasan, Conference on Lasers and Electrooptics, as reported in: Clean cuts for notched hairs, *Science News 123*, 396 (June 18, 1983).
83. R. Srinivasan and B. Braren, Ablative photodecomposition of polymer films by pulsed far-ultraviolet (193 nm) laser radiation: Dependence of etch depth on experimental conditions, *J. Polym. Sci., Polym. Chem. Ed. 22*, 2601-2609 (1984).
84. G. S. Alberts, Process for Etching Organic Coating Layers, U.S. Patent 3,767,490 (Oct. 23, 1973).
85. A. N. Wright, Removal of Organic Polymeric Films from a Substrate, U.S. Patent 3,664,899 (May 23, 1972).
86. L. Robinson, The development of ozone resistant materials for wire and cable, *IEEE Electrical Insulation Mag. 1*, 20-22 (1985).

2

Techniques for Cleaning Liquid Surfaces

JOHN C. SCOTT

1. Introduction

When we speak of "clean liquid surfaces" our meaning is often subtly different from considerations of "clean solid surfaces." In most solid surface cases, what we require is a surface that is "pure", i.e., composed solely of molecules of the desired substance.[1] Although there is a corresponding need for "pure" liquid surfaces in basic research, where delicate measurements of surface tension are related to intermolecular forces, what we require more often is a liquid surface whose fluid mechanics is simply specified or well defined.

The relevance of clean liquid surfaces is twofold, related to chemical engineering processes and to the air–water interface, considered on the environmental scale.

In many chemical engineering processes, principally those involving material exchange between gas and liquid phases, small quantities of certain organic contaminants can have disproportionate effects. In the air–water case it is specifically this interchange of material, and also of wind and solar energy, which determines the influence of the oceans on the atmosphere, and vice versa.

In both cases, dynamically produced surface tension variations on a surface can have marked effects on the turbulence and diffusion processes which bring fresh material to the surface from either of the adjacent bulk phases,[2] and exchanges can thus be largely controlled by contaminant materials. The origin of these dynamic surface tension effects will be described in Section 2.

In general, the techniques which will be considered in this review are rel-

---

JOHN C. SCOTT • Fluid Mechanics Research Institute, University of Essex, Colchester, Essex CO4 3SQ, United Kingdom. *Present address:* Admiralty Research Establishment, Southwell, Portland, Dorset DT5 2JS, United Kingdom.

evant to basic and applied research on liquid surfaces—to ways in which surfaces may be initially cleaned for the better control of research experiments. In the relevant applications, surface cleanliness is determined by other factors, and the techniques described here are usually irrelevant to the chemical engineering or environmental processes which they model. This point will be taken up in Section 6.8, which deals with mechanisms by which surface contamination may be removed or modified in practical applications of clean-surface work.

The major practical distinction between the properties of solid surfaces and liquid surfaces stems simply from the characteristic material distinction between the two phases. A liquid is able to flow under the action of external body forces, and it is unable to sustain shear. Together, these properties allow the physical boundary of the liquid—its free surface—to move relatively easily in response to the forces experienced. Thus, any constraints on such motions which are the result of modifications to the mechanical properties of the surface are enabled to exert a significant influence on the liquid motion.

The present discussion is concerned almost exclusively with the surface of the liquid water, and to some extent also aqueous solutions. Other liquids—organic solvents, for example—are, of course, used extensively in chemical engineering. However, the lower surface tension and intrinsically simpler molecular structure of organic liquids makes surface contamination effects much less significant. Water is not only the commonest liquid in human experience, it also happens to be the most affected—by a large margin—by the phenomenon of surface activity.

The review also avoids consideration of solutions of surface-active materials, concentrating instead on methods of complete removal of such materials. Preparation of a clean surface of an aqueous surfactant solution—effectively restricting the added material to a single species—relies on subtle differences in the chemical or physical nature of the materials concerned, and needs a thorough knowledge of these differences. The reader is directed to a paper by Mysels and Florence[3] which gives an excellent appreciation of such problems.

This discussion will, however, cover the important subject areas concerned with keeping water surfaces clean, including both the choice of materials for water storage and the design of clean-surface apparatus (Section 7), and also the cleaning of the solid surfaces which may come into contact with the clean water (Section 8).

2. The Origin of Dynamic Liquid Surface Phenomena

Surface tension is a force—a measurable physical quantity—which can be observed to act at the surface of a liquid in such a way that it always tends to minimize the surface area of the liquid. It is the force which makes liquid droplets spherical, and which exerts tensile forces on solid boundaries wetted by the liquid.

The force has its origins in the thermodynamic free energy that arises from the unbalanced force field experienced by molecules located near the surface, which tends to attract them towards the bulk liquid phase (Figure 1). The surface region is diffuse compared with surface regions in the solid state, and the transition from liquid phase to vapor phase is necessarily a gradual one, a consequence of the dynamic balance between this attractive force and the turbulent thermal motion of the molecules.

In textbook considerations of surface tension and surface tension phenomena, there has been a tendency to concentrate on the thermodynamic aspects of liquid surfaces, viewing the directly observable manifestations of surface tension as somehow less important. This view, implicit in the textbook of Adam,[4] has tended to efface the true importance of the parameter, and to confuse the general appreciation of what is a real and highly observable physical attribute.

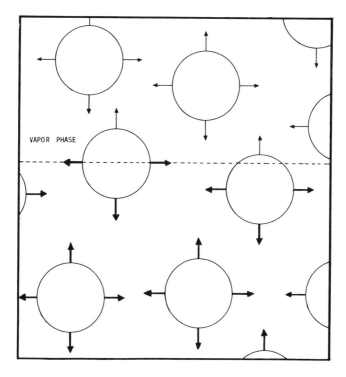

Figure 1. Schematic diagram showing the balance of forces acting on molecules near the surface, which experience a net attraction for the bulk liquid. This is the physical origin of surface tension.

Unfortunately, there is a distinct lack of introductory reading material on this subject, particularly regarding the way it concerns fluid mechanics.

Surface tension effects have, for many centuries, been considered a fascinating part of the properties of the liquid state. The calming effect of oil on water in a storm was known to the Ancients, and references have appeared, increasingly frequently, since Medieval times.[5] The work of Benjamin Franklin[6] can be seen as a major factor in giving the subject the importance it has today, in stimulating the research of his contemporary and subsequent generations.

In general, wave-damping effects arise from wave-induced spatial gradients of the surface tension. These gradients, which arise from compression and extension of the surface as the wave passes, exert tangential stresses on the free surface, acting in a direction which tends to oppose the wave motion. Energy is extracted from the waves via the viscosity of the underlying liquid. The effect may be precisely formulated in mathematical terms as a modification of the surface boundary condition.[7]

Surface tension gradients can also result from spatial variations in temperature and electric field, and these variations can thus also lead to fluid motions. Thermal effects are influential in the well-known phenomenon of Benard convection,[8] gradients arising between rising warmer liquid of lower surface tension and the cooler liquid (higher surface tension) accumulating at the downwards-flowing boundaries of the convection cells. Electric field effects[9] are much less common and do not appear significant for industrial processes.

Surface contamination effects are qualitatively different from other causes of surface tension gradients. The important surface contamination effects are found, almost entirely, in cases where the liquid has an exceptionally high intrinsic surface tension, which the contaminant materials act to reduce.

Apart from water, only liquid metals and molten electrolytes have surface tensions high enough to make them likely candidates for the appearance of interesting surface effects. These other liquids lie outside the scope of the present discussion, and indeed it can be said that very little work has been done on their surface behavior. However, it is certain that scientists considering work in this area should expect practical difficulties similar to those experienced with water, and quite possibly of an even greater magnitude.

Liquid mercury appears to be the only current rival for the attention of scientists. In this case it is common experience that water will not generally spread on a flat mercury surface, forming liquid lenses. It would be expected, however, that water should spread, since it has a much lower surface tension than mercury. The explanation for this observation is that the mercury surface is already contaminated by some material (probably oil or grease) of even lower surface tension.

Contaminant materials are able to reduce the surface tension of a liquid if their molecules, as a result of the detailed force balances at molecular level, find it energetically more acceptable (i.e., having a lower-energy state) to be located in the surface region of the host liquid than to be dispersed homogeneously through the liquid bulk. In this circumstance we say that the dispersed molecules are *positively adsorbed* at the liquid surface, and both the overall free energy of the surface and the observed surface tension are reduced.

In many cases of practical importance, surface-active materials which are completely insoluble in water may be brought into contact with the surface. In such cases, subject to the force balance indicated above, the materials will spread over the surface until the force balance is once more in equilibrium. Gaines[10] has provided a comprehensive study of such cases.

In the case of a single species of spreading molecule, the spreading will cease when the species forms a monomolecular layer—monolayer—on the water surface. Such spread monolayers will show no tendency to become dispersed within the bulk liquid. This may be seen as an extreme case of the (more general) equilibrium between surface and bulk, in which the bulk liquid totally rejects the added material. If the material were thoroughly dispersed within the bulk by mechanical means, it would be expected to gradually reaccumulate at the surface in its adsorbed form, over a time determined by the diffusion process.

It is actually possible for a water-soluble material to behave as an insoluble material in its adsorbed form. Scott and Stephens[11] report the case of poly(ethylene oxide), a high-molecular-weight polymer, soluble in water up to gel-forming concentrations, which does not form a true equilibrium between surface and bulk, at low bulk concentrations. Subject to diffusion, the surface concentration always reaches saturation, as long as there is enough material in the bulk to cover the available surface. This surprising behavior arises from the low probability associated with all of the adsorbed monomer segments of a polymer chain becoming desorbed simultaneously, a necessary step if the molecule is to diffuse into the bulk once again. Scott[12] suggests that this case is almost certainly important for the accumulation of materials of biological origin at the ocean surface.

The magnitude of the reduction in surface tension effected by an adsorbed molecule is, as indicated above, a function of the detailed force balances it finds with the host molecules. However, the observed surface tension at any particular location on the surface will always be lower, the greater the concentration of the adsorbed species on the surface. The basic property of surface tension will, therefore, act to make the surface distribution of contaminant as uniform as possible, and a contaminated liquid at rest will have a constant surface tension, albeit a value somewhat lower than the clean-liquid value.

It is this variation of the surface tension with surface concentration of contaminant which causes the majority of surface tension phenomena. Imposed variations of surface concentration will cause the liquid to move, in a manner which tends to reduce the variations, and, conversely, externally imposed motions of the liquid will tend to cause variations of the surface concentration (and surface tension) which act in a way which resists the motions.

3. The Notion of Surface Cleanliness

The concept of material purity is, in general, a relatively easy one to define in physical situations: it is usually directly quantifiable as a mass or molecular percentage of "foreign" matter incorporated in the bulk of the material concerned. To a certain extent this simplicity also extends directly to considerations of the surface of a solid or a liquid, and we could thus define a "surface purity" in a similarly proportional way.

In fluid mechanical terms, however, the concept of the surface cleanliness of a liquid is rather broader than this. What is found to be important from the fluid dynamics point of view is not the exact composition of the surface. The liquid will still behave as a continuum even when its surface layers contain more than one species. What is important is the way in which the forces at the surface act to modify the flow behavior of the liquid. It is a matter of common experience that while very thin layers of certain organic materials (detergents, oils, etc.) can completely still the ripples on an exposed pond,[6] the presence of much larger fractions of some inorganic materials (i.e., inorganic salts, common salt itself in particular) have little or no observable effect.

4. The History of Clean Surfaces

As was indicated above, surface-tension-controlled motions of liquids are represented among the very oldest of scientific phenomena. Perhaps the oldest of all is the phenomenon of wave-calming using oils, which was certainly known to the Ancient Greeks and Romans and which was probably known to the earliest Mediterranean civilizations.[5] Other effects with a lengthy history include the formation of "teardrops" inside the rim of a glass containing strong alcoholic liquors and surface-tension-driven motions such as the camphor boat, in which a small floating body can be propelled across a water surface by the action of spreading of a small piece of surface-active material (such as solid camphor) attached to one side of it.[13, 14]

Such topics were discussed at considerable length by the forerunners of today's "modern" scientists in the 17th and 18th centuries, and although simple observation and speculation was then usually preferred to direct experiment, it is to these beginnings that we owe the early foundation of modern theories of surface tension and surface fluid mechanics by Carlo Marangoni, Agnes Pockels, Lord Rayleigh, and Horace Lamb (see Reference 15).

One of the problems facing early workers was that they were then largely unaware of the possibility of contamination and its effects. This blissful ignorance is still quite common today, although with rather less excuse. Because we are dealing with surface layer thicknesses measured in molecular dimensions rather than in wavelengths of light, a contaminating layer can be extremely important even when it is so thin as to be completely invisible to the eye. Its effects can, in fact, be greatest when there is insufficient material to form a coherent layer even one molecule thick.

Perhaps the best illustration of this comes from the modern work on the age-old wave-damping problem. Lucassen-Reynders and Lucassen[16] review modern hydrodynamic theory of wave damping, and show that the greatest damping effect generally arises long before enough material is present to form a condensed monolayer. The work of Scott and Stephens[11] supports this conclusion.

It is quite understandable that early workers felt that water surfaces should be naturally in a clean state if they had added nothing to them. Their confidence would be maintained as long as their results were more or less reproducible between individual experiments.

The more careful workers would use distilled water in their experiments, expecting that the distillation process would remove any significant contamination, but as we will describe below, the process of distillation is not always effective for removing the important contaminants. This difficulty would probably have been insignificant had it not been for the fact that there was at that time no reliable means of measuring the surface tension.

Indeed, the science of measuring surface tension (or capillarity, as it was then known) was only then being developed, and the measurement was inextricably bound up with thinking about the concept itself. The lack of appreciation of the effects of contamination, and the consequent variability in contamination of the water being used to test the measurement techniques, gave rise to tremendous confusion about the different results being obtained by different workers using different procedures.

It required the steady persistence of careful workers such as Lord Rayleigh, steadily refining his experimental techniques and always keeping an open mind, to realize that, for water at least, surface contamination was the rule rather than the exception. Rayleigh's conclusions were aided by the vital work of the inventive "kitchen experimentalist" Agnes Pockels, with her open dishes of water and movable surface barriers. Indeed it was Pockels who originated what is still one of the most important experimental techniques in this field.[17]

5. The Nature of Surface Contamination

Water is unique among nonmetallic liquids in having a relatively high surface tension. Its room temperature value of around 72 mN/m is greater by 30 to 50% than the values measured with liquids of organic origin, and the reason for this exceptional behavior lies in the structure produced by the interactions of the component molecules.

The molecular structure of water is such that molecules of other materials can move freely through it, occupying vacancies between molecular groups. If the nature of the foreign molecules is such that they have no specific interactions with water molecules (as would be the case with a nonpolar material such as, for example, methane), then their solubility in water will be slight, and their distribution throughout the bulk liquid would be quite uniform.

If, however, the foreign molecules are able to interact with the water structure in some way, then the situation changes. Ionic inorganic materials, such as sodium chloride, for example, will form transient ionic structures with the H^+ and OH^- ions of the water, and they will feel energetically compatible with the water structure.

If the foreign molecules are neither wholly compatible nor wholly incompatible with the water structure [as would be the case with molecules made up of nonpolar (hydrophobic) sections and also ionic or polar (hydrophilic) sections], then there will probably be some form of mixed behavior, and the competing hydrophilic and hydrophobic sections of the foreign molecules will arrive at some balancing compromise at the water surface.

The stronger the hydrophilic attraction between molecules relative to their hydrophobic repulsion, then the greater will be the bulk solubility of the material, but the important factor from the surface contamination point of view is that the surface region will present an alternative state for the foreign molecules—the state of adsorption. On this simple picture, at least, it is reasonable to suppose that mixed hydrophobic/hydrophilic molecules might, by becoming oriented at the free surface, be able to find lower energy levels than are available in the bulk liquid.

The diagrammatic representation shown in Figure 2 is the conventional way of indicating the molecular alignment at the surface, and indeed it largely fits the way the monolayer is observed to behave. However, it should be remembered that the thermal motion of molecules at the liquid surface is usually at least the same order of magnitude as the molecular sizes.

Even on this simple picture of the origin of surface activity it is possible to deduce the factors which make the effect stronger or weaker. If the hydrophilic part of the contaminating molecule is very dominant, as it would be, say, with ethanol, C_2H_5OH, then the solubility would be expected to be high, and the surface activity of the added molecules would be quite moderate. With ethanol itself this is indeed the case, the two liquids being miscible at all concentrations, and the surface tension reduction of the water by the alcohol is not particularly marked.

If we stay with the example of ethanol-like foreign molecules, and examine the effect of increasing the hydrophobic nature of these molecules, then we will

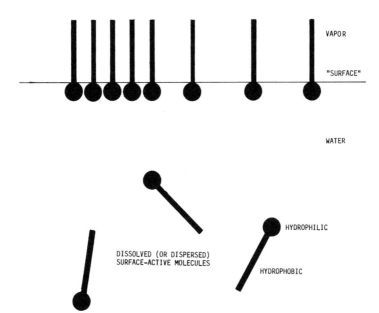

Figure 2. Schematic representation of polar molecules aligned at a water surface. It should be remembered that the thermal motion of the molecules is usually large compared with the molecular dimensions.

see that as we increase the length of the linear hydrophobic paraffin chain C_nH_{2n+1} through the series propanol, butanol, pentanol, hexanol, etc., the water solubility decreases and the reducing effect on the surface tension increases steadily. By the time we get to hexadecanol ($n = 16$, cetyl alcohol), the material is essentially insoluble, and as molecules are added to the surface they form a stable monomolecular layer.[10]

On the basis of this simple outline of the nature of surface activity it is possible to grasp the water contamination problem. Water as supplied to us through the public supply is pure enough for drinking purposes, but before it reaches our taps it has been exposed to biological action in the ground, in streams, and in a reservoir, and even before that it fell as rain through an atmosphere quite rich in organic materials.

We saw above that the sort of materials that are likely to cause surface contamination problems are those that are the least soluble in water, but also saw that the water might nevertheless be saturated with whatever barely perceptible concentration represents complete saturation. The high surface activity of the molecules will result in the stable accumulation of these molecules at the surface as they diffuse there randomly out of the bulk.

The exact nature of tap water contamination is rarely known in practice.

The bulk concentration of contaminant—even in highly industrialized areas—is often so small that analytical methods cannot detect it, much less characterize it. In most cases, however, it is probably reasonable to assume that the contamination consists of several chemical species, and that the mixture varies, perhaps greatly, from day to day or from season to season. Some of the contamination present could well be the result of chemical treatment of the water for drinking purposes.

Historically, the techniques that have been developed for the removal of surface-active contamination were evolved using empirical tests of surface cleanliness such as the presence or absence of particular fluid mechanical effects. One example is the "camphor test" used by Rayleigh,[18] which evaluates a surface by the ease with which a small piece of camphor propels itself over the surface.

Modern techniques of evaluating surfaces, reviewed elsewhere in this treatise,[19] are still largely empirical, although they are more sensitive and less liable to contaminate the surface they are assessing. The consequences of the circular nature of the reasoning being used here should be appreciated fully. The possibility is a real one that we may, in future, discover free-surface phenomena that are sensitive to the contamination levels present in even the cleanest water samples achievable today. However, we are in general confident that the cleaning techniques described in this paper are adequate for the applications currently anticipated.

6. Cleaning Techniques

The techniques used for the removal of surface-active organic materials from water are several. Some workers use combinations of more than one of these techniques, in successive processing steps, and there is considerable variation in current practice. Most workers, however, use an initial stage comprising simple distillation, and this technique will be described first.

The other methods that will be described here may be used in any order or even omitted completely, depending on what any particular researcher finds necessary to obtain water of the necessary cleanliness for a particular application. It is possible that the treatment necessary varies from location to location, and even from season to season, particularly in urban areas, where heavy demands are placed on the supply.

The literature contains scant reference to cleaning techniques, and this review leans heavily on the author's experience over a number of years, together with information on usual practices of other workers in the surface chemistry field.

The suitability of any particular technique must be assessed by the experimenter, being guided as much as possible by the conditions of the proposed fluid mechanical experiments, and it is helpful to have a range of different techniques to evaluate. Evaluation should be made using techniques such as those described in Reference 19.

It should be remembered that direct chemical identification of a contaminant is rarely possible, or indeed worthwhile, since removal will usually be by a physical rather than a chemical process.

6.1. Primary Distillation

Distillation is a well-known and obvious method of removing nonvolatile foreign materials from water. Many of the important surface-active materials from the surface cleanliness point of view are, fortunately, found to be removed by this means. Any inorganic salts present in the water, being nonvolatile, are also eliminated in the process.

The water cleanliness resulting from a single distillation is (regrettably) not usually found to be adequate for fluid mechanics purposes. In the author's experience, some fluid mechanically relevant surface-active contamination will persist, and repetition of the distillation will give no improvement, the contaminating species passing over into the distillate. As will be indicated later, the chemical nature of this residual contamination has not (to the author's knowledge) been determined, and indeed it is likely to vary from time to time, and from place to place. A second type of distillation, or some other process, is usually found necessary to remove it.

The primary distillation step is not absolutely essential, if it is found that either a bubble cleaning (Section 6.3) or solid adsorption (Section 6.5) technique successfully removes whatever contamination is present. However, the author has found that distillation is almost invariably used in practice in clean-water preparation.

The first distillation is conveniently done using a continuous-flow arrangement, operating with mains water and powered either by electricity or by gas combustion. It is usually advisable that workers use an apparatus made completely from Pyrex glass or silica, although it should be noted that some workers seem to experience no difficulty if the electrical element in contact with the water is made of chrome steel. Materials such as plastics and rubbers, and organic or silicone sealants, which might add organic contamination, should obviously be avoided.

To some extent it is true to say for any material which acts as a source of contaminant that either it will eventually become exhausted or its surface will become steadily more blocked by the deposition of inorganic salts. Once again, the only rigid rule which can be imposed is that the resulting water must be clean enough for its intended purpose. It should always be remembered, however, that if the planned experiments should uncover unexpected behavior, critics will always point to what they feel are irregularities in the water cleaning process. This appears to be an unavoidable attribute of the "circular" process of developing tests and preparation techniques, mentioned above. In the absence of more thorough cleaning techniques, or absolute detection methods, we must be satisfied to work to the highest standards that are practicable.

There are many ways of arranging a suitable apparatus, but the example shown in Figure 3, used for many years by the author, is often found adequate.

One of the main functions of the first distillation step is to remove inorganic mineral materials. By themselves these materials are rarely a problem for clean-surface fluid mechanics (although they are of undoubted importance for studies of a more purely surface chemical nature), but they are likely to affect the second distillation, which is more specifically designed to remove the more persistent organics. It is usually found that much of the organic material is removed in the first distillation.

As was indicated above, some respected workers have found that once-distilled water is of adequate quality for their purposes. Lord Rayleigh found this to be so, although he emphasized that water for experiments should always be drawn fresh from a tap at the bottom of the storage bottle. This almost

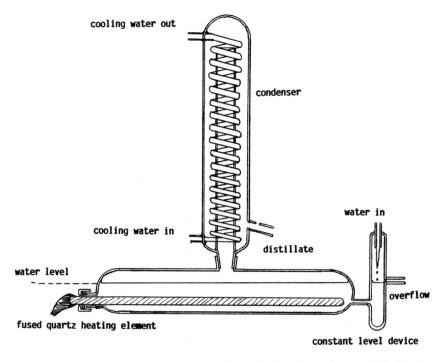

Figure 3. Layout of a typical continuous-flow still suitable for clean-surface work. The heating element (up to 3 kW) is totally enclosed in a silica sheath.

certainly indicates that the contamination continuously accumulating (by diffusion) at the surface was detectable in this case. Interestingly, Rayleigh does indicate[20] that once-distilled water is sometimes less clean than the source tap water.

The wide variations in water cleaning practice found between different workers may be the result of variations in the concentration of organic material in the source water.

No one design of still is unanimously preferred. The all-Pyrex glass construction is very difficult to achieve, unless the heating element is a sealed-in glass tube carrying hot steam. No such apparatus is known to the author in commercially available form. Indeed, such sophistication is probably best reserved for a second distillation, and a simpler construction involving some material compromises is often found adequate, in practice, for primary distillation. Acceptable compromises may be found with commercially available units.

For instance, electrical heating elements of chrome steel or of fused quartz are quite commonly available. In these cases some means has to be found for sealing the heating element into the body of the still in a way that avoids leaks. The most convenient way of doing this sealing uses small flexible washers of some plastic material, and while plastics are generally frowned upon as being likely sources of some of the worst contaminants, they have in practice been found to be adequate by some workers. One plastic material that generally seems to be beyond reproach is PTFE [poly(tetrafluoroethylene), "Teflon"], and this should be used if at all possible.

One possible justification for using "unacceptable" materials, and a possible reason for their apparent acceptability in some cases, is that the offending materials they contain—mold release agents, plasticizers, and other additives—may be gradually leached from them over a considerable period of time.

The construction of the rest of the apparatus—condenser and pipework leading the distilled water to storage—presents no difficulty since Pyrex glass can be used.

One refinement to the apparatus which is often omitted and reserved for a second distillation is the provision of an area of heated glass—a dry zone—between the body of the still and the condenser. This area will prevent the passage of adsorbed liquid-phase contaminant along the thin water film covering the glass, which could conceivably result in appreciable contamination of the final product.

6.2. Further Distillation

Current practice varies from researcher to researcher on the exact procedures to be used for the removal of the last traces of surface-active material, and to some extent this variation may well be associated with variations in the original contaminants encountered by the different researchers. Researchers do, of course, tend to persist in the practices they find adequate for their own particular applications, and much of the variation may therefore be simply due to different ways of approaching the problem.

One of the commonest second distillation techniques is to distill from a strongly oxidizing solution, namely an alkaline solution of potassium permanganate. The present author has always found this to give water of acceptable quality, using a 10-liter batch procedure. The concentration needed for best results should be determined experimentally in each case, but a good starting point is 2 g/liter of potassium permanganate, plus 4 g/liter of potassium hydroxide.

A major disadvantage of permanganate distillation is that it is essentially a batch, rather than a continuous, process. The first fraction of the distillate is often found to remain unacceptably contaminated, and perhaps as much as the first liter must be discarded from a 10-liter batch.

This operational difficulty represents a major limitation on the scale of clean-surface experiments, apparatus requiring more than about 10 liters of water being difficult to keep supplied. The point in a batch at which the contamination becomes acceptably small can be assessed using the bubble persistence test.[19] If about 100 ml of distillate is shaken vigorously in the receiving vessel and bubbles persist at the surface for as long as half a second, then the water should be discarded. As soon as the distillate passes the test, all of the subsequent distillate may be collected and stored for use.

6.3. Bubble Cleaning

In the discussion of surface cleanliness given above, it was made clear that the interference of surface-active materials with fluid mechanical experiments is a result of their adsorption at the water/air interface. Once adsorbed, these materials are able to create surface tension gradients which oppose surface motions.

There is a way of removing surface-active materials which is highly appropriate to the end use, in that it uses the surface activity itself to remove the material. If small bubbles are generated in the water by blowing a clean inert gas into it, then fresh water surface is created continuously. Surface-active material becomes adsorbed onto this fresh, clean free surface, and as the bubbles rise to the water surface, the material is carried on upwards.[21]

The process of material removal by this method is highly complicated. Materials that are soluble in the water form a local dynamic equilibrium between the molecules in the adsorbed and dissolved states. It cannot therefore be assumed that once a molecule has become adsorbed it will stay attached to the bubble all the way to the free surface. Indeed, the turbulent oscillations excited in the bubble as it interacts with other bubbles may well promote desorption.

However, there will always be a net transport of material upwards, and the technique is likely to be more effective, the greater the activity of the contaminants. Indeed, the technique has been used with considerable success.[21] A major difficulty in quantifying the degree of success is that it almost certainly varies from material to material, and certainly varies with the initial bulk concentration. Unpublished experiments by the author using a wide range of artificial contaminants indicated that the final result was apparently independent of the initial concentration, even when this was large.

Another indication of the effectiveness of the technique is how well it works on water that has already been cleaned with other techniques. W. B. Krantz (personal communication) has indicated that reduction of bulk concentration by a factor of three can be obtained with the cleanest available feedwater. Such estimates are necessarily highly specific to the exact nature of the contamination present, and (as has been indicated above) this is rarely well known. These estimates were obtained using a very sensitive measurement technique[22] which is described elsewhere in this treatise.[19]

The bubble cleaning technique is illustrated schematically in Figure 4. An inert gas, such as oxygen-free nitrogen, is filtered, to remove possible oily contamination from its containing cylinder and valve, and is then passed through a circular porous glass membrane sealed into the bottom end of a vertical Pyrex glass tube. Dimensions successfully used by the author have been a 2-m length of 40-mm diameter tube.[21] The tube is replenished at the top with feedwater, and an overflow is available for flushing out the more highly contaminated upper layers of liquid. Clean water is withdrawn slowly through a stopcock near the bottom.

With care it is possible to arrange the gas flow and water input rates so that a continuous flow of "scrubbed" water is produced. Construction of this apparatus is quite easily arranged using acceptable materials (see Section 7). Stopcocks constructed of only Pyrex and PTFE are readily available.

Apart from its use in cleaning pure water, this technique is also extremely valuable for cleaning aqueous solutions (see Section 6.7). It provides an often viable alternative to the uncertain procedure of making up the solution from clean water and clean solute.

In solutions, the technique may in fact be much more effective than in nearly pure water, as an additional factor often exists which gives a significant increase in the efficacy of the method. For example, with sodium chloride solutions, as the solute concentration increases above (say) 2 g/liter, the size of the bubbles produced by the porous glass membrane begins to decrease significantly.[21] Because the bubble surface area associated with a given volume of

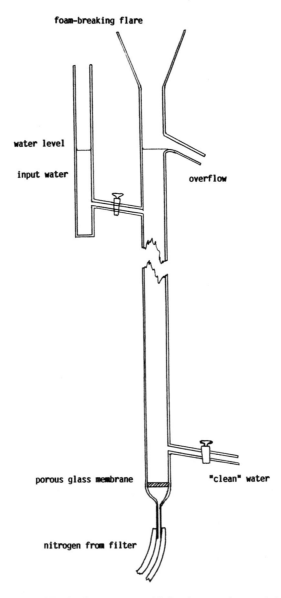

Figure 4. Layout of a bubble cleaning apparatus. Oil-free inert gas is passed through a porous glass membrane into the water or aqueous solution, creating fresh surface area. Contaminating molecules become adsorbed onto this fresh surface and rise to overflow near the top of the column. Continuous or semi-continuous operation of such apparatus is possible.

gas increases as the bubble radius decreases, this effect of the solute can lead to a remarkable increase in the free surface area available for adsorption. Also, the smaller bubbles, being slower moving, are given more opportunity for material adsorption, and it is likely that they experience less desorption-promoting disturbances.

This technique has been used with success by the author for solutions of sodium chloride and of sucrose, for use in measurements of water-wave damping. These materials are good examples of solutes that induce the rapid decrease in bubble size just described, and it should be noted that the effect occurs in these solutions over a particular concentration range. Use of the method with other solutes would require experimentation to determine the minimum concentration for the effect. If the solution concentration required for eventual use is smaller than this optimum value, then clean pure water may be used to dilute the cleaned solution.

This technique may also be useful in experiments involving interior surfaces, such as those of fixed bubbles in coalescence experiments or bubble oscillation observations, in which a volume of water is completely enclosed and inaccessible. Periodic flushing of the water with a cloud of fine bubbles, upwards to some point where water may be withdrawn, may be a useful means of reinforcing or refreshing the cleanliness of the water sample close to the observation point.

6.4. Surface Skimming and Talc Cleaning

In experiments which involve a large open surface it can sometimes be arranged that the contaminated surface layer is physically removed from the bulk. If the water is contained in an open tray it is often possible to skim the accumulated contaminated layer, using a clean, close-fitting solid barrier, to one end of the tray. There it may either be swept quickly over the end or sucked away using a glass or stainless steel capillary attached to a jet pump.

This technique is not ideal, since leakage around the barrier and contamination of the walls are very difficult to prevent. However, the author has used the technique to remove gross contamination in cases where the cleanliness requirements were not stringent. The technique works best when the barrier fits snugly on the sides of the tray and the contacting surfaces are hydrophobic. The surfaces may be made of a hydrophobic material, such as PTFE, or be coated with clean paraffin wax.

A variation of the skimming technique which is possible with relatively narrow surfaces is to blow the contaminant material to one end of the surface with an air jet. If the surface is broad, however, it is not possible to avoid unevenness of the air stream over the surface, and consequent escape of the material around the edges of the jet impingement area.

A related technique involves covering the surface completely with a thin layer of finely powdered talc and removing the talc by suction through a capillary.[23,24] This may cause an unacceptable loss of water from the experiment, but in conjunction with some means of concentrating the film, such as one of those just mentioned, this problem is very much reduced. The talc provides a valuable indicator of the extent of the uncleaned surface.

In some experiments it may be possible to incorporate an overflow for the contaminated surface, but care must be taken to avoid contamination of the solid surfaces all the way along the exit path of the overflow water, because surface tension gradients will tend to spread surface-active material back along the water surface and thus into the experiment.

One technique which has been found useful by the author in experiments in which absolute contaminant levels are rather less critical[25] is to absorb the whole of the liquid surface layer on a thin tissue paper, quickly laying the tissue over the surface and removing it before it has a chance to become saturated. This method may be found useful for removing heavy contaminating layers.

6.5. Solid Adsorption Techniques

Ion-exchange resins, of the type used for deionizing water, are found to be completely unacceptable for clean-surface work, the water being observed to be rather more contaminated after treatment than before (J. F. Padday, private communication). This would appear to follow from the chemical nature of the resins employed in the exchange column, and there seems to be no way of avoiding the problem completely.

However, some workers have had success using adsorption of the offending material on a column of activated charcoal, located between the final distillation vessel and the storage vessel. Back-flow must be prevented with a suitable valve. This is a very convenient water-cleaning step, and it seems to be effective, but it must be remembered that the carbon will eventually become saturated with organic material, and require renewal. Thus, the water being produced must be continually tested.

6.6. Laser Burning

The possibility of using high-power laser beams, focused on the liquid surface, to remove surface contamination was considered by Askar'yan *et al.*[26] These authors observed that a focused high-intensity beam will selectively heat the surface and evaporate material that is preferentially located in the surface layers. Experiments were reported using a 50-W continuous-wave carbon dioxide laser and a pulsed carbon dioxide laser (1 MW, 1 J, 1–2 Hz pulse rate), both operating with 10-μm wavelength radiation. Ejection of contaminating materials to a height of 200–300 mm above the surface was described, and ignition of materials such as kerosene was indicated.

It is considered unlikely that such techniques will ever prove satisfactory for either fluid mechanical work or surface chemical work, but the possibility should be borne in mind in case a suitable situation should ever occur.

6.7. Solution Preparation

In many experimental situations which require the use of water as the observed liquid it can be useful to be able to vary some of the physical properties of the water, without, that is, changing the essentially water-like surface chemical characteristics of the liquid examined. Adding solutes that are either not surface-active or only very weakly so is often a convenient way of doing this. Common inorganic salts are only weakly surface-active, and in a negative sense—increasing the surface tension rather than reducing it—and these solutes provide a useful means of increasing the specific gravity, viscosity, and electrical conductivity of the water, while retaining an essentially aqueous surface chemical behavior. Sucrose is particularly useful for effecting large increases in viscosity.

This review is concerned only with essentially aqueous liquids, containing minimal quantities of surface-active materials, and the surface cleaning of solutions of the surface-active materials themselves will not be considered. Mysels and Florence[3] deal with the purification and assessment of these materials before they are added to clean water, and it is apparent that the techniques required are largely concerned with discriminating between different surface-active materials. Small fractions of a more highly surface-active material can lead to the surface properties being dominated by the impurity. Such discrimination needs considerable knowledge of the specific materials involved and crosses the boundary between the fluid mechanics of surfaces and the surface chemistry of surfaces.

The very existence of even the weakest surface activity—positive or negative—in a solute indicates the need for the utmost caution in the interpretation of experimental results. It is well known[27] that inorganic solutes significantly modify the surface activity of organic materials on water. The solute changes the surface-tension-reducing effect of any given quantity of added material, and hence also the elasticity of the surface, and this effect should be considered when the surface is used for fluid mechanical work.

In experimental situations it is important to measure directly such parameters as the surface tension and the surface dilatational elasticity, and not to assume that the values for an aqueous solution are the same as those for water. The differences are essentially quantitative rather than qualitative, however, even though the surface thermodynamics may be completely changed by the added solute. It is unlikely, therefore, that the associated changes in the surface mechanics are significant compared with the planned effects on the bulk liquid properties.

An example which amply illustrates such effects is the one mentioned above (Section 6.3) of the effect of quite small bulk concentrations of sodium chloride on the coalescence of bubble clouds in water. This effect, although highly relevant to the oceanographical behavior of rising bubble clouds,[28] is little understood, but appears to concern the surface forces that occur when the surface is affected by large shear gradients. The effect is such as to change the whole character of this fluid mechanical situation of the rupture of the thin water film between two approaching gas bubbles in water.[29]

The preparation of clean aqueous solutions is complicated by the need to add the solute after any distillation steps employed to remove surface-active materials from the water. Even the purest available grades of inorganic salts may contain enough surface-active materials—perhaps even as by-products of the manufacturing process—to cause contamination. There are two basic remedies available. The first is to heat the solute to red heat for sufficient time for organic materials to be burnt out, before adding to the freshly cleaned water. The second possibility is to use the bubble cleaning technique (Section 6.3) on the prepared solution. As was indicated above, the solute sometimes actually cooperates in the removal of the surface-active material, by retarding bubble coalescence and thereby making the scavenging process more efficient.[19]

6.8. Surface Cleaning in Engineering Applications

The techniques described above are essentially techniques for cleaning water for use in fluid mechanics research. Although much of this clean-liquid-surface research is ultimately aimed at specific chemical engineering processes (or oceanography, which is a special case of interest to the author), the systems studied and physical scales studied are abstractions from these applications.

Only by researching these abstractions can we hope to understand the more complex situations. Fortunately, however, in some of the more complex situations the problem of surface cleaning is often less problematical than it is in the laboratory. For instance, processes involving material interchange at a gas/liquid interface almost always use turbulent-flow regimes, and surface renewal can be promoted so that surface-active material cannot build up on a surface, in the way that is almost inevitable in the laboratory.

A surface may be renewed in practice simply by inserting a form of sluice, trapping the surface and allowing passage only of the lower layers. The surface layer is thus either removed or (more probably) dispersed within the bulk liquid. Even if the material is still present in the water, in the time taken for it to become adsorbed at the surface once more, the vital part of the interchange may be complete, or well advanced. Laboratory experiments involving surfaces known to be initially clean are necessary here to show the effects of controlled amounts of contamination at the surface, so that their influence in the practical case may be better recognized.

Turbulent flow of either gas or liquid phase will similarly promote surface renewal, the turbulent eddies clearing "bare patches" in whatever surface film is present. For instance, work done by the author in connection with the effect of surface-active films on wind-wave generation[30] on the ocean showed that a film of even an insoluble material could be dispersed within the bulk liquid quickly under the action of a wind of speed 6 m/s. This is therefore an effective means of preventing buildup of contaminant.

7. Materials for Clean-Surface Experiments

7.1. Principles

The requirements for clean-water-surface experiments are that the water, having been prepared to as high a standard of cleanliness as is appropriate, must be protected from the contamination normally present in the atmosphere, and the materials used to contain the water in the apparatus must add no contamination to it. Additionally, the materials must be quite easy to clean using reliable techniques.

As far as atmospheric contamination is concerned, it is a good general rule that materials that give any detectable smell should be avoided. Tobacco smoking should be avoided completely, and organic solvents, glues, etc., should not be used in the vicinity of exposed clean surfaces. Perfumes and strongly perfumed plants should also be avoided.

7.2. Construction of Apparatus

Cleanliness requirements for constructional materials are found, in practice, to be a severe restriction on experiment design. Pyrex glass is acceptable without question, for it may be cleaned readily using aggressive agents such as chromic acid. Stainless steel is found similarly acceptable in many cases. However, the other metals commonly used for construction purposes are usually suspect from an oxidation point of view and due to the possibility that they may be releasing metallic ions into the water.

Only one polymeric material also appears to be completely above suspicion—poly(tetrafluoroethylene), PTFE—as this is highly resistive to chemical attack and is easily cleaned.

With the exception of PTFE, however, plastics in general are highly suspect, and without doubt they are best avoided. However, in practice, bearing in mind the necessary trade-offs between constructional requirements and cleanliness requirements, Perspex [poly(methyl methacrylate)] is found acceptable by some workers in some applications.[31]

Perspex has excellent constructional properties, being strong, transparent, easily machined, reasonably dimensionally stable, and with an optically smooth surface that is easily cleaned using mild detergents (of the washing-up liquid type) and soft cloths. Jointing of the material may be carried out using solutions of the polymer in a volatile solvent. The solvent can be relied on to evaporate from the joint in a matter of days, leaving an acceptably smooth and cleanable finish. Commercially available cements for this material are probably less acceptable, but even so they have been used successfully by the author if a period of two or three days is allowed for volatile active components to disappear.

When a Perspex surface is clean and free of scratches, clean water will run off easily, leaving no streaks behind. Contaminated patches, however small, are betrayed by the reluctance of water to leave them. This is a particularly helpful property of the material, and it appears (from experience) to be a sensitive test. The use of Perspex gives surfaces that are acceptably clean and free from contaminating influences, at least over an experimental period of a day or so.

The principal objections to the use of Perspex center on the possibility that some of the smaller-molecular-weight fraction present in the material, or perhaps plasticizers and mold release agents used in its preparation, might eventually be dissolved into the contained water. The author has found, however, that, while this possibility may be quite irrefutable, in practice there are often no observable effects directly attributable to it, even after lengthy experiments in which water has been in contact with the material for perhaps days. Thus, while experiments made in Perspex apparatus are always open to criticism on these grounds, it is usually acceptable to point out that the experiments were reproducible, and that no test for contamination had disturbing results.

7.3. Water Storage Materials

The only readily available storage medium that is completely acceptable is the Pyrex bottle, fitted with a ground glass stopper. It is quite easily cleaned using the most severe of cleaning materials, chromic acid, it is strong, and, being transparent, it allows a simple, rapid assessment of the cleanliness of the contained water using the shaking technique.[19]

Molded polyethylene bottles seem, by contrast, to be incapable of being adequately cleaned. Chromic acid cannot be used on them, and detergents always seem to leave some residual contamination, perhaps a result of the mold release agents used in their formation.

For glass bottles to be used for storage, a wide neck is preferred, allowing easier pouring, less splashing, and easier stopper removal.

After cleaning with hot chromic acid, Pyrex bottles should be rinsed out several times with clean water. This process gives a wet-draining hydrophilic glass surface which is undoubtedly contaminated with chromate ions, but this is not important for fluid mechanical applications. The ground glass stopper must also be cleaned thoroughly.

Pyrex bottles are commercially available up to 20-liter capacity, although 2-liter and 5-liter sizes are much easier to handle. The larger sizes of bottles are available with stopcocks, and these bottles are very convenient for the collection of water from the final distillation or water cleaning process.

8. Cleaning of Apparatus

8.1. "Hard" Materials—Chromic Acid

One of the traditional cleaning agents in chemical work of many kinds is an infusion of chromium dioxide in concentrated sulfuric acid, commonly known as "chromic acid." This is a highly effective liquid, which acts as a powerful oxidizing agent, and it must be used with great care. In its concentrated form it is probably more dangerous than pure sulfuric acid, with the same property of causing nasty skin burns that is characteristic of that acid. Dilution should be by pouring acid carefully into water, and not vice versa, because the combination of the two liquids can give a dangerous spray of boiling acid.

Protective clothing must be worn when handling chromic acid. Gloves made of polyethylene seem to be the only practical and acceptable covering for the hands, as they resist chemical attack, but these are easily torn, and extreme care must be taken. A nearby bucket of strong sodium bicarbonate solution is advisable. Rubber gloves are quite rapidly attacked by chromic acid, and while they are more robust than polyethylene gloves, they are not acceptable from a surface cleanliness point of view when handling surface-clean apparatus.

Chromic acid is the best way of cleaning glassware and stainless steel. These materials can be left in contact with the acid for long periods. PTFE is resistant to chromic acid, but other plastics, such as Perspex and polyacrylates, should not be brought into contact with it.

Chromic acid can be prepared by adding powdered potassium dichromate to a bottle of concentrated sulfuric acid and shaking the bottle. The conversion to chromic acid takes a few days. The powdered chromate tends to settle to the bottom of the bottle, and will become solidly caked there if left, so the bottle should be shaken every few hours. The acid is ready for use when the milkiness caused by the suspension of small particles has disappeared, leaving a dark brown, clear liquid.

8.2. Perspex—Detergents

Because of their organic nature it is, in general, not advised to use detergents for cleaning clean-surface apparatus, except perhaps as an aid to removing gross soiling before using more refined techniques. However, the author has found that certain detergents, of the washing-up liquid type, may be used to clean smooth Perspex surfaces. If the detergent is rubbed gently into the surface with a soft cloth, it may be removed with a hot water rinse, to give a self-drying surface which is free of both water and detergent. If the hot water fails to recede from the whole surface, leaving streaks or droplets, this is an indication that in these places there is still some contamination, and the cleaning must be repeated. This appears to be quite a sensitive test of the cleaning process, and is, perhaps, in itself a justification for the use of detergents in this case, in that Perspex surfaces cleaned in this way have been used in contact with clean water for many hours without the water showing any detectable contamination. It should be pointed out, however, that the more purist of surface chemical researchers frown on this practice. There appears to be no better way of cleaning Perspex, however.

9. General Design Considerations

Apparatus should be designed bearing in mind the need to clean it frequently, and the most effective way of doing this is to have the parts that come into contact with the clean water easily removable. These parts may then be carried to a cleaning station, and rinsing with clean water may then be accomplished satisfactorily before reassembly. Apparatus should be designed to drain to complete dryness quickly, wherever possible.

Experiments should be designed bearing in mind whatever water volume can be easily prepared, consistent with the needs of the experiment. In practice, this consideration tends to restrict the size of experiments to less than about 10 liters of water.

The avoidance of incidental contamination is also an important factor in design. Measurements should be made using devices which cannot easily come into contact with the surface under observation, as this could lead to complete failure of the experiment. Open surfaces should be covered by clean, dust-free lids where possible. All aromatic substances should be excluded from the laboratory during the cleaning, setting up, and experimental phases, and this includes solvents, perfumes, adhesives, and tobacco smoke.

10. Summary

This chapter has reviewed the principles underlying clean-liquid-surface work, and has outlined the techniques that are available for the successful performance of experiments using clean liquid surfaces, including in some detail the rationale behind the requirements and the means of achieving them.

It is readily seen that the technology has changed little over the years, and differs from that of 100 years ago mainly in the greater understanding we can apply to our requirements. Changes in materials have had little impact, with the (limited) exceptions of Perspex and PTFE. We have learned to take more care with clean-surface preparation and experiment design, and for those who both appreciate the requirements and accept the challenge presented by the experiments, the rewards are considerable. It should be clear, however, that this is by no means an easy experimental field.

The majority of clean-surface work experienced by the author has involved a relatively simple subset of the techniques described here. Typically, the preparation technique has used a simple first distillation, followed by either a second distillation from permanganate or passage through an activated carbon column.

However, as has been made clear above, a technique or set of techniques can only be judged by its results, and since these depend on a wide variety of factors, not least of which is the particular experiment being done, a wide variety of solutions to the problem are needed. Half of the problem is understanding what the problem actually consists of, and if this review has succeeded in illustrating this, then it may prevent the reader from making at least some of the mistakes made by the author.

References

1. K. L. Mittal, Surface contamination: An overview, in: *Surface Contamination: Genesis, Detection and Control* (K. L. Mittal, ed.), Vol. 1, pp. 3-45, Plenum Press, New York (1979).
2. J. T. Davies and J. P. Driscoll, Eddies at free surfaces, simulated by pulses of water, *Ind. Eng. Chem., Fundam. 13*, 105-109 (1974).
3. K. J. Mysels and A. T. Florence, Techniques and criteria in the purification of aqueous surfaces, in: *Clean Surfaces* (G. Goldfinger, ed.), pp. 227-268, Marcel Dekker, New York (1970).
4. N. K. Adam, *The Physics and Chemistry of Surfaces*, 3rd Ed., Dover Publications, Inc., New York (1968).
5. J. C. Scott, The historical development of theories of wave-calming using oil, *Hist. Technol. 3*, 163-186 (1968).
6. B. Franklin, Of the stilling of waves by means of oil, *Phil. Trans. 64*, 445-460 (1774).
7. L. D. Landau and E. M. Lifshitz, The effect of adsorbed films on the motion of a liquid, in: *Fluid Mechanics*, pp. 241-244, Pergamon Press, (1959).
8. E. Palm, Nonlinear thermal convection, *Ann. Rev. Fluid Mech. 7*, 39-61 (1975).
9. C. A. Miller, Wave motion of low-tension interfaces with electrical double layers, *J. Fluid Mech. 55*, 641-657 (1972).
10. G. L. Gaines, *Insoluble Monolayers at Liquid-Gas Interfaces*, Wiley-Interscience, New York (1966).
11. J. C. Scott and R. W. B. Stephens, Use of moire fringes in investigating surface wave propagation in monolayers of soluble polymers, *J. Acoust. Soc. Amer. 52*, 871-878 (1972).
12. J. C. Scott, The effect of organic films on water surface motions, in: *Oceanic Whitecaps and Their Role in Air-Sea Exchange Processes* (E. C. Monahan and G. Mac Niocaill, eds.), 159-165, D. Reidel Publishing Co. (1986).
13. G. van der Mensbrugghe, Sur la tension superficielle des liquides considerée au point de vue

de certains mouvements observés à leur surface, *Mém. couronnes et Mém. des Savants étrangers, de l'Acad. Roy. des Sci., Lett., et des Beaux-Arts de Belgique* 34, 3-67 (1869).
14. L. E. Scriven and C. V. Sternling, The Marangoni effects, *Nature* 187, 186-188 (1960).
15. C. H. Giles, Franklin's teaspoon of oil. Studies in the early history of surface chemistry, *Chem. Ind. 1969*, 1616-1624 (1969). [See also C. H. Giles, Historical aspects of surfactant adsorption at liquid surfaces, in: *Solution Behavior of Surfactants: Theoretical and Applied Aspects* (K. L. Mittal and E. J. Fendler, eds.), Vol. 1, pp. 113-122, Plenum Press, New York (1982).]
16. E. H. Lucassen-Reynders and J. Lucassen, Properties of capillary waves, *Adv. Colloid Interface Sci.* 2, 347-395 (1969).
17. A. Pockels, Surface tension, *Nature* 43, 437-439 (1981).
18. Lord Rayleigh, Measurements of the amount of oil necessary in order to check the motions of camphor upon water, *Proc. Roy. Soc.*, 47, 364-367 (1890).
19. J. C. Scott, Techniques for characterizing surface cleanliness of liquids, in: *Surface Contamination—Genesis, Detection, and Control,* (K. L. Mittal, ed.), Vol. 1, pp. 477-497, Plenum Press, New York (1979).
20. Lord Rayleigh, On the tension of water surfaces, clean and contaminated, investigated by the method of ripples, *Phil. Mag.* 300, 386-400 (1890).
21. J. C. Scott, The preparation of water for surface-clean fluid mechanics, *J. Fluid Mech.* 69, 339-351 (1975).
22. Anon., Device to measure surfactant concentration, *Anal. Chem.* 46, 799A-800A (1974).
23. J. T. Davies and R. W. Vose, On the damping of capillary waves by surface films, *Proc. Roy. Soc. A286*, 218-234 (1965).
24. Lord Rayleigh, Foam, *Proc. Roy. Inst.* 103, 85-97 (1890).
25. J. C. Scott, Flow beneath a stagnant film on water: The Reynolds ridge, *J. Fluid Mech.* 116, 283-296 (1982).
26. G. A. Askar'yan, E. K. Karlova, R. P. Petrov, and V. Studenov, *J.E.T.P. Lett.* 18, 389-390 (1973).
27. I. Langmuir and V. J. Schaefer, The effect of dissolved salts on insoluble monolayers, *J. Am. Chem. Soc.* 59, 2400 (1937).
28. J. C. Scott, The role of salt in whitecap persistence, *Deep-Sea Res.* 22, 653-657 (1975).
29. J. D. Robinson and S. Hartland, The effect of surface active agents on coalescence, *Tenside* 9, 301-308 (1972).
30. J. C. Scott, The influence of surface-active contamination on the initiation of wind waves, *J. Fluid Mech.* 56, 591-606 (1972).
31. J. C. Scott, The preparation of clean water surfaces for fluid mechanics, in: *Surface Contamination: Genesis, Detection, and Control* (K. L. Mittal, ed.), Vol. 1, pp. 477-497, Plenum Press, New York (1979).

3

Hydroson Cleaning of Surfaces

ROBERT WALKER

1. Introduction

Hydroson generators produce sonic waves in solutions, and the system can be fitted to a tank to clean immersed, contaminated articles. It was patented[1] in 1973 and then developed and transferred to the new company Nickerson Ultrason Ltd. in 1976. It is now widely used with over 150 installations employed in Europe, Japan, South Africa, and the USA.

2. The Hydroson System

The word Hydroson is derived from "hydro-," which is a form of the Greek *hudor* meaning water, and "sonic," from the Latin *sonus* meaning sound, which refers to sound or sound waves. Hence the term refers to the production of sonic frequencies in aqueous solutions and the hydromechanical process is used to augment the cleaning reactions.

The principle of the system is that solution is circulated by a multistage vertical pump at pressures up to 1.38 MN/m^2 through pipework outside the tank and into a manifold system fitted with generators (or nozzles). These immersed generators, which have no resonators or other moving parts, rely on the internal contours to convert the velocity of the solution into acoustic energy. The solution is emitted below the liquid level in the tank so that areas of high and low pressure are produced. An analysis of the frequency range formed by a generator indicates a range from 2 Hz to 20 kHz at a reasonably constant pressure level and ranging beyond 50 kHz at reducing pressures, and this range

---

ROBERT WALKER • Department of Materials Science and Engineering, University of Surrey, Guildford, Surrey GU2 5XH, United Kingdom.

of frequencies, or "pseudo-sound," assists accessibility to the sheltered areas, thus reducing the blind areas.[2]

It has been suggested[2] that the cleaning of components immersed in the tank is due to the formation of bubbles during the low-pressure cycle and these produce cavitation. The collapse of the bubbles creates a multidirectional scouring action on immersed surfaces and accelerates cleaning. The possibility of cavitation in the tank is discussed later.

The continuous and rapid circulation of the liquid constantly removes contaminant from dirty surfaces, and this is extracted by filters or strainers in the pumping system which is outside the main tank. The filtration is important and may be a relatively simple dual changeover system or more sophisticated if required. Continuous replenishment occurs because, after filtration, the fresh solution is recirculated through the generators and enters the tank as a high-velocity jet which produces considerable turbulence at the immersed surfaces so that the cleaning process continues. The sound intensity, generated by the individual nozzles, is proportional to the supply pressure to the third power so it is beneficial to use the minimum number of nozzles at the highest practical pressure necessary to give a uniform coverage and cleaning of an article. The number and arrangement of the jets depend upon the size and shape of the tank and the articles being cleaned. One commercial tank which was used to investigate the mechanisms involved, described later, is shown diagrammatically in Figure 1. The strong agitation within the tank also assists emulsification, which is important in cleaning. This process involves the dispersion of one liquid, normally in cleaning a contaminant oil or grease, as fine droplets in another immiscible liquid so that the formation of an oil-in-water emulsion occurs. The high shear forces produced in the liquid by the generators give a mechanical form of emulsification while the chemical reaction is facilitated by the use of surfactants in the solution. Emulsification is particularly intense within the Hydroson generators and tends to minimize subsequent separation of oil on the surface of the cleaning solution in the tank, and it also reduces "drag-through" problems. Another similar mechanism included in cleaning is peptization or deflocculation which involves the maintenance of finely divided solid particles in solution. The action of the Hydroson system helps this process and also tends to prevent any agglomeration of particles which could be redeposited onto the cleaned surfaces. Hence the life of the cleaning fluid can be considerably increased. The use of the Hydroson system to clean components from a variety of companies has been described in an earlier article.[3]

3. Experimental Investigation of Mechanisms in the Tank

The original hypothesis of the manufacturers and the Design Council in Wolverhampton, England was that cavitation occurred within the Hydroson

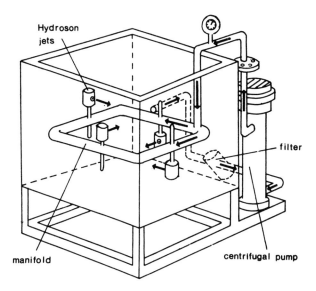

Figure 1. An 80-liter Hydroson tank.

tank. Hence this was investigated by using one of the simplest tests to detect cavitation which is to immerse aluminum foil in water and await perforation. A general rule[4] is that an average of ten holes is produced in an area of 6.45 cm^2 (1 in^2) in 10 seconds in a foil of thickness 2.5×10^{-5} m. With ultrasonic agitation applied to water, perforation of aluminum foil of thickness 2.0×10^{-5} m occurred within 30 seconds. Perforation of the foil did not occur, even after an hour, when the foil was immersed in the Hydroson tank. This test was repeated with water at 50°C and 80°C, because perforation should occur more easily at these temperatures, but nothing was observed.

In an article on wire cleaning, Clough[5] observed that erosion did not occur at Hydroson generators as a direct result of cavitation. This cavitation damage can be a common problem with ultrasound as well as on surfaces of rapidly moving parts such as propellers and pumps and those associated with fast-moving liquids.[6] Hence the fact that cavitation and the consequent damage does not appear to occur in the Hydroson system is particularly beneficial.

Another effect of cavitation which has been observed with ultrasonic agitation is the production of shock waves and surface hardening of immersed metals. In this work annealed copper sheet was found to increase in hardness from 59 ± 3 Hv to 87 ± 5 Hv after being immersed in water in an ultrasonic tank for three hours and similar changes have been recorded elsewhere.[7,8] No hardening, however, was observed over this period of time with the Hydroson tank.

From this observation it was concluded that the very efficient agitation in the Hydroson tank did not give effects which could be attributed to cavitation within the solution. Cavitation may, however, occur in the solution within the generator or in the orifice outlet.

In order to investigate the mechanism, and in particular the geometric distribution of cleaning, a new technique described elsewhere[9,10] was used with glass slides coated with wax and lamp black which were immersed in the Hydroson tank. On surfaces immersed in still or stirred solutions cleaning was observed to commence at the sharp edges where the bonding of the contaminant to the substrate is weakest. This undermining of the surface soil continued around the edges and then eventually spread to the middle of the larger flat faces. In the Hydroson tank, however, the removal started close to, but not at, the focal point of the two Hydroson jets. At this area the high-pressure jets impinged on the surface and created a turbulent flow with a scouring action and shear forces. These effects proved to be very efficient in removing surface contamination. The cleaning action then spread rapidly away from the central areas. The sonic waves produced by the generators could cause the formation of minute bubbles which oscillate on the surface and penetrate between the soil and surface and so assist in the removal of surface films. The values of the cleaning efficiency are given in Table 1 and are discussed in Section 8.

In order to further understand the processes which occur in the Hydroson tank, it was considered necessary to measure the flow rate through a nozzle.[11] A 9×10^{-3}-m^3 volume of solution flowed through a nozzle of diameter 4×10^{-3} m in 60 s so that the velocity was 11.9 m/s. The speed at the point equidistant from the two upper jets, i.e., the region of maximum turbulence, was calculated as 1.79 m/s, at a distance of 0.17 m from the nozzles. This is the velocity of the liquid entering into a tank containing stationary liquid.

In order to confirm the flow rate from the generators, the morphology of zinc deposited onto a rotating surface was compared with that produced in the Hydroson tank. These were similar when the angular rotating velocity was 356 rad/s, which corresponded to a flow rate of 1.16 m/s. Hence the flow rate from this study is considered to be 1.16 m/s. The difference between these two values is attributed to the interaction of the two opposing jets which could reduce the higher value of 1.79 m/s to 1.16 m/s. Other studies on the electrodeposition of zinc from the zincate bath are described in a separate section.

Another basic parameter for immersed surfaces is the thickness of the layer of solution adjacent to the surface through which transport of species to or from the surface occurs by diffusion rather than convection. This is called the diffusion or double layer; see Figure 2. It consists of electrons in the metal surface, a film of adsorbed ions, and an ionic atmosphere in which there is an excess of ionic charge. This results from a concentration gradient at the interface, and the thickness can be calculated from Fick's First Law of Diffusion and the limiting current density, which is determined from electrochemical studies.[12] The value

Table 1. Effect of Agitation on the Cleaning Efficiencies for the Removal of Grease on Glass Slides

Bath agitation	Time (s)	Cleaning efficiency (%)					
		Water		1% Alkali		5% Alkali	
		Grav[a]	Optical	Grav	Optical	Grav	Optical
Still	180	2	0.5	2	1	2	1
	360	3	2	4	2	3	3
	720	6	7	6	6	7	9
	1440	9	13	10	22	8	18
Stirred	180	2	7	0.5	8	2	8
	360	5	21	9	19	13	21
	720	5	26	14	35	18	34
	1440	9	28	24	40	50	64
Ultrasound	0.5	2	11	1	18	2	48
	1	3	72	23	58	31	71
	2	3	81	55	76	64	78
	5	8	93	77	90	90	93
Hydroson	1	2	3	2	2	7	4
	2	12	20	19	28	29	54
	5	25	35	43	44	54	74
	10	54	78	73	96	91	94

[a]Grav = gravimetric.

for the limiting current density in the Hydroson tank was measured as 18.2 A/dm^2 which gives a corresponding double-layer thickness of 13.5 μm; see Table 2. The values for a still and a magnetically stirred solution are 2.5 A/dm^2 and 97 μm and 8.3 A/dm^2 and 29 μm, respectively, so that the Hydroson system has a very significant effect. Another method,[12] involving the cathodic precipitation of magnesium hydroxide, confirmed that the double-layer thickness was about 13.5 μm.

This experimental work, together with the studies on cleaning and electrodeposition, was described in a recent article.[13] This is significant in the cleaning of immersed surfaces and electrochemical reactions such as electrodeposition and phosphating.

4. Commercial Applications of Hydroson Cleaning

The Hydroson system has been particularly successful when introduced into the first-stage soak cleaner for the removal of polishing compounds in a finishing line. The best results are obtained if cleaning occurs before the pol-

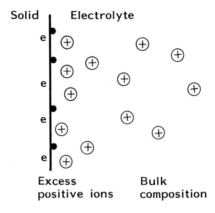

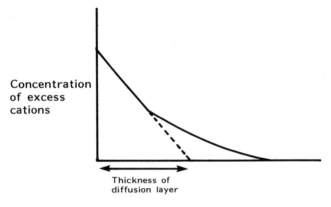

Figure 2. Diagrammatic electrolytic diffusion layer.

Table 2. Limiting Current Density and Diffusion Layer Thickness in Electrodeposition of Zinc

Bath agitation	Limiting current density (A/dm²)	Diffusion layer thickness (μm)
Still	2.5	97
Stirred	8.3	29
Ultrasound (13 kHz)		12.5
Hydroson	18.2	13.5

ishing compound dries. It has been used to remove a wide range of contaminants[3, 14] including polishing compound,[2, 15] oil,[2] grease,[2] swarf, general soils, heat treatment and quench oil deposits, and asbestos fibers. It has been employed to clean zinc-based die castings, brass and aluminum components, rubber, plastic, and asbestos molds, castings prior to assembly and components prior to dismantling, shaped and fine wires, strip and coil, and machine components.

Because the cleaning solution is circulated through strainers, particulate matter is removed. Dual-changeover types of strainer are normally used and fitted with a warning light or switch to indicate the need to change and clean the filter while using the other system. This is beneficial because the constant removal of solid contaminants increases the life of the solution. Also it can be particularly beneficial in the emulsification[16] of oil and mixtures, and improves the efficiency of rinse tanks and spray nozzles in washer systems.[2]

Many applications involve preparation prior to the production of a final coating. Successful applications of the Hydroson system include the cleaning of pressed aluminum caps for perfume bottles prior to anodizing,[17] mild steel components before painting,[18] and light and bathroom fittings and chandeliers contaminated with polishing compounds preceding electroplating.[18] Hydroson cleaning has also been applied to finished articles including aircraft components at Sabena, Belgium, prior to refitting, trays used for food preparation, crown wheels and pinions to remove traces of lapping compounds, and switch components for televisions at Philips, Holland.[18]

5. Safety and Economy

Because one of the most common causes of failure of a coating is due to inadequate surface preparation, it is becoming increasingly important in the finishing industry to be able to consistently produce a sufficiently clean surface to give good adhesion. This should be achieved without risk to health or safety, and it is therefore advisable to avoid the use of dangerous, flammable, toxic chemicals or solvents. The cleaning process also has to be economically viable so very expensive chemicals should not be used, if possible. The Hydroson system can help attain both these aims.[15] Fortunately, it is often possible to fit the system retrospectively into existing tanks with relatively little disruption of production, and this minimizes the expense.

The greater agitation and increased efficiency of the Hydroson system may enhance safety[2] by the use of:

1. Safe rather than toxic chemicals, such as alkaline aqueous solutions instead of toxic organic solvents.
2. Weaker solutions of acids or alkalies for pickling and/or degreasing.

3. Lower operating temperatures.
4. Biodegradable cleaners which minimize effluent problems and costs.

The economic savings can also be considerable and some of the factors are interdependent. Another advantage is that it is not necessary to develop special chemicals because the usual cleaning solutions are normally satisfactory. Because of the considerably increased efficiency it may be possible to either (1) operate with the standard solutions at a faster rate and hence save space and time or (2) maintain the same speed and use

    (a) Lower solution temperatures and save energy.
    (b) Less concentrated solutions and save chemicals.
    (c) The standard solutions, which will have a longer effective life.

A reduction of 50–75% in the time necessary for immersion during phosphating has been reported[16] which is comparable with spray times. A three- to fourfold improvement in the life of an aqueous cleaner in an anodizing line has been recorded,[17] which may be due to the enhanced emulsification of oil and grease. A division of Wilmot-Breeden[15] found that the installation of Hydroson equipment increased the life of chemicals from two to four weeks with a reduction in cost of $2200 in the first year. After four years of successful operation it produces over 250,000 plated die castings per week with a reject rate of less than 0.25%.

Hence the operation of solutions with a lower concentration of chemicals can reduce the installation and operating costs and possible effluent problems and costs, and may even eliminate the necessity of a rinsing stage after cleaning. If this can be accomplished, there will be a saving of floor space and a reduction in both the volume of water consumed and the amount of chemical lost due to drag-out. An increase in the efficiency of a process also means that the rejection rate will be lower and more consistent quality achieved. This can be illustrated by the fact that one plating company[15] has been able to eliminate two presoak hot-alkaline cleaning stages and so saved on space, heating, labor, and chemicals a total of $12,000 pa. The reject rate was also reduced from approximately 2% to almost 0%.

With the Hydroson system it may be possible to produce a saving of energy, water, chemicals, effluent, labor, and time.

6. Recent Developments

6.1. Wire Cleaning

The cleaning of wire has traditionally been carried out by a conventional soak system. For shaped and fine wire, however, more efficient cleaning is required to penetrate into small crevices and to remove contaminants on the surface of the metal. If these impurities are not removed, problems may arise

Hydroson Cleaning

during subsequent heat treatment and other processing. With the high speeds currently being used in the industry, it would be necessary to use long soaking times and very large tanks to produce the high standard of cleanliness required. These big tanks necessitate large floor areas and high volumes of cleaning solution, both of which increase costs. Hence the Hydroson system was considered.

In order to improve the efficiency of cleaning wire and impart optimum energy to the surface, the Hydroson generators have been modified so that each strand of moving wire passes through the center of one or more generators.[5, 18-20] This arrangement can be designed[5] for wire up to a maximum diameter of 5 mm. The generators are arranged in series, with perhaps ten in a manifold box, fitted into a chamber which is flooded with the aqueous cleaning solution. As the wire leaves the orifice the liquid, which is also emitted, produces an intense scrubbing action on the surface. The fact that the generators are immersed in the liquid prevents the transfer of sound to the surroundings. The chamber is fitted with an overflow pipe, and both ends have drains so that the solution is circulated and flows back into a tank below, where it is filtered or strained.

The generators are constructed from tungsten carbide, silver steel, ceramics, or polymers depending upon the type of wire, speed, and cleaning fluid, and the surface finish required. Wires which have been successfully cleaned include low and high carbon, stainless and high-alloy steel, copper, brass, aluminum, and zinc.

In industrial practice the system is used to remove contaminants such as rolling and forming lubricants prior to heat treatment in a furnace. On leaving the furnace, the wire may be quenched in oil, washed, and rinsed prior to final tempering. Hydroson agitation is also usefully employed in these processes.

The speed of the wire cleaning depends upon the particular metal, arrangement of generators, and particular application. Hydroson has been successfully employed[19] for the removal of:

1. Sodium stearate from coppered steel wire at speeds of up to 1000 m/min.
2. Chlorinated drawing oils from finished nickel alloy welding at up to 275 m/min.
3. Zinc phosphate, sodium stearate, calcium stearate, and mineral rolling lubricants from shaped steel wire at up to 33 m/min.
4. "Salt" carrier coating and sodium stearate from nickel alloy wire at up to 350 m/min.

Other contaminants which have been removed include lime and borax coatings, temper oxides and oil, rust, patent oxides, and light scale.

The advantages of using Hydroson have been claimed[5] by a wire manufacturer to include:

1. The wire is kept straight, which minimizes any distortion and breakage.
2. The cleaning and rinsing processes now occupy only 10% of the process area.
3. Emulsification of the oil and water has more than doubled the life of the chemicals used for cleaning.
4. Heating costs have been reduced because of the heat produced during the process.
5. Less toxic chemicals have been used in closed chambers rather than open tanks so that the safety has been improved.

The dies used in drawing the wire can also be cleaned by placing them in a wire basket which is immersed in the cleaning tank. Barrel cleaning is also beneficial.

6.2. Coil Strip Cleaning

Coil strip can be cleaned, degreased, and pretreated using Hydroson generators.[5, 16, 18] The system is flexible with a single or multi-strip layout, a range of line speeds, and the ability to handle solid or perforated strip. Mild steel, high-carbon steel, or nonferrous metal strip may be handled either vertically or horizontally. Narrow strip, up to 1.5 cm wide, can be processed with two generators opposed to each other and suitably machined so that the strip passes between the energy sources.[5] A different arrangement is employed[16] for wider strip, up to 4.5 cm wide and 2.5 cm thick at speeds up to 15 m/min, with the generators placed on both sides of the strip at the optimum distance from the moving surface. As with wire, a wide range of contaminants can be removed and these include rolling oils, anti-rust oils, lanolin-based protective oil, light scale, and smut. An example of this application is at Chloride Alcan, Redditch, England, where two units have been installed to clean narrow strip prior to electroplating. Both tanks have sixteen Hydroson generators which are located below the liquid level.

6.3. Barrel Cleaning and Rinsing

In barrel cleaning[16, 18] the articles to be cleaned are loaded into a perforated or mesh drum and immersed in the appropriate solvent. The drum is rotated and the components, which are normally small, are constantly moved and exposed to fresh solution. The drum is then removed from the solution and the contents allowed to drain. This procedure enables small objects to be cleaned without expensive jigging or manual operations.

The Hydroson system is applied to barrel cleaning by use of a specially designed manifold in the tank. The manifold follows the circumference of the immersed section of the barrel so that the solution can be pumped into the barrel through the perforations. Hence components in the barrel are subjected to the normal tumbling as well as the Hydroson action. The physical emulsification of the oil and water, which minimizes drag-out, is again particularly beneficial and produces an increased efficiency. A similar arrangement can be used for the subsequent rinsing, and better penetration of the water into the barrel is advantageous.

In one commercial line at Hubert Stuken K. G., West Germany, in which barrel cleaning has been used prior to electroplating, the advantages claimed for Hydroson include improved cleaning, less contamination in the process tanks, and a longer life for the soak and electrolytic cleaner as well as the ion-exchange resin in the water treatment plant as a result of the reduction in the oil content in the rinse tank.

6.4. Cleaning Molds in the Glass Industry

The traditional method of cleaning bottle molds is to soak in paraffin or caustic soda solution to soften baked-on mold lubricants. This is followed by brushing or wiping by hand and polishing by prolonged abrasive blasting to remove surface irregularities. The surface produced is then smooth enough for molding bottles.

Unfortunately, the abrasive blasting removes a surface layer as well as irregularities so that the molds quickly become oversized. This problem can be overcome by the application of Hydroson to the cleaning. The molds are placed in a work basket which is lowered into a tank fitted with a manifold of generators and immersed for a suitable period of time. After removal from the tank and manual rinsing, the molds are washed in hot water and given a final, very short dry blast. The advantage of the Hydroson system is that it permits minimum blasting and wear and gives a maximum mold life. Hence this process is now used by glass bottle manufacturers.

6.5. Nuclear Industry

The action in the Hydroson system has been shown to be particularly beneficial in cleaning and decontamination of equipment used in the nuclear industry.[18] Hence a system has been installed in a nuclear power station, and interest has been expressed by other research workers in this area of operation.

6.6. Phosphating

Work has been carried out recently on the effect of Hydroson agitation on phosphating.[16, 18] The results to date look encouraging because the process has been accelerated and the time to produce the required coating has been reduced by 50–75% so that it is similar to the equivalent spray time. The quality of the phosphate layer has also been improved, and the layer has a more closely knit crystalline structure with a better corrosion resistance and surface finish, particularly on thin films. It may also be possible to reduce the operating temperature of the phosphating solution. Because acidic solutions are used in the phosphating process, it is necessary to use materials for the construction of the system which are sufficiently resistant to stop corrosion and chemical attack. Any dissolution of the system could eventually cause perforation and leakage as well as change the effectiveness of the solution. Hence stainless steel or some other corrosion-resistant material should be used for tanks, pipework, and pumps to reduce corrosion and the subsequent contamination of the acid. One satisfactory application of the system has been installed into the phosphating line in the production of brake drum components for the automotive industry.

6.7. Electrodeposition

Electrodeposition of zinc from an alkaline zincate plating bath[21] has been carried out in a Hydroson mild steel tank. This solution was chosen because the steel would be passive, and no corrosion and subsequent contamination would occur. The zinc deposit from the plain bath was brighter with a finer crystallite structure, less porous, and harder than those produced from the same solution without agitation. The reduction in both the cathodic and anodic polarization was beneficial because it permitted an increase in the cathodic current efficiency of about 10%. Hence good-quality deposits could be obtained at higher current densities, i.e., faster deposition, with an improvement in the maximum current density from about 1.2–1.5 A/dm^2 with a still bath to 7.5–8.5 A/dm^2 with Hydroson. At a current density of about 2 A/dm^2, the still bath gave powdery deposits, magnetic stirring gave better deposits of microhardness of 65 ± 5 Hv, and the Hydroson bath was more effective (80 ± 7 Hv). The deposits from a bath containing commercial brighteners did not show the same improvements in the current but were harder (130 Hv) than those from a stationary bath (86 Hv).

The increased agitation with the Hydroson system also raised the efficiency and rate of electrodeposition of zinc powder from dilute solutions by a factor of 3–4 compared with magnetic stirring.[22] Hence it may be economically viable to recover metals as powders from dilute pickling solutions and it may also be possible to recycle the acid or alkali, provided that corrosion-resistant tanks, pipes, etc., are used. The advantages of using the Hydroson system in electrodeposition may thus include:

1. Recovery of valuable metal powders.
2. Regeneration of spent acids and alkalies for pickling solutions.
3. Avoidance of costs involving the disposal of effluent.

7. Size and Cost of Equipment

The size and design of the equipment depends upon the particular application. For wire cleaning, the generator heads are arranged in series and specially produced so that the wire passes through the center of the generator. The more common arrangement is for the generators to be positioned so that the surfaces to be cleaned are subjected to the maximum effect of the sonic waves produced simultaneously by one or more generators. The arrangement used in the cleaning and electrodeposition work described in this chapter is shown in Figure 1. The position and direction of the jets should be carefully controlled so that minimum shielding occurs, which is particularly important for articles with a complex shape. One installation[17] for cleaning aluminum caps for spray bottles prior to anodizing consists of 140 Hydroson generator heads on a manifold driven by two 15-kW pumps. The tank has a volume of approximately 3.6 m^3 (2.4 × 1.2 × 1.2 m). A specially designed machine has been built to clean the bores of pipes up to 12 m long.[3]

The cost of a Hydroson system depends upon the size and number of generators. Generally it is cheaper than an equivalent ultrasonic system in all but very small units.[3] Hence the cost may vary dramatically from plant to plant over a range of $4500 to $90,000 per unit. Other information about the Hydroson equipment is available from articles published in German[23,24] and Dutch.[25]

8. Comparison with Ultrasonic and Megasonic Cleaning

Hydroson, ultrasonic, and megasonic systems all involve the cleaning of articles immersed in solution in tanks and subject to different forms of agitation.

Ultrasonic cleaning uses frequencies of 20 to 80 kHz, which produce cavitation, implosion of bubbles, and a scrubbing action on the surface. Megasonic cleaning is a specialized and relatively new technique, patented[26] in 1975, which utilizes frequencies of 850 to 900 kHz to give short sonic waves. The particles are removed by the very-high-frequency sonic pressure waves generated in the liquid. The main use of this equipment is to remove small particles, 1 μm or less, from the surface of silicon substrates and device wafers, and this type of cleaning can be used in automated production or batch applications. Penetration of the cleaning fluid between these small particles and the surface is necessary and is indicated by good wetting properties. Suitable solutions include hydrogen peroxide–ammonium hydroxide–deionized water and hydrogen peroxide–hydrochloric acid–deionized water. Once removed, the particle must travel sufficiently far from the surface to prevent its recapture by the existing attractive surface forces. Further details of this technique have been given in a recent and very comprehensive paper by Shwartzman et al.,[27] and the advantages and disadvantages are presented in another paper.[28]

In ultrasonic cleaning in water, the removal of the contaminant commences at the areas around the nodes of the standing wave set up in the tank. When an alkaline solution is used, the process starts at the edge nearest to the transducer probe and continues across the surface. This probably results from the fact that the cleaning rate is dependent upon the cavitation intensity, which decreases with distance from the transducer.

There are several mechanisms involved in ultrasonic irradiation. When ultrasound is applied to a liquid, the particles in the medium oscillate and produce a continuous displacement called acoustic streaming. This results in an overall movement of the liquid away from the surface of the vibrating transducer and helps to remove loose contaminant from the dirty surface as well as bringing clean solution from the bulk to the surface. Ultrasound helps in the cleaning procedure because the very small bubbles in the liquid are compressed and expanded, and some collapse and release high local energy and shock waves. This cavitation is important because it helps to remove impurities on the surface by producing an intense scrubbing action on immersed articles. These processes are discussed in more detail elsewhere.[13,29,30] Because cavitation can occur down holes and in recesses, these areas can be cleaned by ultrasound.

In the Hydroson tank, the jets of solution converge on a surface and the high energy results in an intense scrubbing action. This gives a cleaning action which progresses in a radial direction away from the focal point of the jets. Because the scrubbing action occurs on the surface, cleaning at the bottom of holes and recesses may not be very effective and ultrasound can be more beneficial. The advantage of the band of frequencies in the Hydroson system, rather than the fixed value for ultrasonics, is that there should be less "shadow effect" and more cleaning on the blind side of components. The action is described in more detail in Section 3 on mechanisms.

A quantitative evaluation of the Hydroson system has been made under certain very specific conditions. These were the use of water and 1% or 5% solution of a commercial alkali cleaner with glass plates covered in paraffin wax containing lamp black. The cleaning efficiency was assessed gravimetrically and optically, and the values for four different degrees of agitation are given in Table 1. Some of the differences between the optical and gravimetric values for the same time and system are considerable and can be accounted for by the fact that contaminant removed from one area of the glass slide may coagulate with that on an adjoining area. If this occurs, the optical value of cleanliness increases but there would be no difference in the gravimetric value because of the redistribution of the grease.

The degree of cleanliness produced by cleaning a glass slide coated in grease was assessed by measuring the contact angle between the slide and water. Agitation of the water or the 1% alkali or 5% alkali solution had little effect on the contact angle. A cleaning efficiency of 95% is good and means that the majority of the contaminant has been removed but a surface layer still remains. This residual layer has a significant effect on the contact angle. Thus a glass slide, thoroughly cleaned with organic solvents, had a contact angle of 15 ± 1° whereas the best value with the Hydroson system was 18°, which was obtained with a 5% alkali solution. It can therefore be concluded that Hydroson cleaning is good and accelerates the removal of the majority of bulk contaminant but it does leave a residual thin surface film.

It may be concluded from these experiments that both the Hydroson action and ultrasonic irradiation give a very considerable improvement in the cleaning efficiency. More details can be obtained on ultrasonic cleaning from the chapter by O'Donoghue[30] in this treatise. The value for the Hydroson system is the effect of a range of frequencies, whereas the value for ultrasound is more specific, so that a general comparison is difficult to make. Similarly, the reduction in the value of the thickness of the diffusion layer produced in the Hydroson tank is similar to that with ultrasound (frequency 13 kHz); see Table 2.

The following advantages of the Hydroson system over traditional ultrasonic systems have been claimed[31]:

1. Relatively cheap to install.
2. Simple to maintain.
3. Reduced running costs.
4. System has a built-in strainer to remove solid particles.
5. Circulation of solution helps to reach difficult areas.
6. Allows possible use of lower temperatures.
7. Increased life of chemicals.
8. Permits use of less aggressive chemicals.
9. Faster cleaning cycle.

9. Conclusion

The Hydroson system, although relatively new, is now established as an important and widely accepted technique which produces intense agitation in cleaning solutions and a high efficiency. The system is adaptable and the generators, which are the novel aspect and produce sonic waves in an aqueous media, have been modified to permit continuous cleaning of wire and strip while the system can also be used for barrel cleaning. Because very efficient mixing occurs in the process solution, new applications are constantly being found, such as phosphating, and there is considerable potential in electroplating.

References

1. Nickerson Ultrason Ltd., British Patent 1,475,307 (1973).
2. Hydrosonics, a sound approach to surface cleaning, *Product Finishing* 29(10), 11 (1976).
3. Component cleaning using the Hydroson system, *Finishing Industries* 2(12), 23-25 (1978).
4. J. R. Frederick, *Ultrasonic Engineering*, John Wiley and Sons, New York (1965).
5. R. Clough, Wire and strip cleaning by Hydroson, *Wire Industry* 46, 491-492 (1979).
6. D. J. Godfrey, Cavitation damage in: *Corrosion* (L. L. Shreir, ed.), Vol. 1, pp. 8:124-132, Newnes-Butterworth Press, London and Boston (1976).
7. C. T. Walker and R. Walker, New explanation for the hardening effect of ultrasound on electrodeposits, *Nature* 244, 141-142 (1973).
8. R. Walker and C. T. Walker, Hardening of immersed metals by ultrasound, *Nature* 250, 410-411 (1974).
9. R. Walker and N. S. Holt, A quantitative evaluation of the Hydroson cleaning system, *Product Finishing* 33(4), 14-16 (1980).
10. R. Walker, The efficiency of ultrasonic cleaning, *Plating Surface Finishing* 72(1), 63-70 (1985).
11. R. Walker and N. S. Holt, Determination of the flow rate in the Hydroson cleaning system, *Surface Technol.* 17, 147-156 (1982).
12. R. Walker and N. S. Holt, Determination of the Nernst diffusion layer thickness in the Hydroson agitation tank, *Surface Technol.* 22, 165-174 (1984).
13. R. Walker, Electrodeposition from Hydrosonically agitated solutions, *Thin Solid Films* 119, 223-240 (1984).
14. Hydroson Cleaning Systems, Leaflet, Nickerson Ultrason Ltd., Warley, England.
15. I. A. Thomson, The Hydroson system—A sound system for saving, *Product Finishing* 31(3), 29-31 (1978).
16. The Hydroson cleaning system, *Finishing Industries* 4(12), 28-30 (1980).
17. B. Allen, Hydroson improves pre-anodising cleaning efficiency, *Product Finishing* 36(4), 33-34 (1983).
18. I. A. Thomson, Hydroson scores a century, *Product Finishing* 34(10), 54-55 (1981).
19. R. Clough, Progress on in-line wire cleaning, *Wire Industry* 47, 690-691 (1980).
20. I. A. Thomson, In-line wire cleaning with the Hydroson system, Fine Wires Conference, Aachen (Oct 30-31, 1980), International Wire and Machinery Association, Oxted, Surrey, UK (1980).
21. R. Walker and N. S. Holt, The role of Hydrosonic agitation on the electrodeposition of zinc from a zincate bath, *Plating Surface Finishing* 67(5), 92-96 (1980).
22. R. Walker and N. S. Holt, Recovery of zinc powder by electrodeposition from dilute zincate solutions, *Chem. Ind.* 1985 (19), 783-785 (1985).
23. J. Mesmer, Acoustic vibration cleaning system, *Technica (Switzerland)* 30(5), 368-370 (1980).
24. F. L. Rohling, Hydroson: new cleaning technique for metal surfaces, in: Proceedings of the International Congress on Surface Technology, Berlin (1981); World Surface Coatings Abs. 57(492), 3718 (1983).
25. A. J. v. Wessem, Hydroson cleaning process, environmentally acceptable and economical, *Oppervlakechniken. (Surface Techniques)* (Delft) 26(4), 103-107 (1982).

26. A. Mayer and S. Shwartzman, Megasonic cleaning system, U.S. Patent 3,893,769 (July 8, 1975).
27. A. Shwartzman, A. Mayer, and W. Kern, Megasonic particle removal from solid-state wafers, *RCA Review* 46(1), 81–105 (1985).
28. A. Mayer and S. Shwartzman, Megasonic cleaning: A new cleaning and drying system for use in semiconductor processing, *J. Electronic Mater.* 8, 855–864 (1979).
29. C. T. Walker and R. Walker, Hardening effect of ultrasonic agitation on copper electrodeposits, *J. Electrochem. Soc.* 124(5), 661–669 (1977).
30. M. O'Donoghue, Ultrasonic cleaning in this treatise.
31. A Comparison of Cleaning Methods, Hydroson vs. Ultrasonics, Leaflet, Nickerson Ultrason Ltd, Warley, England.

4

Methods of Measurement of Ionic Surface Contamination

JACK BROUS

1. Introduction

Surfaces which are truly free of contamination are, in our everyday experience, exceedingly rare. Such surfaces can be created under very special, carefully controlled laboratory conditions which, however, do not represent the real world of materials fabrication, processing, and storage.

Contamination on surfaces can be of various types and origins.[1] Examples of contamination are:

1. Sorbed gases and vapors: This consists of gases and vapors, including water, which are absorbed or adsorbed on a surface. These may include the normal atmospheric gases but they could also include other gaseous or vapor contaminants which may be present in the environment.

2. Nonionic organic residues: This can include a wide variety of organic materials from many sources. Also included are degradation, oxidation, and polymerization products formed by effects of heat on organic materials residual on a surface.

3. Inorganic materials which are insoluble in water: This could include oxide or sulfide corrosion or oxidation products on a surface, including the oxidation products of the base metallic substrate such as rust on steel, or oxides foreign to the substrate deposited by various means.

4. Ionic materials which are water soluble: Included here are inorganic materials as well as organic acids and salts which are ionic or generate ions in water. They can be present from a variety of sources including chemicals used

---

JACK BROUS • Alpha Metals, Inc., 57 Freeman Street, Newark, New Jersey 07105.

in processes such as plating, etching, pickling, soldering, brazing, and even cleaning.

It is this last category of contaminants that is being addressed in this chapter.

Ions are discrete molecules with the distinguishing feature of an electrical charge. Their mobility in a liquid with a high dielectric constant, such as water, enables them, under an applied electric field, to carry electrical currents. Their presence or absence on a surface can, therefore, be of critical importance if that surface is intended as a nonconductive surface to provide electrical insulation between conductors.

The presence of ions alone, however, is not sufficient to give rise to surface electrical conduction. To conduct electricity, ions must also possess the mobility imparted by a dielectric medium—most commonly water.[2-4] Moisture is therefore another ingredient necessary for conduction by ionic surface contaminants to occur.

Water is abundantly available in the atmosphere and, depending on the temperature and humidity, can be adsorbed or absorbed on a surface. This is especially the case where hygroscopic materials are present. These can include many ionic salts and some nonionic organic materials. In either case, the presence of ions in a sorbed water film can greatly enhance its electrical conductance and contribute to the degradation of insulation characteristics of a surface.[5,6] The measurement and control of surface ionic contamination is, therefore, more than of academic interest to those concerned with processing of electrical and electronic assemblies and devices.

2. Ionic Measurement

The most obvious property of ions lending itself to measurement is their conductance of electrical current in solutions. Although in some special instances ionic currents can be measured directly on a surface,[37,38] it would be virtually impossible to quantify these currents with respect to amounts and locations of ionic materials on the surface. In order to measure such ionic material, it is therefore necessary to extract the ions into solutions in solvents capable of sustaining ionization. Water is the most common solvent and, because of its high polarity and dielectric constant, shows generally good solvency and ionization characteristics for ionic salts extracted from surfaces. The electrical measurements used to detect ions in water can be made extremely sensitively. The detection of very low levels of ions is, however, limited by the intrinsic conductance of the water itself due to a small amount of self-ionization.

$$H_2O \rightleftharpoons H^+ + OH^-$$

At ordinary room temperatures the concentrations of the H^+ and the OH^- ions in pure water are both 10^{-7} moles/liter.

In order to achieve accuracy of measurement, it is important that any conductance changes attributable to dissolved ions be a significant portion of the total measured conductance. For pure water in equilibrium with the atmosphere at 18°C, the specific conductance is typically ~ 1 μsiemens/cm due to dissolved atmospheric carbon dioxide and other ionogenic impurities.[7] Even the purest water which is completely devoid of all extraneous gaseous or ionic contaminants will exhibit specific conductance values of ~ 0.04 μsiemens/cm (~ 25 MΩ-cm specific resistance).[8-10] This means that in pure, static, aqueous media exposed to the atmosphere, low ionic levels giving rise to specific conductance changes less than 1 μsiemens/cm will be measured with increasing inaccuracy as concentrations are decreased.

Theoretical Background

At this point, a brief discussion of ionic conductance and its measurement will be useful.[11-13]

When an electrical current is passed through a conductor, the resistance to its flow is indicated by the ratio of the applied potential to the current flowing through the conductor. This is expressed as Ohm's law:

$$R = \frac{E}{I} \qquad (1)$$

where R is the value of resistance expressed in ohms and E and I are potential and current, respectively, expressed in volts and amperes. The resistance, R, is a measure of resistance of the conductor to the flow of current. Alternatively, the reciprocal of the resistance, the conductance, can be used to express the ability of the conductor to carry current:

$$C = \frac{I}{E} \qquad (2)$$

where C is the conductance expressed in siemens (formerly mhos, reciprocal ohms or ohm^{-1}).

In order to measure the conductance of a solution as a conductor, it is necessary to immerse two electrodes to serve as the electrical interface between the solution and the external measuring circuit (Figure 1).

A voltmeter and an ammeter can be used to measure E and I so that conductance [equation (2)] can be determined for the solution between the electrodes. If the electrolytic solution is homogeneous, the value of the conductance

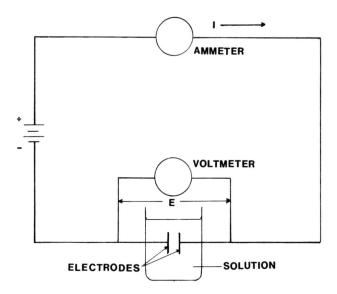

Figure 1. Measurement of solution conductance or resistance.

will depend on the solution, the areas of the opposing electrode faces, and inversely on the distance between them. Thus:

$$C = \frac{LA}{d} \tag{3a}$$

or

$$L = \frac{Cd}{A} \tag{3b}$$

where A is the area of the opposing electrode faces, d is the interelectrode distance, and L is a constant for that solution.

For the particular case where A is 1 cm² and d is 1 cm, then $C = L$. The parameter L is the specific conductance, i.e., the conductance measured across a cube of solution having a volume of 1 cm³.*

Similarly, if we are measuring the resistance of the cell,

$$R = \frac{\rho d}{A} \tag{4a}$$

---

*This is also referred to by many as *specific conductivity* or, simply, *conductivity*. It will, however, be referred to here as *specific conductance* and its analogous reciprocal function, *specific resistance*.

or

$$\rho = \frac{RA}{d} \quad (4b)$$

where ρ (rho) expresses the specific resistance of the centimeter cube.

If one wishes to measure the specific conductance or specific resistance of a solution, it is not necessary, in practice, to measure the actual electrode areas and spacings. The cell can, instead, be calibrated using a solution which has a previously determined specific conductance. This enables the determination, using equation (3b), of the ratio d/A, which is a constant for the cell and can be used to calculate the specific conductance values of solutions subsequently measured in the cell.

In studies of solutions, another value—the equivalent conductance—can be helpful. This is the product of the specific conductance and the volume of solution containing one gram-equivalent of an electrolyte. It can be visualized as the conductance measured in a cell formed by two metal electrodes 1 cm wide and 1 cm apart and infinitely high. When the cell is filled with the solution to a height which includes between the electrodes a volume containing one gram-equivalent of the electrolyte (or $1000/c$ where c is the number of gram-equivalents per liter), then:

$$\Lambda = vL = \frac{1000L}{c} \quad (5)$$

Here Λ (lambda) represents the equivalent conductance and v, the volume in ml containing one gram-equivalent of the electrolyte. If we examine the values of L and Λ for a typical strongly ionized electrolyte, such as sodium chloride, as shown in Table 1, it is seen that Λ approximates a constant value—particularly at very low concentrations—as compared to L, which is very dependent on concentration.

Table 1. Effects of Concentration on Specific and Equivalent Conductances for NaCl

Concentration, c (equivalent/liter)	Equivalent volume, v (ml/equivalent)	Specific conductance, L (μsiemens/cm)	Equivalent conductance, Λ (μsiemens-cm^2/equivalent)
10^{-4}	10^7	10.81	108.1
10^{-3}	10^6	106.5	106.5
10^{-2}	10^5	1,020	102.0
10^{-1}	10^4	9,200	92.0
1	10^3	74,400	74.0

Table 2. Effects of Concentration on Specific and Equivalent Conductances for Acetic Acid

Concentration, c (equivalent/liter)	Equivalent volume, v (ml/equivalent)	Specific conductance, L (μsiemens/cm)	Equivalent conductance, Λ (μsiemens-cm^2/equivalent)
10^{-4}	10^7	10.7	107
10^{-3}	10^6	41.0	41.0
10^{-2}	10^5	143	14.3
10^{-1}	10^4	460	4.6
1	10^3	1,320	1.32

This behavior is typical for all strongly ionized electrolytes. If, however, we look at similar values for weakly ionized electrolytes, such as organic acids, exemplified in Table 2 by acetic acid, Λ shows a much greater drop with increasing concentration.

If we limit our concern only to highly dilute solutions of strong electrolytes, then Λ closely approximates a constant for each material, and from equation (5), L is then proportional to concentration.

$$L = \frac{\Lambda c}{1000} = kc \tag{6}$$

and since the gram-equivalent concentration of a salt in solution is proportional to its mass concentration per unit volume, we obtain:

$$L = \frac{kg}{Mv} \tag{7a}$$

or in a fixed volume,

$$L = Kg \tag{7b}$$

where g is the mass of a salt of molecular weight M in a fixed volume v.

Measurement of the specific conductance, L, can therefore serve as an indication of the total amount of an ionic electrolyte extracted into a specific volume of solvent.

3. Static Extraction Methods

3.1. Egan's Method

Perhaps the simplest method, in principle, for measurement of ionic residue on a surface is to immerse the sample in a static volume of a highly purified, high-polarity liquid, such as water, and, with agitation, extract the ions from the surface. If both the initial and final (after extraction) conductances are measured, then from equation (6),

$$\Delta L = L_f - L_0 = k\Delta c \qquad (8)$$

Since the initial value, L_0, is never equal to zero, it should be subtracted from L_f, the final value, to indicate the specific conductance change, ΔL, directly attributable to the extracted ions. It should be noted that because ΔL is linearly proportional to concentration, each additional increment is linearly additive (remembering that we are discussing here very dilute solutions). The change of L with change in ionic concentration is, therefore, represented by a straight line, with each additional increment of extracted ions contributing to the change in conductance in proportion to its amount. This is shown in Figure 2 (curve A). If, however, specific resistance of the solution is used to measure ionic extraction, the reciprocal curve is obtained (Figure 2, curve B), and the expression for the relationship between ρ_f and the change in concentration, Δc, is considerably complicated:

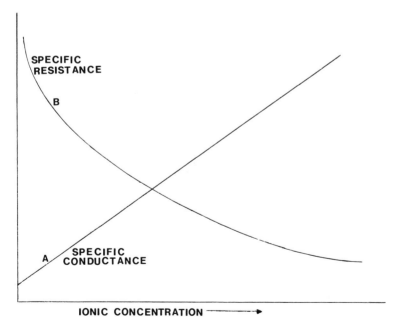

Figure 2. Variation with ionic concentration of (A) specific conductance and (B) specific resistance.

$$\Delta L = L_f - L_0 = k\Delta c = \frac{1}{\rho_f} - \frac{1}{\rho_0} \tag{9}$$

$$L_f = \frac{1}{\rho_f} = k\Delta c + \frac{1}{\rho_0} = \frac{\rho_0 k\Delta c + 1}{\rho_0} \tag{10}$$

$$\rho_f = \frac{\rho_0}{\rho_0 k\Delta c + 1} \tag{11}$$

In addition, the sensitivity of the specific resistance function to concentration, as indicated by the slope of the line, decreases as concentration builds, whereas for specific conductance the slope remains constant. This was pointed out by Egan[7] who also demonstrated the feasibility of using the specific conductance of extracted solutions for the measurement of ionic residues left on plated parts or assemblies after cleaning. By immersing the sample to be measured in a fixed (or static) volume of deionized water (Figure 3), the residual water-soluble ionic contamination can be extracted. This extraction is enhanced by applying agitation of some sort to the extracting solvent. As indicated in equation (8), the specific conductance increases by an amount proportional to the amounts of ions dissolved from the sample.

In this method, Egan measured samples of plated parts using a constant 5 in^2 (32.3 cm^2) of sample area and 100 ml of deionized water which had been equilibrated with the carbon dioxide content of the ambient air. From equations

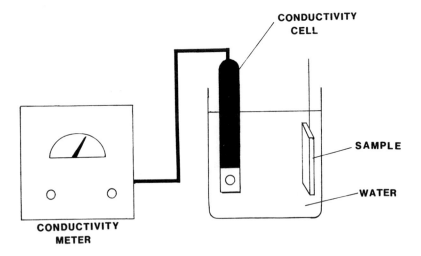

Figure 3. Egan's method for ionic residue measurement.

(7) and (8), the increase in specific conductance resulting from the ionic extraction is represented by:

$$\Delta L = k\Delta c = \frac{k\Delta g}{Mv} = K\Delta g \qquad (12)$$

where Δg is the total weight of an extracted ionic material of molecular weight M and v is the volume of the solution used. For numerous ionic materials extracted from the surface, ΔL will be the sum of the ΔL values for each material. Thus:

$$\Delta L = \Delta L_1 + \Delta L_2 + \cdots + \Delta L_n = K(\Delta g_1 + \Delta g_2 + \cdots + \Delta g_n) \qquad (13)$$

In practice, for simplicity, the total ionic extract level from a sample can be represented, as is often done for water purity measurements, as a specific conductance factor expressed in terms of a single salt, such as sodium chloride. While this does not provide an accurate measure of actual concentrations of the unknown ions, it enables us to represent ionic levels in a situation where many unknown ionic types may be present.

If all of the surface ionic content of a material is extracted, then its Δg is also the product of its original average surface ionic density, D, and the surface area, A, of the sample. Therefore, from equation (12):

$$\Delta L = \frac{kAD}{Mv} \qquad (14)$$

Thus if A, the sample area, and v, the solvent volume, are constant, or held in a constant ratio to each other, then ΔL is indicative of D, the average ionic density which was originally present on the surface of the sample (assuming total extraction of the ionic materials on the surface).

In order to facilitate measurement of ΔL, Egan used a conductivity monitor and a matched pair of conductivity cells. One cell was kept in a small amount of air-equilibrated water, while the other was used to measure the specific conductance of the extracting water in the measuring container. The monitor could be connected to either cell by means of a switch. The parts or assemblies to be measured were then placed in a measuring beaker and agitated for a fixed time. The specific conductance readings of both the extracting solution and the reference air-equilibrated solution could then be easily compared by switching the monitor from one cell to another to obtain L_f and L_0 and hence ΔL.

Advantages of this method are:

1. Measurement is facilitated by a simple reading of ΔL, which serves as an indication of ionic density on samples.
2. Compensation is made for atmospheric CO_2 absorbed into the solution

by allowing the cell to equilibrate with air. This provides a more precise determination of ΔL, assuming complete equilibrium of the water with atmospheric carbon dioxide has been achieved.

3. Specific conductance, rather than resistance, is employed, which facilitates interpretation of results in simple terms, directly proportional to concentrations or quantities of ions extracted from surfaces.

Potential problem areas of this method, however, are:

1. Great care must be exercised in handling containers and electrodes; they must be thoroughly cleaned between sample measurements and extraneous ionic contamination from sources other than samples must be avoided.

2. No effort was made to relate the specific conductance change, ΔL, to some equivalent-acting amounts of ions such as sodium chloride. Results using Egan's method are relative to one another and not to some tangible amounts of a reference electrolyte. It is possible, however, to calibrate the system by measuring the values of ΔL for known amounts of sodium chloride.

3. If nonpolar as well as ionic contaminants are present on the surface, extraction in pure water may not be effective. A good example of this is found where rosin fluxes are used in electronic assemblies. These materials usually contain ionic additives dispersed in a matrix of the nonpolar rosin. Ionic materials embedded in, or under, any rosin flux residues remaining after post-soldering cleaning will stay immobilized in the non-water-soluble, hydrophobic rosin. Other nonpolar materials such as oils and greases can have similar effects in masking the extraction of ions also present on the surface.

The stated objective of Egan's method was measurement of water-soluble residues on parts after plating, and extraction of the ions in pure water would be adequate for that purpose. However, for many other applications of this method where both polar and nonpolar contaminants could be present, a blend such as a water–alcohol mixture should be employed. (See Section 3.2.)

4. Inaccuracies can result from the concentration effects illustrated in Table 2. In a static extraction, such as is used here, measurement of the total extracted material is made when the solution has reached its maximum level of ionic content for that sample measurement. This can be particularly troublesome if the residues contain very weakly ionized materials, since the final reading is taken at the time when maximum concentration is reached, when errors due to concentration effects are greatest.

5. If a surface ionic contaminant is poorly soluble in the extracting solvent, it may not be fully solubilized in the measurement. This and all other static extraction methods can dissolve a residue only up to its solubility limit in the solvent used. Thus, residues such as lead sulfate, lead chloride, zinc carbonate, and many others might not be fully extracted by a limited volume of the solvent.

3.2. Method of Hobson and DeNoon

Another procedure employing extraction of ionic residues by a fixed volume of the solvent was developed by Hobson and DeNoon[14-16] at the Naval Avionics Center, Indianapolis, Indiana. In this method (Figure 4), a stream of the extracting solvent, an isopropanol–water mixture, is directed against the sample to dissolve ions present on the surface. The solution is collected in a graduated cylinder. Here too, the amount of solvent used is chosen so that a fixed ratio of solvent volume/surface area is maintained. As in Egan's method, the ionic concentration—assuming all surface ions are extracted—will bear a relationship to the ionic density of the surface. After collection, a sample of the rinse solution is measured separately for specific resistance.

The use of an alcohol–water mixture represents an important improvement in the effectiveness of the extraction of surface ions. As previously indicated, the use of pure water, alone, for the solvent can result in the masking of the extraction of water-soluble ions by the presence of any water-insoluble contaminants on the sample surface. The addition of alcohol to the solvent system extends its solvency characteristics to enable it to dissolve, to some degree,

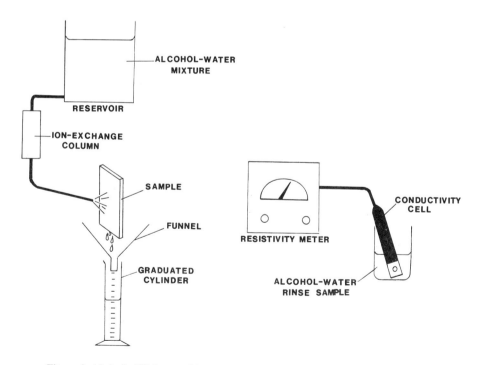

Figure 4. Method of Hobson and DeNoon for extraction of surface ionic contamination.

both types of contaminants. The ionic material is thereby liberated for extraction and measurement. Nonionic material will, however, have virtually no effect on the specific conductance readings, particularly at the very low concentrations in which they are present.

A blend of 75% (by volume) isopropanol, 25% water was chosen to assure good solubilization of rosin residues which may be present, since this procedure was aimed chiefly at measurement of post-soldering contamination on electronic circuits. Some water in the solvent medium is necessary to solubilize the ionic components of the residue and to sustain ionization of the more weakly ionizable materials. The sensitivity of response of a conductivity cell is very dependent on the amount of water, i.e., the alcohol/water ratio of the solvent system, as shown in Table 3. Very careful control of this ratio is, therefore, required in order to obtain consistent results in these measurements.

This procedure was included in U.S. Military specifications, such as MIL-P-28809, to serve as the basis of a means of monitoring ionic cleanliness of electronic circuit assemblies manufactured under government contracts.[17] This test specifies 10 ml of the 75% isopropanol solution to be used for each square inch of circuit area—counting both sides of the printed wiring assembly. The initial specific resistance specified for the solution must be greater than 6 MΩ-cm. In order to pass the specified test, the value of the final specific resistance after extraction must be greater than 2 MΩ-cm.

The advantages of this method include the use of alcohol–water mixtures to solubilize the surface residues and the simplicity and economy of the equipment required. Since this method involves extraction into a static volume of solvent solution, it has associated with it the same concern common to all static extraction processes, as indicated under Egan's method. A further difficulty is brought about by the use of specific resistance rather than a conductance function, which could simplify calculations and understanding of the extraction process.

This procedure can also be very "operator sensitive," varying considerably with operator technique or skill in flushing the surface to extract ionic

Table 3. Measured Increase of Specific Conductances for Injection of 100 μg of Sodium Chloride into 100 ml of Solvent Mixture[a]

Ratio of isopropanol/H$_2$O	%H$_2$O	Specific conductance increase
3/1	25	3.1
2/1	33.3	5.3
1/1	50	7.7
1/2	66.7	10.3

[a]Measurements by the author.

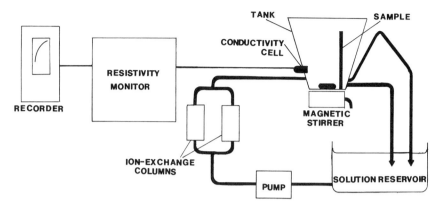

Figure 5. Schematic representation of the Omega Meter system.

material. The exposure of the surface to the alcohol–water mixture used to flush the surface contaminants is very brief, extending, perhaps, to one or two minutes. It is likely that some of the contaminant material will not be dissolved in the brief exposure to the dissolving mixture. This can be particularly true for electronic assemblies with flat components—such as integrated circuits—which are mounted on a circuit board with very little clearance to allow flushing of contaminants from under components. This would then result in some very significant errors, particularly since variable fractions of the total ionic residues can be dissolved in each test run.

3.3. Omega Meter

Another variation of static extraction is employed in the commercially available Omega Meter® (Kenco Industries, Inc., Atlanta, Georgia).[15, 18-20] This is a semi-automated system utilizing extraction into a volume of solvent solution which is in a fixed ratio with respect to the sample area, for a fixed time of immersion in the agitated solution (Figure 5). Alcohol water mixtures— most commonly 75% (by volume) isopropanol, 25% water—are used.

After pumping the alcohol–water mixture through mixed-bed ion-exchange columns, the deionized mixture is fed into the test tank. The sample is then introduced into the tank with agitation of the solution to accelerate solubilization of the surface ionic contamination. In various models, this agitation has been done in different ways, e.g., use of magnetic stirrers at the bottom of the tank or simple recirculation of the solution through the tank utilizing the pumping system which was used to fill the tank initially. By appropriate automated adjustment of valves, the solution can be recirculated without passing through the ion-exchange columns, thereby providing agitation and extraction into a fixed volume of the solvent mixture.

The values of solution specific resistances are measured with a conductivity cell and are indicated both on a recorder and a meter. The specific resistance, after a pre-established period of time of extraction, is thereby determined. The effective amounts and surface density of ions, in terms of sodium chloride, can be calculated using a special slide rule device calibrated for that purpose. On completion of the run, the alcohol-water mixture is recycled to the reservoir where it is held until needed for subsequent measurements.

More recently, a newer model of the Omega Meter was developed which uses a microprocessor to perform all of the necessary conversions and calculations, thus eliminating the use of the special slide rule to calculate the effective ionic density extracted from the surface of the sample. The system, shown in Figure 6, can also be preprogrammed with up to 30 values representing the areas of different standard sample types. At the start of a measurement run, the sample number is designated and the system will automatically fill the tank with a volume of solvent mixture to provide a volume/area ratio of 35 ml/in^2. During the extraction, it will print the values of ionic density removed from the surface of the sample at any time up to the termination time of the run. A printout is then delivered indicating, in addition to the calculated ionic density, the final resistance, alcohol percentage, board area, volume of solvent used, and run time.

The system also utilizes various sizes of interchangeable tanks to permit measurement of many sample sizes and shapes. Like the earlier models, the test cycle time can be specified by the operator; however, the microprocessor-controlled system can also terminate the run automatically when the amount of contamination detected in a two-minute period is less than 5% of the total contamination measured.

Advantages of this method are:

1. In contrast to the previously described extraction process, this procedure is not sensitive to the operational technique of the individual operator. Since the process is automatic, it is not necessary for the operator to clean containers with each sample measurement or to be concerned with handling the sample beyond its initial immersion into the tank.

2. Alcohol-water mixture is recycled automatically after each measurement, thereby conserving solution.

3. Time specified for each test is reasonably short—usually ten minutes or less—thereby allowing high productivity rates of sample measurement.

Problem areas with measurements using this system are:

1. The specific resistance function is used with its associated limitations, as described previously.

2. Running an extraction on a time-limited basis can give rise to basic errors since the extraction, depending on the nature of the surface and its con-

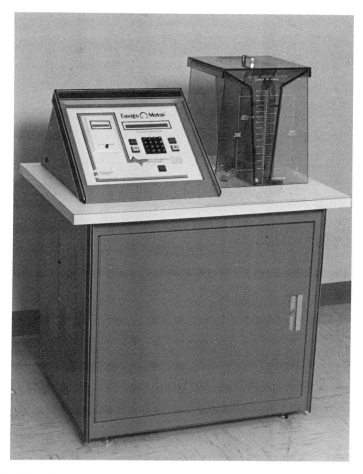

Figure 6. Omega Meter Model 600 with microprocessor control and calculation. (Courtesy of Kenco Industries, Inc.)

taminants, may not be complete in the limited time allotted. This can be particularly troublesome when the specific resistance function is used, since the sensitivity of the measurement is least near the end point of extraction, when the ionic concentration in solution is greatest (Figure 2, curve B). Solution of further quantities of ions will, therefore, not be sensitively indicated.

3. This system is subject to the same basic concentration considerations mentioned previously for static extractions such as deviations from true measurement at the higher ionic concentrations and limitation of complete extrac-

tion in a static solution due to solubility saturation effects of weakly soluble ionic materials.

4. Any carbon dioxide dissolving in the solution during the measurement period will also contribute to the reading and be read as if it were ionic material extracted from the sample.

3.4. Ion Chaser

Another system, called the "Ion Chaser,"[15,21] utilizes a 1-hour static extraction in an alcohol–water mixture (Figure 7). At the end of a static soaking period, the sample is removed and the solvent pumped through a dynamic conductivity monitoring system. (This will be described in detail later in Section 4 under the dynamic extraction method Ionograph.) To facilitate the continuity in measurements, an array of six tanks is utilized. By manipulating a series of valves, any single tank can be selected for measurement. Sample measurement in individual tanks can be staggered so that a different measurement can be made as often as every ten to fifteen minutes, although a full hour is actually allowed for each extraction.

The entire ionic content of all of the liquid in the tank is measured. Ionic density can then be calculated by dividing by the sample surface area. Readings made by this method tend to indicate somewhat higher values of extracted ionic content, not only because of the longer extraction time of the sample, but be-

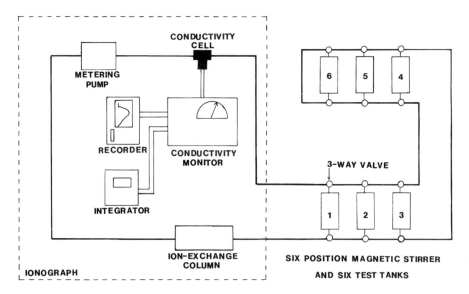

Figure 7. Schematic representation of the Ion Chaser system.

cause of greater ionic contributions by the atmosphere and the system during this longer extraction time.

The chief advantage of this system is that it allows a longer time for each static ionic extraction than most other techniques, thereby allowing for a more thorough solution and diffusion of ionic material from surface capillaries into the extracting solvent. If many samples are being measured, this is done in an effectively short time per sample, since the extraction times are overlapped.

This method, however, retains most of the fundamental difficulties associated with a static extraction process, i.e., the errors attributable to the absorption of atmospheric carbon dioxide into the solution and limitation of solution of weakly soluble contaminants on the sample surface into a limited volume of solvent mixture. This system is not commercially available.

3.5. Contaminometer

Another system for cleanliness measurement is the Contaminometer (Protonique, S. A., Romanel, Switzerland).[22] This system is a modular assembly comprised of the basic pumping/control unit together with separate commercially available components such as the conductivity measurement apparatus and, if used, a calculator or a microcomputer and associated software.

The system, which is available primarily in Europe, is in many respects similar to the Omega Meter which was previously described. It utilizes, however, a fixed volume of solvent (alcohol/water mixture) for all measurements in a tank regardless of the sample area. Thus there is no predetermined volume/area ratio maintained.

Initially the solvent mixture is deionized by pumping and recirculating it through a mixed-bed ion-exchange resin column. When a suitably low value of specific conductance is reached, the resin column is bypassed and a very vigorous flow of the solvent is established through the tank. This solvent is recirculated without deionization. The sample is then immersed into the tank which has been initially filled to an overflow level. The displaced liquid which then overflows the tank is collected and measured in a calibrated overflow compartment. An estimate of the entire sample surface area is then made based upon the measured volume of displaced liquid. As the system pumps, the extracting solvent increases in ionic content until a relatively stable specific conductance level is reached. The extraction is usually terminated within two to four minutes.

Two basic modes of contamination measurement are offered—manual and automatic. The manual method requires draining of a small quantity of the solvent from the system at various times and measuring its specific conductance separately from the tank. For this purpose a separate conductivity meter and cell are used. The results can be interpreted using either a nomogram or a programmable calculator with special software to calculate the results. The values fed into the calculator are: initial and final specific conductances, sample dimensions, volume of displaced liquid, and solution composition. The calculator then determines the average ionic density which was removed from the sample—most commonly an electronic circuit board.

The automatic method for contamination measurement utilizes an in-line conductivity cell to enable continuous sampling of specific conductance values. These values are fed to a microcomputer which, when programmed with special software, can continuously monitor the readings and plot a curve of extracted contamination vs. time. It will then print out the final contamination calculation.

A newer, updated version of the computer-controlled system has also been developed in which the entire modular system is mounted in one rack (Figure 8). This system offers flexibility by making available a variety of different tank sizes which can be changed, if needed, to best fit the dimensions of the intended sample sizes. This enables the establishment of an optimal sensitivity of response for samples of various sizes. Changing tanks, however, requires about one half hour.

Another version of this system is also available for measurement of smaller components and assemblies. This system, the Microcontaminometer, is designed with smaller tanks and pumping capacity to more sensitively accommodate samples with smaller surface areas. A range of eight tanks are available which can provide an optimum fit for most smaller parts or assemblies (up to 6 in^2).

The system which is sold under the name of Contaminometer represents a progression of ideas and different models. Starting with a simple measurement of specific conductance increase in one model, the means of calculation progressed to nomographs, calculators, and microcomputers with increasingly elaborate software. Control of the process and compensation of the readings were also added as well as analytical commentary in the case of printed circuit board contamination measurements.

One further improvement of the system's performance which has been made at the software level is compensation for carbon dioxide absorption from the air. Correction values are estimated by the computer based on the conditions of the run. These amounts are subtracted from the total ionic measurements so that the readings which are plotted and printed represent corrected values.

Advantages of this system are:

1. The time required for leveling off of the specific conductance increase, and hence the termination of the run, is relatively short—usually a few minutes.

2. A measurement is made of the solvent displaced from the tank, from which an estimate is made of the sample area. To do this, however, some assumptions must be made about the sample geometry since it is the total surface

Figure 8. Contaminometer, modular system with microcomputer. (Courtesy of Protonique, S.A.)

area which is required for the calculation of average surface ionic density rather than the volume of the sample. Nevertheless, this method for determining a value for the sample surface area is probably not less accurate than merely guessing at an estimated value, as is often done by many using the other procedures.

3. The modular assembly offers a range of choices of measurement techniques, e.g., manual or automatic control and analysis by computer.

4. The computer can, at least partially, compensate for errors ordinarily contributed by atmospheric carbon dioxide absorption. This, however, applies only to those automatic systems which utilize the computers and the software.

5. Specific conductance, rather than specific resistance, is measured. As

described previously, this provides a constant sensitivity of response to ionic concentration changes over a broad range of concentrations.

Disadvantages of this method are:

1. During the extraction process, the circulation and agitation of the static volume of solvent is high. This will accelerate the rate of absorption of carbon dioxide and, although the run time is generally a few minutes, can contribute significantly to the errors from this source. Those systems which do not have computer control with software for carbon dioxide compensation will be affected most severely by this problem. Even in cases where the computer factors in a value for atmospheric carbon dioxide correction, this is only an estimate based on assumptions programmed into the software, and the results could still be in error with respect to this factor.

2. As with all previously described static extraction methods, the Contaminometer takes its final reading at the level of maximum concentration of ions in solution, where deviations from true indication of ionic concentrations are greatest. Also limited solubilities of slightly soluble ionic residues can result in incomplete extraction while indication is given by the system that no further extraction of ions is taking place.

3. There is no attempt to maintain a constant ratio of solvent volume to sample area. This can result in a wide variation of sensitivity and accuracy of the readings for samples whose areas are greatly differing. In those systems where a different size tank can be interchanged, this problem will be alleviated somewhat.

4. Although the specific conductance appears to have leveled off within a few minutes in most cases, there will still be ionic material remaining to be extracted over longer periods of time. This is due to slower diffusion rates of ions out of surface porosity or capillary spaces, e.g., from under electronic components mounted close to a circuit board. It can also be caused by slower rates of solution of some contaminants which might be present. The need for considerably longer extraction times to allow for more thorough extraction has been shown by Duyck[34] to be crucial to measurement accuracy. While it may not be possible to extract 100% of all surface ionic contaminants, limiting the time to only several minutes will result in greatly reduced efficiency of ionic removal.

4. Dynamic Extraction Method—Ionograph

In all of the static extraction methods discussed thus far, the ionic material is extracted from the sample surface into a fixed volume of solvent, generally with the aid of agitation to accelerate the extraction. The ionic detection is then made on the final extract using either resistance or conductance functions.

Another method and system for measuring the extractable ions from a sample employs a theoretically unlimited supply of freshly deionized extraction solvent mixture. This method, represented by the Ionograph® (Alpha Metals, Inc., Jersey City, New Jersey), uses a dynamic extraction of ions into an alcohol–water mixture and continuous measurement of specific conductance throughout the extraction cycle.[15,23-25]

In this system (Figure 9), the alcohol–water mixture is pumped continuously by a metering pump through a mixed-bed ion-exchange column, then into a tank which contains the sample. The liquid is then passed through a conductivity cell and back to the pump to complete the loop. The measurement system continuously monitors the specific conductance of the solution and provides a DC voltage which is linearly proportional to the ionic content of the solution in the cell at any instant. This voltage is fed to the meter and a chart recorder which indicate the changes of specific conductance of the tank effluent solution over a period of time. The signal is also fed to an electronic integrator which generates a reading proportional to the time-integral of the specific conductance.

With the system filled with an alcohol–water mixture and the pump in operation, the recirculation of the solvent through the mixed-bed ion-exchange column will purge it of all ionic salt content. With no sample in the tank, the specific conductance falls until it reaches a very small constant value. This background ionic level corresponds to the "baseline" of the recorder curves

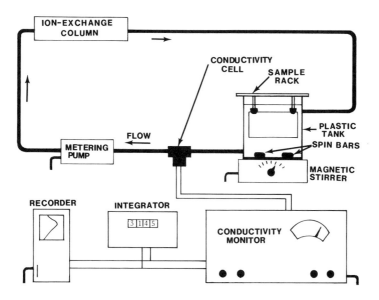

Figure 9. Schematic representation of the Ionograph system. (Courtesy of Alpha Metals, Inc.)

represented in Figures 10A and 10B. This baseline level represents the sum of conductance values contributed by three principal sources:

1. The intrinsic ionization of water determined by its ionization constant in the specific alcohol–water mixture.
2. Ionized carbonic acid formed by a dynamic equilibrium between the solution of atmospheric CO_2 at the surface of the liquid in the tank and the purified mixture flowing through the tank.
3. Traces of ionic materials extracted at a steady rate from the materials of construction of the system, e.g., tubings, valves, tanks, etc.

When an ionically contaminated sample is introduced into the tank, extraction of ions will cause the solution conductance to rise rapidly (Figure 10A). This will be indicated by the conductivity monitor and integrator as the solution is pumped through the cell. As the ions are stripped from the sample and purged through the system, the specific conductance will fall, gradually returning to the original baseline. When the conductance has reached this baseline, it can be concluded that all of the extractable ionic material has been removed from the sample and tank. The response of the instrument at this point is identical to the original response without any sample in the tank. The completion of ionic

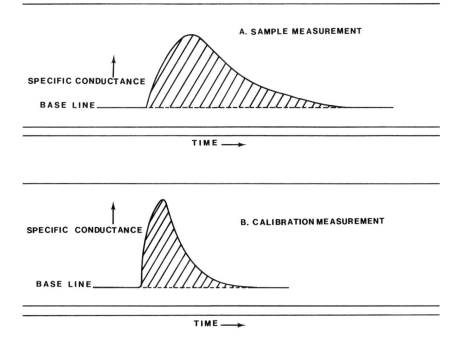

Figure 10. Ionograph response curves: (A) sample measurement; (B) calibration.

removal is very sensitively indicated since the measurement near the end point (return to the baseline) is made at the lowest concentration of ions, so that the slightest additional ionic content is easily discernible relative to the low background level.

The entire amount of ionic material removed from the sample can be related to an integration of the specific conductance function over the time of extraction, i.e., the area of the curve above the baseline. At any time, t, the number of moles of ionic material within the conductivity cell, n_t, is $n_t = v_c c_t$, where c_t is the concentration of ions and v_c the volume of the cell, a constant. Over an infinite amount of time, the total number of moles of ions, N, passing through the cell will be:

$$N = \int_0^\infty n_t \, dt = v_c \int_0^\infty c_t \, dt \qquad (15)$$

Since we are dealing with very low ionic concentrations ($<10^{-4}$ M), we can assume complete ionization and that L is proportional to c [equation (6)] (assuming only one salt is present for purposes of clarity). Substituting for c in the equation above, it can be seen that the total amount of extracted salts is proportional to the integral of the specific conductance readings above the baseline:

$$N = \frac{v_c}{k} \int L_t \, dt \qquad (16)$$

If the conductivity monitor and recorder responses are linear with respect to L and the pumping rate is constant, the area under the specific conductance vs. time curve is a function of the total amount of ions removed from the sample. The area above the baseline can be integrated using the electronic integrator. The instrument can then be calibrated by measuring the response obtained by injecting into the tank a known amount of a standard sodium chloride solution.

Figure 10B represents the curve obtained in such a calibration. It shows the buildup of the specific conductance to a peak value followed by an exponential decline as the salt is purged through the system. Comparison of this curve with Figure 10A shows a more rapid buildup of the conductance to the peak and a more rapid decline to the baseline. In the case of the calibration curve, the entire ionic content had been injected into the solution at the start of the measurement. There is no dependence on rates of solution or diffusion of ionic materials out of surface capillaries or from under components as in the sample measurement. The declining part of the curve of Figure 10B is then determined simply by the pumping rate and dilution kinetics in the volume of the tank.

A sample measurement can be related to the overall response of the system to a known quantity of ions of the standard calibrated salt solution. A response factor can be determined which, in turn, can be employed to determine the effective amounts of ions extracted from subsequent unknown samples:

$$F = \frac{g_c}{Q_c} \qquad (17)$$

where F is a response factor, g_c is the mass of NaCl used in the calibration, and Q_c is the integrator reading for the calibration.

In any subsequent measurement of an unknown sample, the final integrator reading for that sample is then multiplied by F to indicate the total amount of extracted ions expressed as NaCl:

$$Q_s \times F = g_s \qquad (18)$$

where Q_s is the integrator reading of the measured sample and g_s is the total mass of NaCl (equivalent) extracted from the sample.

If the average ionic density, D, of the ions originally present on the surface of the sample is to be calculated, it is necessary only to divide g_s by the sample surface area, A_s:

$$D = \frac{g_s}{A_s} \qquad (19)$$

An important benefit of the dynamic measurement method is derived from the fact that all measurements are made only of the integrated signal above the baseline. Since the initial nulling and balancing of the integrator prior to sample insertion is against a baseline representing the existing dynamic equilibrium with atmospheric carbon dioxide, the carbonic acid contribution is below the baseline. The reading of the integrator above the baseline, therefore, represents only the contribution by the ions extracted from the sample. In a like manner, any steady-state evolution of low levels of ions from the system materials or atmospheric dust fallout will also be subtracted out with the baseline.

There are only two factors which affect the sensitivity function, F, in equation (18):

1. Pumping speed—Since the calibration is related to the time integral of the specific conductance, increasing the flow rate of solvent will decrease the time over which the ions will be purged through the system, thereby decreasing the area under the curve which is integrated. Thus, in effect, the sensitivity of the instrument will be decreased (F increased) by an increase in pumping speed.

2. Alcohol–water ratio—This can affect both the degree of ionization of weakly ionized materials and the activity coefficients of ions in a conductance measurement. Increasing the water content of the two-component alcohol–water solvent system raises its dielectric constant, which directly affects the activity

coefficients of the ions in solution according to classical Debye–Hückel theory[11–13] of ions in solution.

Other factors such as the tank volume will affect the shape of the curve but not the actual area under the curve and above the baseline. Thus a larger tank—all other conditions being the same—will yield a longer (in time) and flatter (because of dilution) curve than a smaller tank. The total area under the curve above the baseline will, however, be the same for equal amounts of ionic materials measured in each tank.

In 1982, a newer model Ionograph was developed which utilizes a microprocessor to compensate for changes or variations of the flow rate and alcohol–water ratio (Figure 11). Flow rate is detected by a flow sensor in the closed loop and fed to the microprocessor which compensates the calibration factor continuously throughout the measurement. Also the specific gravity and temperature of the solution in the tank can be keyed into the system and converted to an alcohol–water ratio value which is then used to internally compensate for solution composition. This can eliminate the need for any direct day-to-day

Figure 11. Ionograph with microprocessor for calculation and compensation. (Courtesy of Alpha Metals, Inc.)

calibration using standard NaCl solutions since the microprocessor is programmed to compensate for changes in the two variables affecting calibration. The system can thus be factory-calibrated.

The microprocessor is also utilized for other timesaving chores such as storing values of sample areas, calculating total extracted NaCl equivalent, and computing the mean ionic density extracted from the surface and comparing it to a predetermined pass/fail limit to indicate inspection status. A printout record is also provided which includes the conductance/time curve and the measured and calculated results for that run.

Advantages of the dynamic extraction method are:

1. Contributions to conductance by dissolved atmospheric CO_2 or by ionic dissolution from equipment materials are below the baseline and thus are automatically subtracted from the measured values.

2. Limited equilibrium dissolution of partially soluble contaminants is avoided by the dynamic extraction into a theoretically infinite volume of deionized solvent.

3. Completion of extraction is self-indicating. In any extraction method, the diffusion rates of ions from surface porosity and the finite rates of solution of various surface contaminants limit amounts of ions which can be removed in a finite period of time. In the dynamic extraction method, completion of the extraction is demonstrated by the return of the conductance to the original baseline value. This indicates that at that point the equilibrium ionic level in the tank is the same as before the sample was immersed and no additional ionic material is being extracted from the sample.

4. During the period of extraction, the tank is continuously purged by freshly deionized solution. The ionic concentrations are thus not allowed to build up continuously as is the case in the static extraction. This, then, will minimize the measurement errors due to higher concentrations of ions in solution. This will be particularly applicable, as indicated in Table 2, if the extracted materials contain weak electrolytes.

5. The system is calibrated directly against sodium chloride. The total ionic contribution to the conductance of the unknown sample is, therefore, expressed as an amount of sodium chloride capable of contributing equally to the solvent conductance. This presents a great convenience in conceptualizing the ionic extraction in terms of quantities of ions rather than its dependent variables—the resistance or conductance of the solution.

The major difficulties encountered with this method are:

1. In order to obtain the most accurate results, it is necessary for the system and sample to purge themselves of ions completely to enable the conductance to fully return to the baseline value. Both the finite rates of surface ionic extraction and the pumping dynamics of the system can serve to limit the speed

at which the measurement is completed. A typical measurement run to a full return to the baseline may, therefore, require a longer extraction time than an arbitrarily terminated static measurement.

2. Since all dynamic conductance readings are referred to a baseline value which is established prior to the introduction of the sample into the tank, accuracy of the results is very dependent upon the initial accuracy of establishing this baseline value. It is, therefore, necessary, for best results, to allow sufficient time for the system to establish a fully equilibrated baseline prior to the introduction of each sample to be measured. This function is done automatically in the newer, microprocessor-controlled model of the Ionograph.

3. For best results, an assortment of tank sizes and geometries is needed to handle a variety of sample sizes and shapes. Optimum results are obtained when the tank volume can be the minimum necessary to completely submerge the sample in solvent. For different tank volumes, the pumping rate should then also be adjusted to provide the same relative turnover rate of the solvent in the tank. This, then, requires adjustments to be made to the system in changing, in any major way, the sample size or shape. It also requires, for optimum performance, the availability of several tanks to handle a variety of sample sizes and geometries.

5. Applications of Ionic Contamination Measurements

As previously mentioned, the chief application of surface ionic contamination measurements, at the present time, is for electronic circuits and components. The methods of measurement, described in this chapter, are used for controlling quality of cleaning processes for both military electronics and non-military commercial applications. Some of the specified methods for measurement as well as their acceptable limits are indicated in MIL-P-28809. It should be remembered that all of these methods indicate ionic extraction from the total surface of each sample. The total ionic material extracted is thus averaged over the entire surface area. The results, as applied in quality control of electronic assemblies, can therefore differ in terms of any resulting effects of observed ionic levels on a circuit's performance.

It is possible, for example, to have a sample with an excessively high concentration of ionic contamination, such as a fingerprint,[23,36] in a localized but critical area of the surface, e.g., between two conductive tracks in a high-impedance portion of a circuit. This could then give rise to performance problems even though the averaged ionic density readings obtained by these extraction methods indicate "safe" levels of contamination.

In addition to quality control, the measurement techniques can be used to determine the effects of various cleaning process variables on the final cleanliness of an electronic circuit. Process variables such as cleaning method, handling, equipment, chemicals, time, temperature, agitation, storage, and others can be studied in terms of their effects on the ultimate cleanliness level.[23, 26-36] These methods of measurement can, therefore, provide much of the information needed to develop the most effective cleaning materials and processes for highest levels of reliability of electronic products.

6. Future Prospects

Looking toward the future, a number of further innovations in this area would appear likely—at least if necessity dictates the course of future developments. Most of these applications are projected in the rapidly changing area of electronics.

At the present time, there are major changes taking place in the technology of circuit assembly. Throughout the world there has been a growing trend toward circuit miniaturization by the use of flush mounting of components directly to the metallic circuitry on the surface of a circuit board. Typically, smaller, unleaded components are used which are mounted and soldered nearly flush against the circuit board, with the circuitry also scaled down in size. The very small clearances (often 0.1 mm or less) of the components from the circuit board can make cleaning of ionic flux residues from under flat, flush-mounted components very difficult and questionable.

The same problem of extracting these ionic materials is encountered in any attempt to measure them. Of course, if they are not removed into the extracting solvent, they will not be measured. Further techniques and processes are needed, therefore, to enhance the extraction of ionic materials from very narrow spacings. New methods employing techniques such as high-velocity directed jets of the solvent, ultrasonic agitation, or elevated-temperature ionic extraction utilizing solvent systems other than the conventional alcohol/water mixtures will, in all probability, be considered in future systems to address this problem.

Another need is for methods to measure ionic material within limited localized areas of a larger surface rather than from the entire surface immersed in the extracting solvent. This would enable measurements to be confined to the most critical regions of an electronic circuit, which are most sensitive to the presence of ions and related electrical leakage currents.

Finally, a most sophisticated and potentially useful development would be an extension of the ion extraction process to include identification and measurement of specific ionic types rather than a mere measurement of a total amount of unidentified ions, as is presently done. Such analyses could provide extremely valuable clues to the sources of the contamination resulting, potentially, in more effective prevention or removal of the contamination in a given situation. Some work in this area has already been done by Wargotz,[36] who used ion chromatographic techniques to identify and quantify extracted ionic types. Further development of this technique in the future is a possibility.

7. Summary

Ionic materials contaminating a surface can be detected and sensitivity measured by taking advantage of the contributions of ions in solution in a polar liquid to the electrical specific conductance of that solution. For that purpose, the ions must first be extracted into the solvent medium to make them available for measurement.

There are, fundamentally, two basic processes for ionic extraction—the static and the dynamic extraction methods. In the static method, ions are extracted into a fixed or limited volume of the solvent, and the final solution with the total accumulated ionic material is measured. The dynamic method extracts the ions into a theoretically infinite volume of the extracting solvent over a period of time. The conductivity is measured continuously during the time of the run and the quantity of ionic material is determined from the time integral of the conductivity function.

Various schemes and equipment employing each of these extraction principles have been described in this chapter along with their advantages and potential difficulties. Many of these methods and systems have found important applications in areas such as processing of electronic assemblies where freedom from ionic contamination is often a major requirement.

References

1. K. L. Mittal, Surface contamination: An overview, in: *Surface Contamination: Genesis, Detection and Control* (K. L. Mittal, ed.), Vol. 1, pp. 3–45, Plenum Press, New York (1979).
2. W. A. Yager and S. O. Morgan, Surface leakage of pyrex glass, *J. Phys. Chem.* 35, 2026–2042 (1931).
3. R. F. Field, Formation of ionized water films on dielectric under conditions of high humidity, *J. Appl. Phys.* 17, 318–325 (1946).
4. S. W. Chaikin and F. M. Church, Character of insulator surface leakage at high humidity, *IRE Transactions of the Professional Group on Component Parts*, Vol. CP-5, No. 4, 153–156 (1958).
5. S. W. Chaikin, C. W. McClelland, J. Janney, and S. Landsman, Contamination and electrical leakage in printed wiring, *Ind. Eng. Chem.* 51, 305–308, (1959).
6. W. B. Wargotz, Quantification of contaminant effects upon electrical behavior of printed wiring, *IPC Tech. Rev.*, 9–15 (January, 1978). Also *Circuits Manufacturing* 18(2), 42–49 (1978).
7. T. F. Egan, Determination of plating salt residues, *Plating* 50(4), 350–354 (1973).
8. N. E. Dorsey, *Properties of Ordinary Water Substance*, pp. 374–375, 380, Reinhold Publishing Corp., New York (1940).
9. F. N. Alquist, Preparation and maintenance of high purity water, in: Symposium on High

Purity Water Corrosion, 58th Annual Meeting of the American Society for Testing and Materials (June 28, 1955), Publication No. 179, ASTM, Philadelphia, Pa.
10. D. Eisenberg and W. Kauzmann, *The Structure and Properties of Water*, Oxford University Press, New York and Oxford (1969).
11. H. S. Harned and B. B. Owen, *The Physical Chemistry of Electrolytic Solutions*, 3rd Ed., ACS Monograph No. 137, Reinhold Publishing Corp., New York (1958).
12. F. Daniels and R. A. Alberty, *Physical Chemistry*, 2nd Ed., John Wiley and Sons, New York, London (1961).
13. J. O'M. Bockris and A. K. N. Reddy, *Modern Electrochemistry*, Plenum Press, New York (1970).
14. W. T. Hobson and R. J. DeNoon, Printed Wiring Assemblies, Detection of Ionic Contaminants On, Materials Research Report 3-72 (1972), Naval Avionics Center, Indianapolis, Ind.
15. W. T. Hobson and R. J. DeNoon, Review of Data Generated with Instruments Used to Detect and Measure Ionic Contaminants on Printed Wiring Assemblies, Materials Research Report 3-78 (1978), Naval Avionics Center, Indianapolis, Ind.
16. W. T. Hobson, Testing for ionic contaminants, *Contamination Control* 3(1), 9 (March, 1981).
17. MIL-P-28809, Military Specification, Printed Wiring Assemblies.
18. E. W. Wolfgram, Means and Method for Measuring Levels of Ionic Contamination, U.S. Patent 4,023,931 (May 17, 1977).
19. I/C Staff, Measure ionic contamination level on the production line, *Insulation/Circuits*, 21(13) (1975).
20. Omega Meter II, Bulletin No. 379, Kenco Industries, Inc., Atlanta, Ga.
21. W. Kenyon, Fundamentals of Ionic Contamination Measurement Using Solvent Extract Methods, IPC-TP-177, The Institute for Interconnecting and Packaging Electronic Circuits, Evanston, Ill. (September, 1977).
22. B. N. Ellis, *Handbook of Contamination*, Protonique, S.A., Romanel, Switzerland (English ed., 1981).
23. J. Brous, Evaluation of post-solder flux removal, *Welding Journal, Research Supplement*, 444s–448s (December, 1975).
24. J. Brous, Self-Purging Apparatus for Determining the Quantitative Presence of Derived Ions, U.S. Patent 3,973,572 (Aug. 10, 1976).
25. J. Brous, Extraction methods for measurement of ionic surface contamination, in: *Surface Contamination: Genesis, Detection and Control* (K. L. Mittal, ed.), pp. 843–855, Plenum Press, New York (1979).
26. R. Woodgate, Use of resistivity meters in pc cleaning systems, *Electronic Production* 10(2), 26 (1981).
27. P. R. Taylor, P. L. Altavilla, and G. J. Simonds, Solvent Extract Resistivity—Is It Meaningful?, IPC Technical Report #TR-41, IPC, Evanston, Ill.
28. H. E. Phillips, Ionic residue removal: Which solvent is best, *Elect. Packag. Prod.* 13(9), 177–180 (1973).
29. W. G. Kenyon, How to use the solvent extract method to measure ionic contamination of printed wiring assemblies, *Insulation/Circuits* 27(3), 47–49 (1981).
30. P. Altavilla, Analysis of chloride contaminants on pcb's reveals causes, suggests preventative action, *Insulation/Circuits* 20(10), 46–48 (1974).
31. D. F. Ball, Equipment parameters for aqueous flux removal after IR reflow, *Insulation/Circuits* 26(11), 55–60 (1980).
32. J. W. Dennison, Jr., Cleaning of printed circuits boards to remove ionic soils, *Materials Performance* 14(3), 36–40 (1975).
33. D. Howell, Make pc boards come clean, *Electronic Products Magazine*, 27–34 (June 1976).
34. T. O. Duyck, Testing large printed circuit boards for cleanliness, *Insulation/Circuits* 24(11), 38–41 (1978).

35. C. J. Tautscher, New generation of board and assembly cleanliness tests, *Insulation/Circuits,* 25(13), 15–19 (1979).
36. W. B. Wargotz, Ion Chromatography Quantification of Contaminant Ions in Water Extracts of Printed Wiring, IPC Technical Paper #TP-248, IPC, Evanston, Ill. (1978).
37. C. W. Jennings, Filament formation on printed wiring boards, in: How to Avoid Metallic Growth Problems (IPC Blue Ribbon Committee Report), IPC-TR-476, Institute of Printed Circuits, Evanston, Ill. (September, 1976) pp. 25–32.
38. W. G. Kenyon, The Evaluation of Commercial Flux/Solder Deflux Systems, Part I: The Effect of Incoming Printed Wiring Board Contamination on Post-Assembly Test Results, DuPont Technical Report RP-11, E. I. DuPont de Nemours & Co., Wilmington, Del., pp. 14–15 (1982).

5

Characterization of Surface Contaminants by Luminescence Using Ultraviolet Excitation

Tuan Vo-Dinh

1. Introduction

Surface detection and characterization of organic contaminants have recently received considerable interest.[1] This is due to the increasing awareness that contamination of skin and other surfaces by organic pollutants is a serious problem in many industries.[2] Organic compounds on surfaces can be monitored by a variety of analytical techniques such as photoacoustic, Raman, multi-reflection infrared, and luminescence spectroscopies. Among these spectroscopic tools, luminescence spectroscopy, which generally uses ultraviolet (UV) radiation for excitation, is the most sensitive method of detection for most organic compounds, especially for materials containing polycyclic aromatic hydrocarbons (PAH). Many PAH compounds are present in various chemicals, environmental samples, and coal tar materials, as well as in many oils and greases used in industry. These PAH compounds are also produced by incomplete combustion of organic substances in many industrial and residential activities. Since most PAH compounds absorb UV radiation and many of them are strongly luminescent, the luminescence technique provides an extremely useful tool for detecting surface contamination by these materials.[3] This chapter reviews the

---

Tuan Vo-Dinh • Advanced Monitoring Development Group, Health and Safety Research Division, Oak Ridge National Laboratory, Oak Ridge, Tennessee 37831. Research sponsored by the Office of Health and Environmental Research, U.S. Department of Energy, under contract DE-AC05-84OR21400 with Martin Marietta Energy Systems, Inc. By acceptance of this article, the publisher or recipient acknowledges the U.S. Government's right to retain a nonexclusive, royalty-free license in and to any copyright covering the article.

basic methodology, instrumentation, and applications of the luminescence technique using UV-radiation excitation to detect surface contaminants.

2. The Luminescence Technique for Surface Detection

Luminescence is the general term referring to the process of light emission from a molecule that has been elevated to an excited electronic state by absorption of energy. Although a variety of energy sources can be used, excitation is most conveniently achieved by irradiation of the molecule with UV light. In general, "near-UV" radiation (>260 nm) is used because high-energy radiation can photodegrade the organic compounds. The luminescence process can either be fluorescence or phosphorescence. Fluorescence is detected when the molecule returns to the ground state by emitting light without a change in the spin quantum number of the molecule. If the electronic transition responsible for the observed emission occurs with a change in the spin state of the molecule, the emission is termed phosphorescence.

Measurement of the fluorescence (or phosphorescence) intensity permits identification and quantification of many organic materials at ultratrace levels.[4] The compounds may be characterized by their emission and/or excitation spectra. The emission spectrum is obtained by recording the fluorescence intensity as a function of the emission wavelength. The excitation spectrum is recorded by varying the wavelength of the excitation radiation while maintaining a fixed observation wavelength. A third type of measurement, synchronous excitation spectroscopy, involves scanning both excitation and emission wavelengths simultaneously while keeping a constant wavelength interval between them. This novel approach for monitoring organic pollutants has been recently reviewed in the literature.[5]

The foundations of quantitative luminescence spectrometry for optically homogeneous samples such as dilute liquid solutions are based upon the Lambert–Beer law of absorption.

The luminescence intensity, I, can be expressed by the following basic equation:

$$I = k\Phi(I_0 - I_T) \tag{1}$$

where Φ = luminescence quantum efficiency, I_0 = incident light intensity, I_T = transmitted light intensity, and k = experimental parameter.

For optically homogeneous and perfectly transparent samples, the fundamental expressions of quantitative luminescence spectrometry can be derived from the well-known Lambert–Beer law:

$$I_T = I_0 k 10^{\epsilon c x} \tag{2}$$

where ϵ = molar extinction coefficient, c = analyte concentration, and x = thickness of the sample.

It is obvious from equation (1) that the analyte must exhibit a reasonably large fluorescence quantum efficiency in order for fluorescence to be a useful analytical method for that analyte.

Equation (1) can be written as

$$I = I_0 k \Phi (1 - 10^{\epsilon c x}) \qquad (3)$$

The quantity $1 - 10^{\epsilon c x}$ can be expressed by a power series approximation:

$$1 - 10^{\epsilon c x} = \epsilon c x - \frac{(\epsilon c x)^2}{2!} + \frac{(\epsilon c x)^3}{3!} + \cdots \qquad (4)$$

For small absorbances ($\epsilon c x < 0.15$ at the exciting wavelength), the higher-order terms in equation (4) are negligible with respect to $\epsilon c x$. In that case equation (3) simplifies to

$$I = I_0 k \Phi \epsilon c x \qquad (5)$$

As already mentioned, the above equation is only applicable to luminescence measurements of homogeneous liquid solutions. The Lambert–Beer law, however, cannot be applied to surface luminescence since the conditions assumed for the Lambert–Beer law are not valid for surface detection where the sample support is neither optically homogeneous nor transparent. Kubelka and Munk derived the differential equations and developed a set of equations that provide the foundations for many quantitative studies of absorption, reflection, fluorescence, and phosphorescence processes in diffuse scattering media.[6] These equations which have been used extensively by the paper and paint industries, are often known as the Kubelka–Munk (K–M) equations.

The K–M differential equations are given by:

$$-\frac{di}{dx} = -(S + K)i + Sj \qquad (6)$$

$$\frac{dj}{dx} = -(S + K)j + Si \qquad (7)$$

where i = intensity of light propagating inside the sample in the transmitted direction, j = intensity of light propagating inside the sample in the back-scattered direction, S = scatter coefficient per unit thickness, K = absorption coefficient per unit thickness, and x = distance from the nonilluminated side.

Equations (6) and (7) simply describe the fact that the light beam traveling in the transmitted direction (i) decreases in intensity due to the absorption (K) and scattering (S) processes and gains intensity from the scattering process that occurs to the beam coming from the other direction (j).

The basic assumptions in the K–M model are that the sample is an ideal diffuser, planar, homogeneous, and illuminated on one side with diffuse monochromatic light that strikes perpendicularly to the surface. It is also assumed that the reflection and absorption processes occur at infinitesimal distances and are constant over the area under illumination and over the thickness of the sample.

It is also implicitly assumed in the one-dimensional approximation of the K–M model that the reflected light beam is normal to the sample surface.

Various researchers have solved the Kubelka–Munk equations to obtain the luminescence intensity expressions for surface luminescence.[7,8] Zweidinger and Winefordner[7] derived the expressions for the phosphorescence intensities emitted in optically inhomogenous and nontransparent media such as snowlike matrices by integrating the K–M differential equations.[6] These equations would find useful applications in luminescence analysis of solid substrates.

Another derivation of the K–M differential equations described by Goldman for fluorescence densitometry should also be applicable to surface luminescence analysis.[8] In that study, Goldman took into consideration both the excitation and luminescence radiation explicitly in the differential equations. Goldman derived two sets of differential equations for the illuminating light beams (i, j) and fluorescence light beams (I, J), respectively. Goldman used one value for the absorption coefficient K but introduced two different scattering coefficients: s for the illuminating (excitation) beam and S for the luminescence beam.

All these theoretical derivations lead to one conclusion of practical importance to analytical applications. Quantification is always possible below a certain concentration level because a simple relationship exists between the luminescence intensity and the contaminant concentration. At sufficiently low analyte concentrations, the slope of the analytical curve ($\log I$ vs. $\log C$) is 1. The corresponding intensity is a linear function of the concentration. In the example of a coal liquid spotted on filter paper shown in Figure 1, this linear relationship occurs usually at about or under 30 ng/cm^2. The analytical curve reaches a maximum at higher concentration.

It should be noted that above the concentration level where the luminescence intensity is not directly proportional to analyte concentration, quantification is still possible if there is a reproducible one-to-one relationship between intensity and analyte concentration.

At high concentrations, prediction of the actual shape of the analytical curve may be complicated by instrumental factors and photophysical interferences, such as filter effect, quenching, and energy transfer. These effects may be so great that the luminescence intensity may even decrease as the concentration becomes too high. In addition, the contaminants may be distributed in several molecular layers and, hence, may be incompletely illuminated.

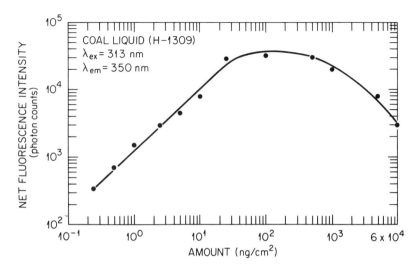

Figure 1. Fluorescence analytical curve (log intensity vs. log concentration) of a coal liquid sample spotted on filter paper.

3. Applications

3.1. The Use of UV "Black Light" for Surface Detection

Detection of contamination on skin and other surfaces using induced fluorescence is not a new concept. Until recently, this method has only been performed in a coarse manner by simple visual examination. In this method, a hand-held "black light" lamp is used to illuminate the body parts of a person and the induced fluorescence is examined with the naked eye in a dark room. The contaminants are detected when they exhibit a fluorescence emission brighter than that of the clean surface or of the skin.

This method has several disadvantages. First, detection has to rely upon the sensitivity and discriminating ability of the operator's eye. Second, quantitative information cannot be obtained from subjective human elements. Third, accidental illumination or reflection of the UV radiation may damage the eye. Finally, the intensity of the hand-held black light lamp is such that it might cause concern about the phototoxicity of UV radiation and the synergistic effects of light and chemicals. Several studies have indicated that UV radiation enhances the activity of chemical carcinogens.[9,10] Ultraviolet irradiation of anthracene, a noncarcinogenic compound, has been shown to induce carcinomas in mice.[11] The photocarcinogenicity of anthracene may be related to its photoinduced covalent bonding to DNA.[12] Although the actual effects of UV radiation remain to be further investigated, the intensity of the UV excitation light has to be selected with care. The threshold limit value (TLV) for occupational exposure to near-UV radiation (320–400 nm) has been set by the American Conference of Government Industrial Hygienists at 10^7 ergs/cm^2 for an exposure period less than 1000 s.[13] This value does not apply to individuals concomitantly exposed to photosensitizing chemicals. For these individuals, it has been suggested that only a UV radiation level equivalent to or less than sunlight may be considered safe. Because of their high levels of UV radiation, the hand-held black lights are presently not recommended for routine industrial hygiene practices on workers.

3.2. Study of Workers' Skin Contamination by the "Skin-Wash" Method

Keenan and Cole described a "skin-wash" procedure to assess the level of skin contamination of workers by coal liquid products.[14] In the skin-wash method, a spraying cup is pressed against the skin and approximately 10 ml of cyclohexane is sprayed onto the skin. The contaminants are removed into the rinsing solvent and drained into a glass vial. Sampling is terminated when the level of the cyclohexane solvent in the glass vial reaches a predetermined level. The glass vial is then removed, sealed, and sent to the laboratory for fluorimetric analysis. The skin-wash sampling device is shown in Figure 2. The skin-wash method was able to provide quantitative determination of heavy distillate fuel oil at submicrogram levels. As discussed by Keenan and Cole, the use of cyclohexane sprayed onto the skin may facilitate penetration of the contaminants into the skin because organic solvents are known to penetrate the skin of

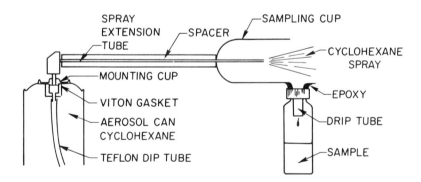

Figure 2. Skin-wash sampling apparatus. Reprinted with permission from R. R. Keenan and S. B. Cole, *American Industrial Hygiene Association Journal 43*, 473 (1982). Copyright 1982 by the American Industrial Hygiene Association.

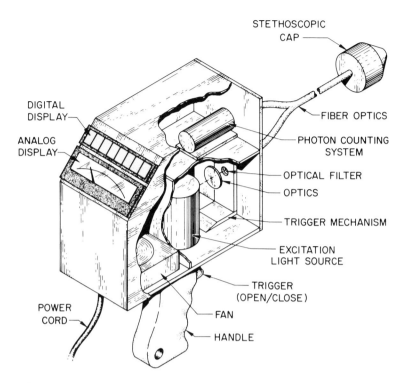

Figure 3. Schematic design of the fiberoptic luminoscope. Reprinted with permission from T. Vo-Dinh and R. B. Gammage, *American Industrial Hygiene Association Journal 42*, 112 (1981). Copyright 1981 by the American Industrial Hygiene Association. A field portable instrument is commercially available from Environmental Systems Corporation, Knoxville, Tennessee 37912.

man and animals. As a result, the skin-wash method should be used with extreme care because it might increase the exposure dose by enhancing the penetration of contaminants into the skin.

3.3. The Fiberoptic Luminoscope for Monitoring Occupational Skin Contamination

Because of the importance of the risk associated with skin contamination and the lack of adequate monitoring instruments, a portable instrument has been developed for detecting organic contaminants on the skin.[15] This instrument, the luminoscope, is based upon a bifurcated fiberoptic lightpipe that transmits the ultraviolet (UV) excitation radiation onto the surface area being monitored and conveys the fluorescence emission back onto the detector (Figure 3).

The radiation from a 125-watt mercury lamp is focused onto the excitation entrance of the waveguide. A set of interference filters transmitting light from 400 to 700 nm are used to select the spectral regions at which the fluorescence emission is to be detected. The top of the common leg of the bifurcated waveguide is mounted into a stethoscopic cap. An optical lens mounted in the stethoscopic cap serves to focus the excitation light from the lightpipe tip onto the area being monitored and to transmit the luminescence from the same surface back into the fiberoptics tip. During the measurement, the open end of the stethoscopic cap is pressed against the targeted area of the skin. The single-photon counting technique, which offers excellent sensitivity for low intensity levels,[16] is used for detection.

The UV light source is enclosed within a compartment designed to shield the photon counter and its associated electronic circuits from the heat generated by the lamp. A small fan flushes cool air into the upper compartment that houses the detection system. The trigger mounted at the instrument handle is used to operate a dual shutter that opens and closes the excitation and emission apertures simultaneously. The hand-held instrument is designed to be low cost and simple to operate. A photograph of the first prototype version of the instrument being field-tested at a synfuel pilot plant is shown in Figure 4.

Figure 4. Use of the luminoscope at a synfuel pilot plant.

Measurements of oils and tars from a variety of processes have been made with the lightpipe luminoscope.[17] The instrument is capable of detecting trace amounts of various coal tars at submicrogram levels. Coal distillates and recycle solvents from liquefaction processes can be detected at concentrations of a few nl/cm^2. The measurements of these materials have also been carried out using hamster skin.

Several field tests were conducted with the luminoscope at various synfuel production plants to evaluate the performance of the prototype apparatus in a real-life workplace environment and to test the applicability of the instrumental concept in actual measurements.[17] The body parts of the workers at this coal gasifier most likely to be directly exposed to coal and tar are the areas that are not protected by clothing, e.g., hands, arms, and faces. The measurements performed during this field trip demonstrated the ability of the luminoscope to detect trace contaminants of coal products on workers' skin. The fiberoptics luminoscope is manufactured and commercially available for industrial hygiene and environmental control applications.[18]

3.4. Detection of Surface Contamination with the Spill Spotter

A spill spotter has been developed for monitoring surface contamination caused by spill or leakage of hazardous materials in coal conversion facilities.[19] The prototype instrument consists of an optical unit that contains the excitation and detection systems and is connected by an umbilical cable to a battery-powered electronics unit. An illumination beam is produced by a high-pressure mercury arc lamp and modulated by an electromagnetic, tuning-fork chopper. A dichromatic beam splitter reflects the excitation radiation to the telephoto output lens and transmits the longer-wavelength fluorescence to the photomultiplier tube. The filters in both units are mounted in thumbwheels allowing for filter selection while the unit is in use. The telephoto lens is adjusted so as to focus the illumination beam 40 cm from the spotter. The UV beam and, consequently, the induced fluorescence are modulated at 1 kHz. Since the electronic frequency spectrum of fluorescent or incandescent room lighting consists of a 120-Hz fundamental and its harmonics superimposed upon white noise, it is possible to separate the 1-kHz fluorescence signal from the background signal by electronic filtering. Demodulating and low-pass filtering of this signal enables detection of fluorescence signals from the surface contaminants. The spill spotter has been evaluated in the laboratory and field-tested at various synfuel production plants.

3.5. Remote Sensing with Laser-Based Fluorosensors

The use of remote sensing via laser excitation has been investigated by a number of researchers.[20,21] In the remote-sensing technique, an airborne laser irradiates a defined area on the ground, and fluorescence induced by the laser light is detected through a telescope by an optical receiver located on the aircraft. Two fluorosensors using a pulsed nitrogen laser and a pulsed krypton fluoride laser as excitation sources have been tested in laboratory conditions to demonstrate the feasibility of remote detection for organic effluents associated with coal conversion.[20] The results of this study showed that the krypton fluoride system offers good sensitivity for both daylight and nighttime operation.

Another airborne laser fluorosensor was successfully tested under field conditions.[21] The excitation radiation was provided by a pulsed nitrogen laser. The fluorescence signal was collected by a telescope that receives light from the area where the airborne laser strikes the surface. The fluorescence radiation was spectrally dispersed by a spectrometer coupled with a multichannel fiber-optics image splitter. The fourteen channels provides a resolution of 20 nm over a spectral range from 400 to 660 nm. The receiver was electronically gated so that the detectors received mainly the photon pulses from the fluorescence induced by the laser excitation. This gated detection mode allowed the sensor to operate in daylight. The background signal was recorded between these laser pulses and subtracted from the laser-induced fluorescence signal. All the spectral information was displayed in real-time on a video monitor and stored in a tape for post-flight analysis. This airborne laser-based remote monitor has been successfully operated in field trials to characterize oil and dye spills, using a correlation technique that allowed different classes of fluorescent contaminants to be distinguished. Although oils are complex mixtures exhibiting structureless fluorescence spectra, the correlation method was able to differentiate different classes of crude oil from the background ocean and from ships. Figure 5 illustrates typical responses of the various detection channels of the system when the aircraft flew over an oil spill.

3.6. A Fluorescent Tracer Detection Technique

Dermal adsorption of pesticides has been a topic of great concern for agricultural workers. Methods developed for monitoring occupational exposure to pesticides include a direct technique of attaching cloth or paper collection devices to various body parts[22] and an indirect technique of measuring pesticide metabolites in the urine.[23] Recently, a fluorescent tracer detection technique has been developed to monitor directly the exposure of field workers to pesticides.[24] In this technique, a known amount of a tracer is added to the pesticide spray tank. The tracer is usually a strongly fluorescent agent that can be detected at trace levels much lower than the pesticide compounds. After a work shift, the field worker is illuminated by UV light and the body parts scanned by a video camera. Several sources of variability and limitations associated with this technique include the changes in the distance of the subject from the camera, nonuniform illumination of the subject, and curvatures of the different body parts. Addition of a chemical agent may also increase the health risk because

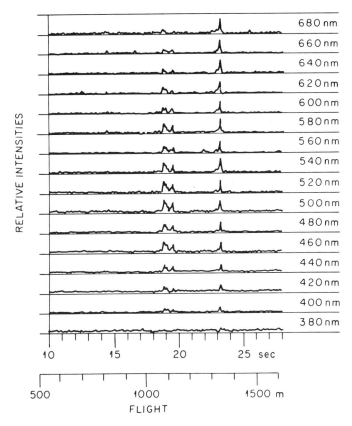

Figure 5. Typical data collected by the laser-based airborne fluorosensor over a crude oil spill. Reprinted with permission from R. A. O'Neil, L. Buja-Bijunas, and D. M. Rayner, *Applied Optics* *19*, 836 (1980). Copyright 1980 by the Optical Society of America.

of synergistic effects between the agent, pesticide, and UV light. At the present time, the tracer technique utilizes elaborate computer data treatment and a sophisticated monitoring device especially developed for laboratory studies of skin contamination in agricultural workers. Further developments are needed to demonstrate the potential of this technique for routine applications.

3.7. Studies of Absorption of Carcinogenic Materials into Mouse Skin

One application of surface fluorescence measurement of great interest to chemists and biologists as well as to industrial hygienists is the determination of the rate of absorption of toxic contaminants into the skin. The luminoscope was used to monitor the fluorescence from various coal liquids spotted on ani-

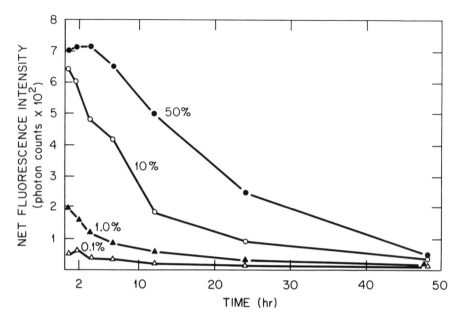

Figure 6. Temporal behavior of the fluorescence signals from coal liquid products spotted on mouse skin. The coal liquids were diluted (0.1%, 1%, 10%, and 50%; vol/vol) in ethanol.

mal skin. The measurements allowed the study of the temporal behavior of the fluorescence intensity of the a coal liquid spotted on skin of a hairless mouse.[25] The intensities of the fluorescence can be used to follow the processes of solvation and absorption of the coal liquid into the skin of the animal. Samples of the coal liquid at various concentrations (0.1, 1, 10, and 50%) in ethanol were spotted on a mouse and the fluorescence intensities of the various sample spots were monitored for 50 hours (Figure 6). Except for the 50% concentration sample spot, the fluorescence intensities of all the spots exhibited a gradual decrease with time. This temporal decrease provided an indication of the rate of absorption of the contaminant into the animal skin. The slight increase in fluorescence of the concentrated spot (50% coal liquid) during the first three hours indicated that the initial concentration of the spot was too high, causing quenching of the fluorescence. Because of gradual absorption of the coal liquid into the animal skin, less and less material was left on the skin surface. As a result, the quenching process decreased and a slight increase in the fluorescence signal was observed for the 50% coal liquid spot.

3.8. Chromogenic and Fluorogenic Spot Test Techniques

Weeks et al. described a chemical spot test procedure to detect several potentially carcinogenic contaminants on a variety of surfaces, including filter paper, metal, and painted and concreted surfaces.[26] These investigators formed chromogenic and fluorogenic derivatives using various chemical reactions: (1) p-dimethylaminobenzaldehyde (Ehrlich's reagent) with primary amines, (2) p-dimethylaminocinnamaldehyde with primary amines, (3) chloranil with primary aryl amines, (4) fluorescamine with primary amines, and (5) o-phthalaldehyde with primary aromatic amines. Several sensitive detection methods were found among these reactions for many cancer-suspect compounds investigated. The chromogenic spot tests were readily recognized by the naked eye, whereas the fluorescent derivatives were made visible by irradiation with a UV light source. Table 1 gives the limits of optical detection for several aromatic amines

Table 1. Detection Limit Values (ng/cm^2) for Aromatic Amines Visualized on Filter Paper[a] via Fluorogenic Derivatization[b]

	Fluorogenic visualization reagent			
		Isomeric phthalaldehydes		
Compound studied	Fluorescamine	o-	m-	p-
Aniline	3	30	n.d.[c]	n.d.
o-Chloroaniline	30	n.d.	—	—
m-Chloroaniline	3	30	—	—
p-Chloroaniline	3	30	n.d.	n.d.
o-Toluidine	3	800	n.d.	n.d.
m-Toluidine	3	30	—	—
p-Toluidine	3	30	—	—
o-Toluidine	5	—	—	—
4,4'-Methylenedianiline	3	30	800	800
4,4'-Methylenebis(2-chloroaniline) (MOCA)	3	800	800	800
Benzidine	3	3	80	80
3,3'-Dichlorobenzidine	3	800	150	80
α-Naphthylamine	30	80	80	80
β-Naphthylamine	3	3	80	80
4-Aminobiphenyl	3	3	80	80

[a]Whatman 42 filter paper.
[b]Reference 25.
[c]n.d.: Not detected at the level of 800 ng/cm^2.

observed on Whatman 42 filter paper using the fluorogenic derivatization method. The fluorogenic method was found generally more sensitive than the procedure relying upon colored derivatives. The limits of detection were usually less than 200 ng of material per cm^2 for most of the compounds and surface studies.

3.9. Sensitized Fluorescence Spot Tests

Another spot test procedure uses the principle of sensitized fluorescence. In the sensitized fluorescence technique, a compound, known as the sensitizer, is used to absorb the UV excitation energy. When the analyte is added to the sensitizer, electronic interaction can occur between the excited state of the sensitizer (or donor) and the analyte (or acceptor) compounds. This interaction is of long-range type and results in resonant energy transfer from the sensitizer to the analyte molecule, followed by radiative emission from the analyte. An important type of electronic energy transfer, investigated by Forster,[27] may be represented by the equation:

$$D^* + A \rightarrow D + A^* \tag{8}$$

where D is the singlet ground state of the donor molecule, D^* is the singlet excited state of the donor molecule, A is the singlet ground state of the acceptor molecule, and A^* is the singlet excited state of the acceptor molecule.

The efficiency of energy transfer is increased by (i) a large overlap between the first absorption band of the acceptor molecule and the emission band of the donor molecule and (ii) a high fluorescence quantum yield of the donor molecule. Figure 7 illustrates the principle of this energy transfer process.

Smith and Levins investigated the use of naphthalene as the sensitizer for detecting polyaromatic hydrocarbons in environmental samples.[28] Experimentally, the test is conducted by marking three circles on a high-grade, ashless filter paper. Sample solution is spotted to two circles (with and without sensitizer), and the sensitizer solution is applied to the third circle. The three spots are then irradiated with UV radiation and the differences in intensity between the analyte spots with and without sensitizer are visualized with the naked eye. Johnson et al. conducted further investigation to evaluate the efficacy of the spot test procedure using naphthalene.[29] These investigators reported that polyaromatic hydrocarbons having a molecular weight greater than 200 have a lower detection limit than those with a lower molecular weight.

Further investigations have found that anthracene is also a very effective sensitizer for the detection of high-boiling PNA species in complex mixtures.[30] Table 2 illustrates the use of anthracene as the sensitizer to increase the fluorescence of coal liquid products. One major advantage associated with the use of anthracene is the low volatility of this compound, which makes this sensitizer

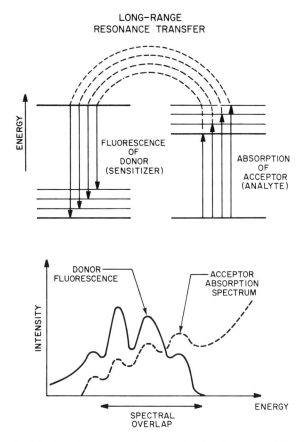

Figure 7. Principle of energy transfer process involved in sensitized fluorimetry.

stay longer on the sample spot and, therefore, improves the reproducibility of the measurements. Figure 8 illustrates the use of anthracene as a sensitizer that increases the fluorescence intensity of the samples of a solvent-refined coal product.

3.10. Surface Detection by Room Temperature Phosphorimetry

One important technique for the detection of contaminants on surfaces is the room temperature phosphorescence (RTP) technique.[31-33] The RTP technique is based upon the detection of the phosphorescence emission of organic compounds adsorbed onto solid surfaces such as filter paper, silica gel, sodium

Table 2. Detection of Sensitized Fluorescence by the Luminoscope Using Anthracene as the Sensitizer[a]

Sample[c]	Net fluorescence signal[b] (photons/s)	
	Without sensitizer	With sensitizer
SRC-II fuel oil blend	1,600	4,000
H-Coal liquid	2,700	4,500
Centrifuged shale oil	0	700
$ZnCl_2$ distillate	19,300	25,000
Signode oil	900	3,200

[a]Excitation = 365 nm; emission = 462 nm.
[b]All samples (3 µl) are spotted on filter paper.
[c]All samples are 100-fold diluted in ethanol.

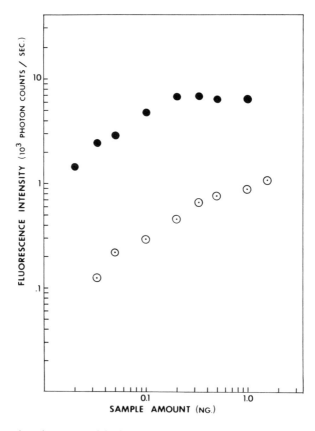

Figure 8. Intensity enhancement of the fluorescence signals from various coal liquid samples with the use of anthracene as the sensitizer (open circles: samples without anthracene; filled circles: samples with anthracene).

acetate, etc. Unlike conventional phosphorimetry, RTP does not require low-temperature refrigerant and cryogenic equipment. In general, three microliters of sample solution are spotted on filter paper. The sample is then dried under an infrared heating lamp for about five minutes. Upon drying, the sample is transferred to a spectrophotometer for phosphorescence analysis. The entire experimental procedure generally takes less than ten minutes. A particular process that is an invaluable aid to RTP analysis is the external heavy-atom effect. The presence of heavy-atom salts, such as lead acetate and thallium acetate, on the sample substrate increases the RTP emission of weakly phosphorescent compounds.[34] The external heavy-atom effect has been used to improve both the sensitivity and the specificity of RTP analysis of organic pollutants in various coal liquids[35,36] and in atmospheric samples[37] and on a passive dosimeter badge.[38] The versatility of the sampling procedure using filter paper is one of the main attributes of the RTP technique. The method can be used to detect not only organic pollutants adsorbed onto the paper but also contaminants trans-

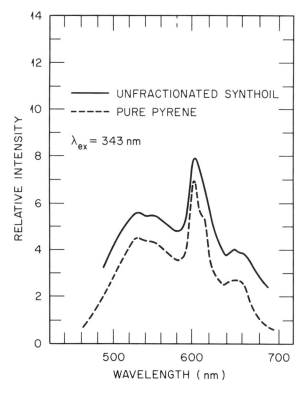

Figure 9. Room temperature phosphorescence detection of pyrene in an unfractionated sample of synthoil. Reprinted with permission from T. Vo-Dinh, R. B. Gammage, and P. R. Martinez, *Analytica Chimica Acta 118*, 313 (1980). Copyright 1980 by Elsevier Scientific Publishing Company, Amsterdam.

Table 3. Limits of Detection (LOD) for Several PNA Compounds by Room Temperature Phosphorescence[a]

Compound	λ_{ex}[b] (nm)	λ_{em}[c] (nm)	LOD (ng)
Homocyclics			
Benzo[*a*]pyrene	395	698	0.07
Benzo[*e*]pyrene	335	543	0.001
2,3-Benzofluorene	343	505	0.025
Benzo[*ghi*]perylene	398	626	0.6
Chrysene	330	518	0.03
1,2,3,4-Dibenzanthracene	295	567	0.08
1,2,5,6-Dibenzanthracene	305	555	0.005
Fluoranthene	365	545	0.05
Fluorene	270	428	0.2
Phenanthrene	295	474	0.007
Pyrene	343	595	0.1
N-Heterocyclics			
Acridine	360	660	0.4
5,6-Benzoquinoline	355	502	0.06
7,8-Benzoquinoline	353	509	0.04
Carbazole	296	415	0.005
Dibenzocarbazole	295	475	0.002
Quinoline	315	505	0.1

[a]Reference 35.
[b]λ_{ex} = excitation wavelength.
[c]λ_{em} = emission wavelength.

ferred onto paper from other surfaces by leaching and swiping techniques. These latter approaches could be used as an alternate method for quantifying surface contamination. Figure 9 illustrates the identification of pyrene detected in a 3-μg sample of synthoil on the surface of a filter paper. The sensitivity of the RTP technique is excellent for many polyaromatic pollutants, the detection limits being in the subnanogram range (Table 3). The RTP technique can be combined with a swipe procedure using filter paper to provide a sensitive and practical method for monitoring organic contaminants on surfaces.

4. Conclusion

The variety of applications discussed in this chapter demonstrates the usefulness and wide applicability of surface luminescence detection using UV radiation as excitation sources. By far the most important advantage of luminescence over absorption methods is that for substances that have significant luminescence quantum efficiencies, luminescence techniques can achieve much better sensitivity than absorption techniques. Another important attribute of surface luminescence techniques is the fact that the optical excitation and detection allow the analyses to be conducted directly on the surface areas being monitored, without the need for desorption, removal, or extraction of the materials of interest. With the rapid development of powerful UV radiation sources such as lasers, low-cost holographic gratings, and sensitive semiconductor sensors, luminescence spectroscopy will undoubtedly receive a renewed interest in the critical area of monitoring contamination of skin and other surfaces in the workplace.

References

1. K. L. Mittal, ed., *Surface Contamination: Genesis, Detection and Control,* Vols. 1 and 2, Plenum Press, New York (1979).
2. T. Vo-Dinh and A. R. Hawthorne, Monitoring instrumentation, in: Proceedings of the DOE-OHER Workshop on Monitoring and Dosimetry, CONF-8403150, pp. 43–55, National Technical Information Service, Springfield, Va. (1984).
3. T. Vo-Dinh, Luminescence spectroscopy, in *Analytical Measurements and Instrumentation for Process and Pollution Control* (P. N. Cheremisinoff and H. J. Perlis, eds.), pp. 47–80, Ann Arbor Science, Ann Arbor, Mich. (1981).
4. J. D. Winefordner, T. C. O'Haver, and S. G. Schulman, *Luminescence Spectrometry in Analytical Chemistry,* Wiley-Interscience, New York (1972).
5. T. Vo-Dinh, in: *Synchronous Excitation Spectroscopy, Modern Fluorescence Spectroscopy,* Vol. 4 (E. L. Wehry, ed.), pp. 167–191, Plenum Press, New York (1981).
6. P. Kubelka and F. Munk, *Z. Tech. Phys. 12,* 59 (1931).
7. R. A. Zweidinger and J. D. Winefordner, *Anal. Chem. 42,* 639 (1970).
8. J. Goldman, *J. Chromatog. 78,* 7 (1972).
9. W. M. Baird, *Int. J. Cancer 22,* 292 (1978).
10. E. Cavalieri and M. Calvin, *Photochem. Photobiol. 14,* 641 (1971).
11. W. Heller, *Strahlentherapie 81,* 529 (1950).
12. G. M. Blackburn and P. E. Taussig, *Biochem. J. 149,* 289 (1975).
13. Threshold Limit Values for Chemical Substances and Physical Agents in the Workroom Environment, *Amer. Conf. Government Ind. Hyg.* (1983).
14. R. R. Keenan and S. B. Cole, *Am. Ind. Hyg. Assoc. J. 43,* 473 (1982).
15. T. Vo-Dinh and R. B. Gammage, *Am. Ind. Hyg. Assoc. J. 42,* 112 (1981).
16. J. D. Ingle, Jr., and S. R. Crouch, *Anal. Chem. 44,* 785 (1972).
17. T. Vo-Dinh and R. B. Gammage, in: *Chemical Hazards in the Workplace—Measurement and Control* (G. Choudhary, ed.), ACS Symposium Series No. 149, pp. 270–281, American Chemical Society, Washington, D.C. (1982).
18. The U.S. Department of Energy has given approval for Environmental Systems Corporation (ESC) to acquire exclusive rights to manufacture and distribute the luminoscope; Environmental Systems Corporation, Knoxville, Tennessee 37912 (1985). Further information on the commercial availability of the luminoscope should be requested from Environmental Systems Corporation, 200 Tech Center Drive, Knoxville, TN 37912.
19. D. D. Schuresko, *Anal. Chem. 52,* 311 (1980).
20. G. A. Capelle and L. A. Franks, *Appl. Opt. 18,* 3579 (1979).
21. R. A. O'Neil, L. Buja-Bijunas, and D. M. Rayner, *Appl. Opt. 19,* 863 (1980).

22. J. E. Davis, *Residue Review 75*, 33 (1980).
23. D. P. Morgan, J. L. Hetzler, E. F. Slach, and L. I. Lin, *Arch. Environ. Contam. Toxicol. 6*, 159 (1973).
24. R. A. Fenske, J. T. Leffingwell, and R. C. Spear, Paper presented at the 185th National Meeting of the American Chemical Society, Seattle, Wash. (March 20–25, 1983).
25. T. Vo-Dinh, R. B. Gammage, and G. D. Miller, in: Proceedings of the American Industrial Hygiene Conference, Philadelphia, Pa. (May 22–27, 1983).
26. R. W. Weeks, Jr., B. J. Dean, and S. K. Yasuda, *Anal. Chem. 48*, 2227 (1976).
27. T. Forster, *Discussion Faraday Soc. 27*, 7 (1959).
28. E. M. Smith and P. L. Levins, in: *Polynuclear Aromatic Hydrocarbons* (A. Bjorseth and A. Dennis, eds.), p. 973, Battelle Press, Columbus, Ohio (1980).
29. L. D. Johnson, R. E. Luce, and R. G. Merrill, in: *Polynuclear Aromatic Hydrocarbons* (M. Cooke and A. J. Dennis, eds.), p. 119, Battelle Press, Columbus, Ohio (1980).
30. T. Vo-Dinh and A. White, *Anal. Chem. 58*, 1128 (1986).
31. T. Vo-Dinh and J. D. Winefordner, *Appl. Spectrosc. Rev. 13*, 261 (1977).
32. R. T. Parker, R. S. Freedlander, and R. B. Dunlap, *Anal. Chem. Acta. 119*, 189 (1980); *ibid. 120*, 1 (1980).
33. T. Vo-Dinh, *Room Temperature Phosphorimetry for Chemical Analysis*, Wiley-Interscience, New York (1984).
34. T. Vo-Dinh and J. R. Hooyman, *Anal. Chem. 51*, 9115 (1979).
35. T. Vo-Dinh and P. R. Martinez, *Anal. Chim. Acta 125*, 313 (1981).
36. T. Vo-Dinh, R. B. Gammage, and P. R. Martinez, *Anal. Chim. Acta 118*, 313 (1980).
37. T. Vo-Dinh, R. B. Gammage, and P. R. Martinez, *Anal. Chem. 55*, 253 (1981).
38. T. Vo-Dinh, in: *Identification and Analysis of Organic Pollutants in Air* (L. H. Keith, ed.), Chapter 16, pp. 257–270, Butterworth Publishers, Woburn, Mass. (1983).

6

Particulate Surface Contamination and Device Failures

Joseph R. Monkowski

1. Introduction

An interesting situation exists in the microelectronics field with regard to particulate contamination. A large number of vendors have as their primary purpose the supplying of clean rooms and clean-room accessories to the semiconductor industry. Integrated circuit manufacturers expend a substantial amount of both time and money in order to maintain a production facility as free as possible of particulate contamination. In fact, the initial capital investment for the clean room is from several hundred to more than one thousand dollars per square foot.[1,2] Furthermore, chemical companies are now making special efforts to remove insoluble particulate contamination from their electronic-grade chemicals.[3-9] However, in comparison to this overwhelming attention to the control of particulate contamination and the overall attitude that this control is entirely necessary, very little has been reported on the exact role of particulate contamination in device failures.

Predictably, most of the work done concerning the role of particles has been related to their mechanical aspects. For example, particles present on photomasks block the ultraviolet light used to expose the photoresist.[10,11] Particles present in the photoresist film or in any thin film give rise to nonuniform coverage of the film, and can even cause breaks or shorts in the film if the particle is sufficiently large.[12,13]

In contrast to these mechanical defects, those situations in which the chemical nature of the particle plays a large role have been far less extensively reported. Consider, for example, low-field breakdown in MOS (metal-oxide-semiconductor) integrated circuits. This type of breakdown is the most preva-

Joseph R. Monkowski • MRI, 9250 Trade Place, Suite 500, San Diego, California 92126.

lent failure mode for these circuits; however, only recently and in very few reports have particles been identified as the primary cause of these low-field breakdowns.[14-16]

Undoubtedly, one of the major problems encountered in attempting to assess the significance of particulate contamination in device failures is that with an average size of approximately one micrometer or less and a typical density of only one to several per square inch, these particles are very difficult to locate on the silicon wafer. As Gordon Moore of Intel Corporation recently pointed out, the problem is "like looking for a single gum wrapper in a six-acre field."

However, the work that is reported in the literature does indeed show that the attention paid to particulate control is justified. There are a number of ways in which particles can lead to device failure, and these are each addressed in the various sections of this chapter. Most of the work covered in this chapter deals with silicon technology, and before discussing the actual device failures, a brief introduction into the various sources of the particulate contamination is given.

2. Sources of Particulate Contamination

Almost every aspect of device processing is a potential source of particulate contamination. The source that is given by far the most attention in any processing facility is the air.[1,17-22] However, the air in most processing facilities has been made so clean that now other sources such as the gases[23-25] and chemicals[3-9,17,26] are being recognized as potential problems. In addition, the general handling of the silicon wafers generates a substantial amount of contamination.

2.1. Air

Typically, most semiconductor processing is done in Class 100 clean rooms.[1,17,19] The standardization of definitions and air cleanliness for clean rooms is given in Federal Standard No. 209B, and is briefly summarized in Table 1. The classification is based on the maximum allowable number of particles, 0.5 μm (micrometer) and larger, per unit volume. Thus, a Class 100 clean room contains 100 or fewer particles, 0.5 μm or greater in diameter, per cubic foot of air.

Many people in the industry feel that FED STD 209B is out of date for the current microelectronics technology.[17,19-22] They feel that the minimum particle size of 0.5 μm should be pushed down to 0.3 μm or even 0.1 μm. At the present time, HEPA (high-efficiency particulate absolute) air filters are capable of removing better than 99.99% of all particles 0.3 μm or larger. Furthermore, particle measurement instruments based upon light scattering techniques can

Table 1. Air Cleanliness Classes (from FED STD 209B)

Class	Maximum number of particles/ft³ (liter) 0.5 μm and larger	Maximum number of particles/ft³ (liter) 5.0 μm and larger
100	100 (3.5)	
1,000	1,000 (35)	
10,000	10,000 (350)	65 (2.3)
100,000	100,000 (3,500)	700 (25)

detect particles as small as 0.1 μm.[17,20] Most significant, however, is that particles as small as 0.3 μm or less are believed to be deleterious to presently manufactured integrated circuits.[25]

Most of the particulate contamination inside of clean rooms is generated by personnel movement,[1,17] although at least one investigation has shown that the HEPA filters used to clean the air can themselves be a source of contamination.[18] Table 2 lists a number of typical sources of particles, along with the typical sizes of the particles generated. Note that in most of the cases the particle size is quite large.

The large particles are typically not the most troublesome, though. As shown in Figure 1, for a particular Class rating, the concentration of particles falls off very rapidly as a function of size, and consequently, there are typically very few large particles in a clean-room facility. Another factor, perhaps even more important, is the tenacity with which the different particles adhere to a surface such as a photomask or a silicon wafer. Table 3 shows the relative adhesional forces for particles of different sizes.[27] The larger particles, such as those several micrometers or larger, can be quite readily removed in various cleaning and rinsing operations. The smaller particles, however, especially those of submicrometer size, can only be removed in scrubbing or very high pressure rinsing operations.[28-30]

Table 2. Typical Sources of Particles[a]

Source	Particle size (μm)
Rubbing painted surface	90
Crumpling or folding paper	65
Rubbing an epoxy-coated surface	40
Writing with a ballpoint pen on ordinary paper	20
Rubbing the skin	4

[a]From Reference 1.

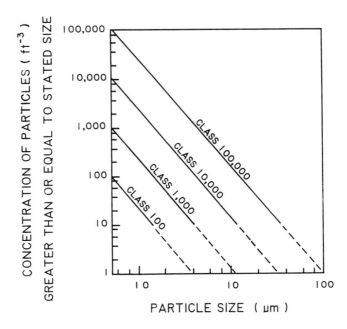

Figure 1. Particle size distributions for different classifications of clean rooms (from FED STD 209B).

Figure 2 shows a distribution of particle sizes as observed in the gate region of MOS integrated circuits.[31] These data clearly show that the particle size corresponding to the most prevalent contamination on the silicon or oxide surface is in the range of one micrometer; very few particles larger than 2 μm are observed. As pointed out in an investigation reported below, particles as small as 0.1 μm appear to play a large role in low-field gate failures.

Table 3. Relative Adhesional Force (Force/Particle Weight) for Quartz Bead on Glass Slide in Air at $T = 25°C$, R.H. = 95%[a]

Bead size (μm)	Relative adhesional force (gravity units)
100	510
50	2,159
10	57,716
1	674,600
0.1	749,552,300

[a] From Reference 27.

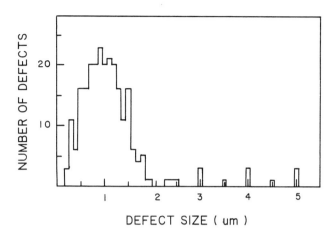

Figure 2. Size distribution of defects observed in gate region of MOS transistors (from Reference 31).

2.2. Chemicals

Until recently, very little attention was paid to particulate contamination in the processing chemicals and deionized water. Most of the effort with regard to these materials went into being certain that soluble contamination, such as sodium and metallic ions, was in the ppm (parts per million) range or lower.

The measurement of dissolved contamination uses techniques that, although sophisticated, are relatively well known.[4] These include atomic absorption,[32] inductively coupled plasma atomic emission spectroscopy,[33] gas chromatography,[34] infrared and ultraviolet spectrophotometry,[35] and neutron activation analysis.[36]

The particulate contamination, on the other hand, is much more difficult to define and identify. First, the composition varies greatly.[3] The particles can be simple elements, complex compounds, organic matter, or mixtures. Second, the detection of the particles and their classification according to size is not as straightforward as in the case of air. The liquids can contain bubbles that interfere with the light scattering techniques, or they may be opaque, which rules out the light scattering techniques entirely.

Another standard technique is the filtration of a particular volume of the liquid through a clean membrane, with subsequent microscopic examination of the membrane.[37] Two drawbacks of this technique are that it is limited to relatively large particles of at least 5-μm size and that it is subject to extraneous contamination. In general, the main problem is that, as mentioned above, the large particles are not the worst offenders, and there are few techniques for detection of micrometer-sized particles in liquids.

Recently, Tolliver et al.[9] have developed an acoustic technique that uses

Table 4. Concentrations of Particles 0.8 μm and Larger
for Three HF Samples before and after Filtration[a]

Sample	Particles per liter	
	Upstream of filter	Downstream of filter
HF "A"	>100,000	47,948
HF "B"	>100,000	35,897
HF "C"	>100,000	38,461

[a]From Reference 9.

the principle of ultrasonic sound scattering by the particles in the liquid. They have carried out experiments showing that submicrometer particles can be detected in a real-time system.

Some of the results obtained via this acoustic technique are very interesting. Table 4 lists particle counts determined with the acoustic technique for hydrofluoric acid both before and after filtration. Note the relative effectiveness of the filtration system for the various sizes of particles. Although the researchers used a standard submicrometer chemical filter, the filtration of particles at the 0.8-μm level was very inefficient. As one of their conclusions, Tolliver et al. state that "single pass filtration with submicron filters provides little control of submicron particles in many acids and chemicals presently in use."

Over the past several years, the chemical suppliers have become more aware of the stringent needs brought about by the move of the microelectronics industry to very large, high-density circuits, and as a result some of these suppliers now offer special product lines of low-particulate chemicals. In general, the level of particulate contamination in these specially processed chemicals is approximately two orders of magnitude lower than in the standard electronic-grade or MOS-grade chemicals. A typical rating is fewer than 10 particles of size 1 μm or greater per ml of the chemical.

2.3. Gases

Gases are prone not only to the inherent contamination introduced during manufacture or separation,[24] but can also pick up a large amount of contamination as they flow through the plumbing and various pieces of equipment.[5,23] Particles can be released into the gas stream from movement of parts such as regulators, valves, and flowmeters, and also from the plumbing, furnaces, and other deposition and etching equipment.

Ironically, particles can be introduced by filters intended to remove particulate contamination.[5,25] This enhanced contamination frequently occurs with the use of glass-fiber filters. In addition to the particles, these filters can also release ionic and metallic impurities. The membrane filters are much preferable for gas filtration.

Many times the reactivity of a gas enhances the probability of particulate contamination. For example, a major fraction of MOS IC manufacturers incorporate some form of chlorine into their oxidation process.[38–40] Often the chlorine is introduced in the form of HCl. If the HCl has even a trace of moisture in it, it will readily corrode any metallic parts with which it comes into contact. Quite frequently, stainless steel is used for the plumbing, and after some time the inside of the plumbing can become severely corroded. From time to time, pieces of the rust are picked up by the gas and carried into the furnace where they can drastically damage the devices being processed.

Often, particles are generated as a result of gas-phase reactions. When silane is used for silicon epitaxy or chemical vapor deposition, if there are any traces of oxygen in the system the silane will react to form particles of SiO_2.[41] These particles can cause failure of regulators and flowmeters and, more significantly, are often incorporated into the growing films, producing defects in these films. The presence of particulate contamination is particularly deleterious in the case of epitaxial growth.

2.4. Wafer Handling

This is the broadest of the categories of particle sources, and since there are so many facets to this category, and most of them involve personnel, this source may be the hardest to control.[1,17] Wafer handling typically involves movement by personnel in some manner, whether it is a matter of carrying wafers from one location to another or handling the wafers in a particular work station. As is shown in Table 5, almost all activities by personnel are accompanied by an increase in the ambient contamination level in the vicinity.

Table 5. Increase of Contamination by Personnel[a]

Activity	Times increase in ambient level (particles, 0.2 to 50 μm)
Normal walking	1.2–2
Sitting quietly	1–1.2
Hands inside laminar flow station	1.01
Brushing sleeve of uniform	1.5–3
Stamping on floor with shoe covering	1.5–3
Stamping on floor without shoe covering	10–50
Normal breathing	No increase
Breathing of smoker up to 20 minutes after smoking	2–5
Sneezing	5–20
Rubbing skin	1–2

[a] From Reference 1.

Handling of wafers with metal tweezers can generate a large amount of damage in the silicon, not only from the mechanical aspects, but also from particles of metal rubbed onto the wafer as a result of the motion and friction. Stacy et al.[42,43] have shown via transmission electron microscopy (TEM) that where the tweezers contact the silicon, remnants of stainless steel are left behind, and the nickel from this stainless steel diffuses into the silicon wafer during subsequent high-temperature processing, ultimately forming Ni-decorated stacking faults. On a macroscopic scale, these decorated stacking faults comprise what is referred to as "haze." In regions of the wafer removed from the tweezer contact, the haze is much reduced.

In a recent investigation, Baginski et al.[44] were studying the effect of the presence of HCl in the processing ambient on the gettering of metallic impurities from silicon. Their approach was to analyze the wafers after processing by etching off several micrometers of silicon at a time using ultrapure acids, followed by inductively coupled plasma (ICP) atomic emission spectroscopy of the acid solution. In the case of copper and gold, they were able to obtain reproducible results down to the ppb (parts per billion) level in the solution. In the case of iron, however, the scatter in the results was at the 100-ppb level. Since they had taken special effort to be certain that the wafers contacted only high-purity plastic from the time of processing up to and including the measurement, their general conclusion was that iron is a very ubiquitous impurity probably introduced as particles originating from metal cabinets, clean hoods, or other equipment.[45]

Another source of contamination associated with wafer handling is the friction between the furnace process tubes and the boats or paddles used to insert the wafers into the furnace. This friction produces particles which can be picked up by the gas stream and carried to the silicon surface. These particles are difficult to avoid since their deposition occurs immediately prior to the high-temperature processing. To circumvent this problem, techniques aimed at avoiding the contact between the boat and the tube are being introduced.

In general, contamination from wafer handling is the most difficult to categorize and to avoid, primarily because it depends so much upon personnel behavior. For example, in a study of defects in photomasks, Angel et al.[10] noted that almost half of the anomalies on the masks were attributed to handling, with the greatest part of these resulting from contamination. Analysis of the contamination using gas chromatography, infrared spectroscopy, and Auger spectroscopy showed the majority of the contamination to be condensate of the mask inspector's breath.

In order to reduce the amount of contamination related to personnel, many semiconductor manufacturers are introducing training programs for their cleanroom personnel.[1,17] Many of these programs are conducted by special outside firms, but a number of the large manufacturers conduct their own in-house programs. Typical topics covered in these programs include movement under laminar flow, personal hygiene, clean-room garments, the cleaning of work stations, and the general role of contamination in device failures.

Another general approach taken by semiconductor manufacturers is to incorporate more automation into the processing facility. This automation includes wafer handling and transport systems as well as automated cassette transport systems. A major consideration, however, is whether the automated equipment actually contributes less contamination than the personnel it replaces. This has not always been found to be true.[13]

An interesting alternative approach to automation is the "standard mechanical interface" (SMIF).[13] In this approach, the operators carry the wafers from machine to machine; however, the wafers are held in special containers that connect, via special noncontaminating interfaces, to the various processing machines. In this way, the wafers are kept in an environment totally separate from the clean-room personnel.

3. Effects on Device Performance

The most obvious manner in which particles can degrade the performance of integrated circuits and their semiconductor devices is through the mechanical aspects of their presence. For example, particles adhering to the wafer during metallization or chemical vapor deposition can cause breaks, pinholes, or other structural defects in the thin films. Particles adhering to photomasks can cause pinholes which ultimately lead to shorts through insulating regions such as gate or field oxides. A relatively large amount of literature exists on these subjects,[9-13, 19, 25] and the literature is generally quite consistent.

In contrast to the mechanical aspects, the chemical aspects have been much less studied. One of the main problems encountered in trying to pinpoint the chemical effects of particles is their wide diversity in composition. A particle containing only silicon and oxygen behaves very differently from one containing iron or sodium. In addition, the chemical features of particulate contamination are considerably more difficult to study than the mechanical aspects and the results are generally less conclusive.

In the first part of this section, the role of particulate contamination on photomasks is examined. This is followed by a review of crystalline defects in epitaxial films as brought about by unwanted particles. The latter two parts of the section deal with the chemical effects of particles on both silicon and silicon dioxide.

3.1. Particulate Contamination on Photomasks

The photomask contains the pattern that is transferred via a photolithographic process into a thin film on the silicon wafer.[46] For example, one mask defines where the gate regions of MOS transistors will be located; another mask defines where the interconnections will be made between components. Typically, up to approximately a dozen different masks can be used for a particular integrated circuit.

Obviously, if any particles adhere to the surface of the mask, these particles will behave very much like opaque patterns in the mask, and will be printed onto the wafer during the photolithographic process.[10,11] In fact, random defects on photomasks, of which particulate contamination accounts for approximately 50%, is the major source of yield loss in semiconductor devices, particularly for LSI (large-scale-integrated) and VLSI (very-large-scale-integrated) circuits.

Not all particles give rise to defects. Three main considerations are the size of the defect, the size of the IC components, and whether the photoresist used in transferring the pattern to the wafer is of the positive or negative type.

Consider Figure 3a. If the mask is used with negative photoresist, the result of the particle would be a pinhole. Pinholes are typically deleterious, regardless of their size, and particularly when they occur in an insulating film. Figure 4a shows the effect of a pinhole formed in the field oxide over the base

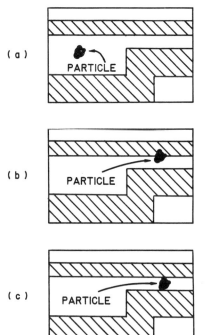

Figure 3. Interference of particles with photomask patterns. Depending upon whether the photoresist is negative or positive, and whether the particle is in an open field (a), on a pattern border (b), or is bridging across an entire area (c), the impact on device performance can vary widely.

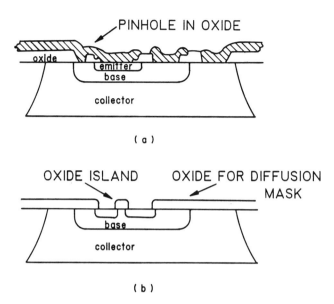

Figure 4. Results of particulate contamination present during photomask exposure. In (a), the particle has led to the formation of a pinhole in the field oxide; in (b), the particle has led to the formation of an oxide island, which has subsequently blocked the emitter diffusion. In both cases, emitter-base shorts are created.

region during the photolithographic step dealing with opening of the contact windows. As shown in Figure 4a, if the metal line contacting the emitter runs over the pinhole, the emitter and base become shorted together. In an analogous manner, pinholes in MOS gate oxides allow a shorting of the gate electrode to the underlying substrate. In all of these cases, the defect is catastrophic.

If the mask shown in Figure 3a is used with positive resist, the result is an island of photoresist, which would ultimately leave an island of the thin film. If the film being patterned is, for example, metal being used for interconnections, there is little problem with an isolated island.

If the film is an oxide layer being used for masking of a subsequent diffusion, the relative sizes of the island and the device dimensions become important. If the diffusion step is that of an emitter, and if the island is smaller than the final emitter junction depth, the dopant will merely diffuse around and under the island of oxide, and the final device will be relatively unaffected. If, however, the island is larger than the final emitter junction depth, the dopant will be unable to diffuse a sufficient lateral distance under the island, and there will be a small region in the middle of the emitter where the base contacts the silicon surface. This is shown in Figure 4b. The final result is that after metallization of the emitter region, the base and emitter will be shorted together.

Figure 3b shows a particle located on the border between the transparent and opaque regions of a mask. If the transparent region corresponds to a metal pattern (negative photoresist) or a window for diffusion (positive photoresist), the severity of the defect would depend upon the size of the particle. For example, if the metal line were substantially constricted, electromigration could eventually lead to a break in the line.[47] In the case of a diffused region, the resistance of the region could be significantly altered.

Figure 3c shows a situation that clearly would be catastrophic. Here the particle is so large as compared to the device dimensions that two opaque regions have been connected. For positive photoresist, this would be a short between two metal lines; for negative photoresist, it would be a break. If this were the masking layer for a diffusion, negative photoresist would yield a short; positive photoresist would produce a break.

Assuming that mask defects are the exclusive cause of device failures, Angel et al.[10] have shown that the average number of good devices resulting from a particular process is given by

$$\bar{X} = Ne^{-M/N} \qquad (1)$$

where N is the total number of possible good devices in the subject area size and M is the total number of defects from all critical mask levels.

For example, consider a 0.2-inch (0.5-cm) square device that is produced through a process that contains six critical mask levels. If each mask has only one defect per square inch (6.45 cm^2), the number of good devices per square inch is given by $\bar{X} = 25e^{-6/25} = 20$ mask-limited good devices/inch^2. If the number of defects goes up to five per square inch, by the same calculation, $\bar{X} = 25e^{-30/25} = 7$ mask-limited good devices/inch^2.

Obviously, with even very few defects, the yield can be severely limited for large integrated circuits. Angel et al. estimate that in order to be cost competitive within the industry, the aggregate defect density on a completed, high-complexity NMOS (n-channel MOS) technology wafer cannot exceed 20/inch^2 (3/cm^2).

As pointed out above, approximately one-half of all mask defects arise from contamination. A breakdown of the various sources is given in Figure 5. Note that handling and personnel account for the major amount of the contamination. Smaller, but approximately equal amounts of contamination are contributed by the working environment and the raw materials, the latter including particulate contamination in the photoresist.

Photomask inspection is prevalent within the microelectronics industry. Traditionally, the inspection has been done by an operator using an optical microscope. However, studies comparing this mode of inspection with that of automatic inspection have clearly shown the superiority of automatic inspection.[10,11,31,48] The automatic technique often finds up to ten times as many defects as the manual mode.

Particulate Surface Contamination

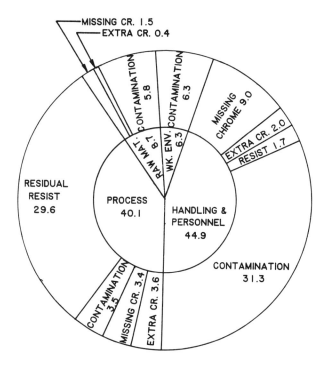

Figure 5. Probable sources of photomask defects (from Reference 10).

The general concept upon which the automatic techniques are based is comparison of one die against another on the same mask. For example, in one instrument, two objectives are placed over adjacent die. As the mask moves under the objectives, the patterns of each die are digitized. A built-in microprocessor-based computer system compares the two patterns and accounts for discrepancies caused by mask runout, operator alignment, and other mechanical errors. After this pattern fitting, any differences between the two patterns are considered defects. The size and location of these defects are then stored in memory. These automatic inspection techniques can be used to detect defects down to below one micrometer in size.[11]

3.2. Epitaxial Growth

Epitaxial layers have been used prominently in bipolar integrated circuits, where they form the uniformly doped collector regions of the transistors.[49] Now, as a result of a number of advantages, they are being used in many MOS integrated circuits.[50,51] In all cases, they need to be as free of impurities and crystalline defects as possible.

One of the ways in which particulate contamination can degrade the quality of epitaxial layers is through interference with the growing film. To obtain a high-quality film, the epitaxial layer must be deposited onto a defect-free substrate. Any defects on the surface of the substrate, including contamination, will be propagated into the growing film.

As a result of the strong influence of these defects on device performance, much research has been concerned with understanding and preventing these epitaxial defects.[52-56] Surprisingly, however, no detailed work has been concerned with the effect of particulate contamination on the substrate surface.

Obviously, if a microscopic particle is situated on the surface of the substrate, the atoms of silicon being deposited from the vapor phase will not be able to accurately follow the appropriate crystalline structure. This misplacement will give rise to a crystalline defect, probably a stacking fault. Although epitaxial films are often found to contain stacking faults, circumstantial evidence suggests that these faults nucleate on crystalline defects at the substrate surface rather than on particles of contamination.

As pointed out by a large number of researchers, epitaxial stacking faults can be prevented to a large extent through standard gettering techniques.[52,54,56] These techniques include ion implantation, back-surface damage, silicon nitride film deposition on the back surface, and heavy diffusions on either the front or back. Since these techniques are carried out prior to epitaxial growth, and particularly since many of them take place on the back surface of the substrate, it is very unlikely that they have any influence on particles located on the front surface of the silicon.

One of the reasons that particles do not appear to play a large role in the formation of epitaxial defects may be that the influence from other sources is yet too overwhelming. Whereas defect densities of tens to hundreds per square centimeter are tolerated at the present time, VLSI circuit requirements for the near future will dictate that the epitaxial film contain fewer than one defect per square centimeter.[51] At that time, the influence of particulate contamination may dominate.[57]

Another way in which particles can influence epitaxial films is through the metallic impurities that they release. In a striking demonstration, Pearce and McMahon[55] showed that metal particles can severely degrade the quality of epitaxial films. Prior to their epitaxial growth, they lightly dragged their stainless steel tweezers across the susceptor in the pattern of the letter "W". They then placed their silicon wafers on the susceptor and deposited an epitaxial film. After preferential etching of the defects in the epitaxial layer surface, one could clearly see the "W" delineated in the epitaxial film. Further investigation confirmed that heavy metals, particularly iron, chromium, and nickel, had diffused through the wafer and had given rise to metal precipitates which ultimately formed dislocation loops in the epitaxial silicon.

3.3. Failure Mechanism in MOS Gate Oxides

One of the most prevalent failure modes in MOS (metal-oxide-semiconductor) transistors is low-field breakdown of the gate insulator.[31,58-63] In fact, time-dependent dielectric breakdown (TDDB) has been reported as the primary failure mode of present MOS memory devices,[58,59] and the consensus is that as more emphasis is put on thin gate oxides for VLSI circuits, this failure mode will become even more important.[63]

Much research has been carried out in order to obtain a theoretical model to accurately describe the dependence of breakdown on time, voltage, and temperature.[58-67] These models for the most part are phenomenological, and their primary purpose has been to serve as reliability screens for MOS devices.[59,64]

In general, the occurrence of TDDB has been observed to follow a lognormal distribution; i.e., the occurrences of breakdown are normally distributed with the logarithm of time.[59] In addition, there is an exponential dependence upon temperature and upon electric field, for which the dependencies are described by thermal and electric-field acceleration factors.

As pointed out by Anolick and Nelson,[64] the origin of these dielectric breakdowns has been attributed to a number of causes, including electrostatic discharge, sodium localization, microcracks, pinholes, and particulate contamination. Compared to the large amount of interest in this failure, there has been relatively little research on the mechanisms of low-field breakdowns, and the conclusions that have been reached appear to not be in total agreement. However, closer inspection of the research findings reveals that, in general, there is agreement concerning the results of the various experiments, and when all of the results are taken into consideration, the role of particulate contamination appears to be very important.

A number of researchers have found that the sites of failure are randomly distributed over an area.[31,64] Figure 6 shows the locations of failure sites in an FET (field-effect-transistor) gate region as found by Anolick and Nelson.[64] Note that there appears to be no strong preference for either the diffused regions or the oxide edges.

From a study of the phenomenological behavior, several researchers have attributed the breakdown to ionic contamination.[61,65,66,68-70] In particular, they have attributed the breakdown to sodium emission from the metal/oxide interface and subsequent clustering of the sodium at the oxide/silicon interface.

Using scanning internal photoemission (SIP), DiStefano[68,69] and Williams and Woods[70] correlated sites of particularly high sodium contamination with oxide breakdown. The technique consists of scanning an MOS structure with a small light spot, typically from 5 to 100 μm in diameter. The light has an energy that produces very little photocurrent in an ideal MOS structure. However, if for any reason the interface barrier is lowered, the photocurrent is

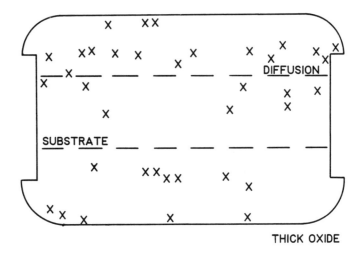

Figure 6. Location of failure sites in an FET gate region showing randomness of sites (from Reference 64).

significantly enhanced. As determined by these researchers, sodium located at the oxide/silicon interface does indeed lower the barrier, and consequently, the photoemission technique is an ideal means to investigate the spatial distribution of interfacial sodium.

Figure 7 shows an SIP map of an MOS device that contains no intentional contamination. After stressing the device for 40 hours at a field of 10^6 V/cm at 150°C, several sharp peaks have developed. Subsequent testing showed that dielectric breakdown occurred at two points corresponding to enhanced photocurrent. The breakdown occurred at a field of approximately 3×10^6 V/cm. DiStefano further points out that in all of the samples he tested, SIP maps predicted the location at which breakdown occurred.

Another nondestructive technique useful for detecting defective regions in SiO_2 films is the liquid crystal approach used by Keen[71] and Zakzouk.[61,62] In this technique, a nematic liquid crystal is put in contact with the SiO_2, and an electric field is applied. Conducting regions in the SiO_2 film cause a turbulence in the liquid crystal which can be seen under an optical microscope. The technique is nondestructive, however, because the liquid crystal limits the current to levels below that required to destroy the oxide film.

Using this technique, Zakzouk found that the defective regions were typically submicrometer in size. From the field and time dependence, he concluded that the defects result from migration of sodium to the oxide/silicon interface, with consequent formation of sodium clusters. As DiStefano and Williams and Woods had concluded, Zakzouk reported that such a clustering would result in enhanced electron emission, which would ultimately lead to breakdown.

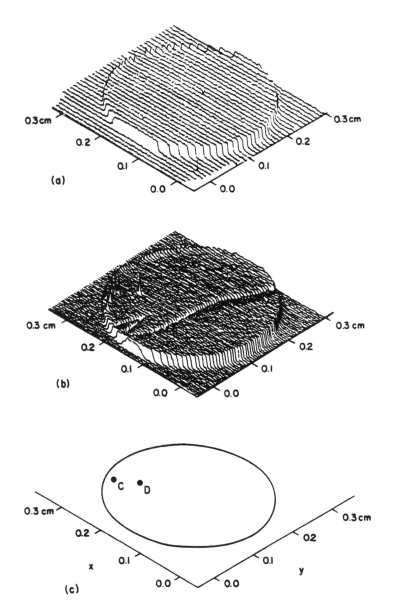

Figure 7. Scanning internal photoemission map of an MOS device. (a) Initial map. (b) Map after drifting sodium ions to oxide/silicon interface under applied field at 150°C. Line is a result of scratch in electrode. (c) Sites of subsequent dielectric breakdown (from Reference 69).

Although the idea of ionic contamination clustering fits very well with the reported data, one of the main difficulties that the researchers faced was explaining the agglomeration of the ions in the oxide. As DiStefano pointed out, radiotracer studies had indeed shown that sodium was inhomogeneously distributed at the interface, but the reason for the nonuniformity was unknown. He suggested that it may result from fast diffusion of sodium through "pipes" in the SiO_2, by a lateral diffusion of sodium along the interface to areas where accumulation is favored, or by an inhomogeneity in the initial sodium distribution.[69]

Although Revesz and Schaeffer[72] have suggested the existence of pipes, or "channels" as they called them, in the SiO_2 film, it is unlikely that these channels could lead to micrometer-sized clusters of sodium separated by distances typically in the 10–100 μm range. Furthermore, the data of Stagg and Boudry[73] concerning lateral diffusion of sodium indicate that diffusion over the distances involved in the breakdown experiments is unlikely. The suggestion of an inhomogeneity in the initial sodium distribution appears to be the most likely explanation, and moreover, as pointed out by several researchers, this initial inhomogeneity appears to result from particulate contamination incorporated into the growing oxide film.

Some of the first work relating dielectric breakdown to particulate contamination was done by Berenbaum,[14] who used transmission electron microscopy in conjunction with X-ray microanalysis to investigate defects in thin dielectric films. He found that the particles, typically submicrometer in size, became incorporated into the oxide during growth, and some of the particles actually reacted with the oxide to form a new structure. He noted that during oxidation, a radial diffusion front grows outward from the particle, and may grow to 25 times the diameter of the original particle.

In order to verify that the incorporation of submicrometer particles causes time-dependent dielectric breakdown, Berenbaum carefully examined and mapped wafers after oxidation. They were then further processed, packaged, and tested. The results showed that some of the failures could be traced, one-to-one, to the submicrometer particle sites previously mapped.

Further work dealing with particulate contamination was carried out by Monkowski and Zahour.[15] Using optical microscopy, they found ramdomly distributed defects, typically micrometer-sized, in the SiO_2 films. These defects show up as slight perturbations in the background color of the oxide film. For example, defects in a deep-blue 1000-Å film show up as a lighter blue, indicating a thicker oxide growth.

A thicker oxide in the defective regions is consistent with the mounds and bumps that Bolin[31] observed using optical and scanning electron microscopy (SEM). Furthermore, the sizes of the defects observed by Monkowski and Za-

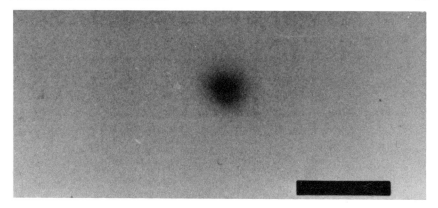

Figure 8. Defect in SiO$_2$ observed with SEM (from Reference 15). Scale bar: 1 μm.

hour agree well with the distribution of sizes observed by Bolin (Figure 2).

To further study the defects in the SiO$_2$ films, Monkowski and Zahour used SEM in conjunction with X-ray microanalysis. Figure 8 shows one of the defects as observed in the SEM. For this sample, no conductive coating was put onto the oxide film as is typically done to prevent charging in the SEM. As can be seen in the micrograph, the defect is devoid of any apparent structure. In fact, the only reason that one obtains contrast formation with these defects is that the defect is more highly conducting than the remainder of the film, and consequently charges less, appearing darker in the SEM. A number of these defects were observed, and the submicrometer size is typical.

X-ray microanalysis of the defect shown in Figure 8 results in the spectrum shown in Figure 9. In addition to the strong silicon background, peaks for sodium, sulfur, and calcium can be seen. Other defects were similar, but some contained potassium while others contained iron. In all cases, X-ray microanalysis just outside of the defect site showed only silicon, indicating the high degree of localization of the contamination.

Monkowski and Zahour[15] attribute the contamination present in these defect sites to particles that have become incorporated into the oxide during oxide growth, similar to the suggestion of Berenbaum.[14] In Figure 10 is shown an SEM micrograph of a particle in an MOS device. In this case, the particle was coated with aluminum. This particle is unusually large, but serves to point out the proposed origin of the defect formation. During oxidation, the particle fluxes or reacts with the growing SiO$_2$, forming a highly contaminated region around

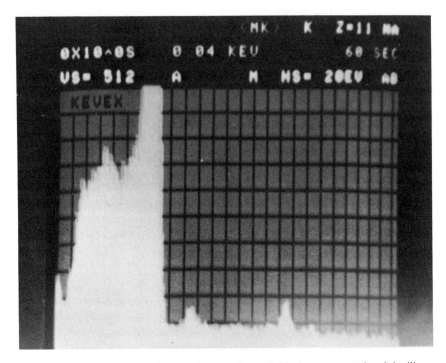

Figure 9. X-ray microanalysis of defect shown in Figure 8. The large, truncated peak is silicon, the peak to its left is sodium, and the two peaks to the right of silicon are sulfur and calcium.

the particle. The contamination, which in this case was found to be sodium, calcium, and sulfur, lowers the viscosity of the SiO_2, allowing a more rapid growth. In fact, a mound of approximately 30-μm diameter can be seen around the particle. The researchers go on to explain that if the particle were small enough, it could be entirely assimilated before a defective region even as large as 1 μm had formed, such as in Figure 8.

To investigate the effect of particulate contamination on yield and reliability, split lots of wafers were run, in which one lot was cleaned in electronic-grade chemicals and the other lot was cleaned in low-particulate chemicals. The wafers were then oxidized together and MOS capacitors were formed.

Figure 11 shows the results of the yield test, which consisted of ramping a voltage across the capacitor and monitoring the voltage at which the first self-healing breakdown occurred. Three sets of capacitors with different oxide thicknesses (200 Å, 500 Å, and 1000 Å) were tested. Each data point corresponds to a minimum of 400 tests, and the bars indicate ± 1 standard deviation. The differences are significant at better than the 99% confidence level.

Particulate Surface Contamination 143

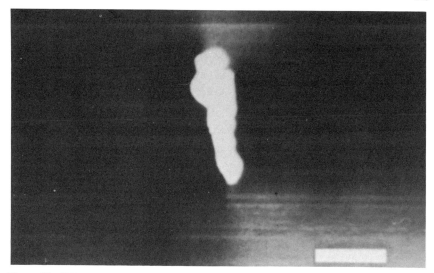

Figure 10. SEM micrograph of particle in an MOS device, showing mound in oxide surrounding particle (from Reference 15). Scale bar: 10 μm.

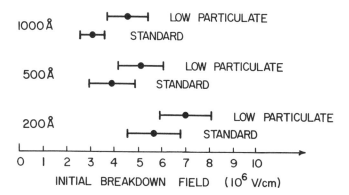

Figure 11. Average initial breakdown fields (±1 standard deviation) for MOS devices processed in standard MOS-grade and in low-particulate chemicals (from Reference 15).

As can be seen, the concentration of particles in the chemicals plays a significant role in dielectric breakdown strength. Of particular importance is that the particles adhere to the silicon surface quite tenaciously; if this were not the case, the final rinse in deionized water just prior to oxidation would have washed off the particles introduced by the chemicals, so that there would not have been any differences in the results obtained for the two cleaning treatments.

Figure 12 shows reliability data for MOS capacitors that have been fabricated from wafers processed in a split lot as described above. In general, the time-dependent dielectric breakdown is lognormal as anticipated.[59] Furthermore, there is a considerable improvement in the time to failure for those devices fabricated from wafers cleaned in the low-particulate chemicals.

3.4. Impurity Contamination in Silicon

Just as particles can introduce impurity contamination into SiO_2 films, they can also introduce impurities into the silicon wafer. However, since impurity contamination in silicon can arise from a large number of sources,[52] very little attention has been paid to the specific role of particulate contamination in the introduction of metallic impurities into the silicon wafer.

Whenever the silicon wafer is in contact with some form of impurity, though, and is undergoing high-temperature processing, that impurity will find its way into the silicon material.[74] Once in the silicon, the impurities serve to enhance junction leakage currents, reduce transistor gain, and reduce junction breakdown voltages.[52]

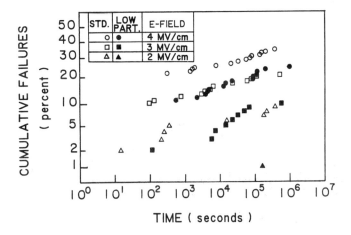

Figure 12. Reliability data for MOS devices processed in standard MOS-grade and in low-particulate chemicals.

A specific case in which particles were shown to be the source of contamination was that of haze formation as caused by particles from stainless steel tweezers.[43] This case was described above in Section 2.4.

Overall, since the diffusivities of metals in silicon are typically high, and since extremely small amounts of metallic impurities can severely degrade device performance,[75] the avoidance of any kind of metallic impurity source—including particulate contamination—is of utmost importance.

4. Summary

The investigation, measurement, and control of particulate contamination is becoming increasingly important in the semiconductor industry. As other types of contamination are being better understood and better controlled, the role of particulate contamination is becoming a more significant part of overall contamination control. Unfortunately, though, as a result of its wide diversity in structure, composition, and properties, particulate contamination is clearly the most difficult type of contamination to define, measure, and control.

This chapter began by reviewing the various sources of particulate contamination, including the air, the processing chemicals, the processing gases, and the overall handling of the silicon wafers. Clearly, the most attention has been paid to the air; however, the industry now recognizes that the other sources are becoming increasingly important. As a result, low-particulate chemicals and high-purity gases are now available. With regard to wafer handling, a major emphasis is being placed on education of personnel in the field of clean-room procedures.

One of the primary means by which particulate contamination degrades device yield is through obstruction of the patterns on photomasks. In this regard, the severity of the degradation depends upon the size of the photomask features, the size of the particle, and whether positive or negative photoresist is being used.

The other major failure mode in which particles play a role is dielectric breakdown of MOS gate oxides. Although the available literature is not in obvious agreement concerning the role of particles in this type of failure, close inspection of the data and conclusions does indeed show that particulate contamination is the most probable cause of these failures and is essentially consistent with all of the data.

References

1. T. G. O'Neill, Clean room efficiency: A combination of design and operation, *Semiconductor International* 3(11), 49–62 (1980).
2. VLSI Facilities Overview, ICE Corporation Publication 16-1781-02, Figure 7337C (1985).

3. D. LaFeuille, D. Roche, and E. M. Juleff, Purity of chemicals for semiconductor processing, *Solid State Technol. 18*(1), 43-48 (1975).
4. E. M. Juleff, W. J. McCleod, E. A. Hulse, and S. Fawcett, Advances in contamination control of processing chemicals in VLSI, *Solid State Technol. 25*(9), 82-86 (1982).
5. A. Weiss, Particulate filtration of chemicals, gases, and photoresist, *Semiconductor International 5*(7), 55-64 (1982).
6. C. M. Juleff, Verifying purity of chemicals, *Electron. Prod. Methods Equip. 4*(1), 34-38 (1975).
7. J. T. Przybytek and K. L. Calabrese, Measuring low level particle counts in solvents, *Microcontamination 3*(6), 51-54 (1985).
8. P. Burggraaf, Applied wet chemical microfiltration, *Semiconductor International 8*(3), 86-89 (1985).
9. D. L. Tolliver, N. Davenport, and L. R. Abts, The detection of microcontaminants in semiconductor process fluids using an acoustic technology, *Solid State Technol. 25*(9), 116-123 (1982).
10. D. Angel, P. H. Johnson, and M. B. Vye, Automatic defect inspection of chromium-on-glass photolithographic masks, *Semiconductor International 1*(1), 100-109 (1978).
11. P. S. Burggraaf, The value of photomask inspection, *Semiconductor International 3*(2), 33-46 (1980).
12. W. Kern, Characterization of localized defects in dielectric films for electron devices, *Solid State Technol. 17*(3), 35-42 (1974).
13. S. Gunawardena, U. Kaempf, B. Tullis, and J. Vietor, SMIF and its impact on cleanroom automation, *Microcontamination 3*(9), 55-62, 108 (1985).
14. L. Berenbaum, The effect of submicron particulate contamination on the properties of thin dielectric films, Abstract #63, 147th Meeting of the Electrochemical Society, Toronto, Canada (1975).
15. J. R. Monkowski and R. T. Zahour, Failure mechanism in MOS gates resulting from particulate contamination, in: Proceedings of the IEEE Reliability Physics Symposium (1982), pp. 244-248.
16. E. S. Anolick, Area and electrode effects on time dependent breakdowns through Si_3N_4 and SiO_2 double layers, Abstract #60, 147th Meeting of the Electrochemical Society, Toronto, Canada (1975).
17. P. S. Burggraaf, Airborne-particle monitoring know-how, *Semiconductor International 5*(7), 35-50 (1982).
18. C. M. Davis, G. Bergeron, R. LaCourse, and G. Trombley, HEPA filters as a contamination source, *J. Environ. Sci. 24*, 27-35 (1981).
19. K. H. Stokes, Class 10—Can we do it?, *Microcontamination 1*(4), 12-14 (1984).
20. R. P. Donovan, B. R. Locke, D. S. Ensor, and C. M. Osburn, The case for incorporating condensation nuclei counters into a standard for air quality, *Microcontamination 2*(6), 39-44 (1984).
21. J. Burnett, Class 1 cleanroom specifications, *Microcontaminations 3*(6), 21-24 (1985).
22. J. McDonough, Status and update report: IES recommended practice program and Federal Standard 209B revision, *Microcontamination 3*(9), 47-52 and 104-106 (1985).
23. J. A. DeNicola and R. D. Mastropiero, Design and maintenance of high purity gas handling systems, *Solid State Technol. 15*(2), 51-54 (1972).
24. H. Boyd, Gases for semiconductor production, *Microelectronic Manufact. Testing*, 80-83 (May, 1981).
25. R. L. Duffin, Process gas filtration in integrated circuit production, *Microcontamination 1*(4), 35-38 (1984).
26. T. G. O'Neill, Ultra-pure water update, *Semiconductor International 4*(7), 55-69 (1981).

27. M. Corn, The adhesion of solid particles to solid surfaces, *J. Air Pollution Control Assoc.* 2, 567 (1961).
28. P. S. Burggraaf, Wafer cleaning, *Semiconductor International* 4(7), 71–100 (1981).
29. S. Bhattacharya and K. L. Mittal, Mechanics of removing glass particulates from a solid surface, *Surface Technol.* 7, 413–425 (1978).
30. J. M. Duffalo and J. R. Monkowski, Particulate contamination and device performance, *Solid State Technol.* 27(3), 109–114 (1984).
31. H. R. Bolin, Process defects and effects on MOSFET gate reliability, in: Proceedings of the IEEE Reliability Physics Symposium (1980), pp. 252–254.
32. H. H. Willard, L. L. Merritt, Jr., and J. A. Dean, *Instrumental Methods of Analysis*, 5th Ed., pp. 350–389, D. Van Nostrand Co., New York (1974).
33. J. Reednick, A unique approach to atomic spectroscopy, *Am. Lab.* 11(3), 53–61 (1979).
34. H. H. Willard, L. L. Merritt, Jr., and J. A. Dean, *Instrumental Methods of Analysis*, 5th Ed., pp. 522–560, D. Van Nostrand Co., New York (1974).
35. H. H. Willard, L. L. Merritt, Jr., and J. A. Dean, *Instrumental Methods of Analysis*, 5th Ed., pp. 150–188, D. Van Nostrand Co., New York (1974).
36. H. H. Willard, L. L. Merritt, Jr., and J. A. Dean, *Instrumental Methods of Analysis*, 5th Ed., pp. 328–331, D. Van Nostrand Co., New York (1974).
37. A.S.T.M. F-312-80 Specification.
38. J. R. Monkowski, The role of chlorine in silicon oxidation, *Solid State Technol.* 22(7), 58–61; *ibid.* 22(8), 113–119 (1979).
39. J. Steinberg, Dual HCl thin gate oxidation process, *J. Electrochem. Soc.* 129, 1778–1782 (1982).
40. C. Hashimoto, S. Muramoto, N. Shiono, and O. Nakajima, A method of forming thin and highly reliable gate oxides, *J. Electrochem. Soc.* 127, 129–135 (1980).
41. W. A. Brown and T. I. Kamins, An analysis of LPCVD system parameters for polysilicon, silicon nitride and silicon dioxide deposition, *Solid State Technol.* 22(7), 51–57 (1979).
42. W. T. Stacy, D. F. Allison, and T. C. Wu, The role of metallic impurities in the formation of haze defects, in: *Semiconductor Silicon 1981*, pp. 344–353, The Electrochemical Soc., Inc., Pennington, N.J. (1981).
43. W. T. Stacy, D. F. Allison, and T. C. Wu, Metal decorated defects in heat-treated silicon wafers, *J. Electrochem. Soc.* 129, 1128–1133 (1982).
44. T. Baginski, J. R. Monkowski, and I. S. T. Tsong, The role of chlorine in the gettering of metallic impurities from silicon, Abstract #399, 160th Meeting of the Electrochemical Society, Denver Colo. (1981).
45. T. Baginski, private communication (1983).
46. See, for example, R. A. Colclaser, *Microelectronics: Processing and Device Design*, pp. 22–52, John Wiley and Sons, New York (1980).
47. A. B. Glaser and G. E. Subak-Sharpe, *Integrated Circuit Engineering*, pp. 751–753, Addison-Wesley Publishing Co., Reading, Mass. (1977).
48. M. Martin and H. Williams, Optical scanning of silicon wafers for surface contaminants, *Electro-opt. Syst. Des.* 12(9), 45–49 (1980).
49. M. L. Hammond, Silicon epitaxy, *Solid State Technol.* 21(11), 68–75 (1978).
50. P. Burggraaf, High resistivity epi may solve MOS problems, *Semiconductor International* 3(4), 71–75 (1980).
51. G. R. Srinivasan, Silicon epitaxy for high performance integrated circuits, *Solid State Technol.* 24(11), 101–110 (1981).
52. J. R. Monkowski, Gettering processes for defect control, *Solid State Technol.* 24(7), 44–51 (1981).
53. G. B. Larrabee and J. A. Keenan, Neutron activation analysis of epitaxial silicon, *J. Electrochem. Soc.* 118, 1351–1355 (1971).

54. G. A. Rozgonyi, R. P. Deysher, and C. W. Pearce, The identification, annihilation, and suppression of nucleation sites responsible for silicon epitaxial stacking faults, *J. Electrochem. Soc. 123*, 1910–1915 (1976).
55. C. W. Pearce and R. G. McMahon, Role of metallic contamination in the formation of "saucer" pit defects in epitaxial silicon, *J. Vac. Sci. Technol. 14*, 40–43 (1977).
56. M. C. Chen and V. J. Silvestri, Pre- and post-epitaxial gettering of oxidation and epitaxial stacking faults in silicon, *J. Electrochem. Soc. 128*, 389–395 (1981).
57. R. E. Logar and J. O. Borland, Silicon epitaxial processing techniques for ultra-low defect densities, *Solid State Technol. 28*(6), 133–136 (1985).
58. C. R. Barrett and R. C. Smith, Failure modes and reliability of dynamic RAMS, paper presented at the International Electronic Devices Meeting, Washington, D.C. (1976).
59. D. L. Crook, Method of determining reliability screens for time dependent dielectric breakdown, in: Proceedings of the IEEE Reliability Physics Symposium, (1979), pp. 1–7.
60. C. M. Osburn and D. W. Ormond, Dielectric breakdown in silicon dioxide films on silicon, *J. Electrochem Soc. 119*, 591–602 (1972).
61. A. K. M. Zakzouk, Time dependent MOS gate oxide defects using liquid crystals, *J. Electrochem. Soc. 127*, 932–936 (1980).
62. A. K. M. Zakzouk, The dependence of the SiO_2 defect density on both the applied electric field and the oxide thickness, *J. Electrochem. Soc. 126*, 1771–1779 (1979).
63. R. A. Williams and M. M. Beguwala, Reliability concerns for small geometry MOSFETs, *Solid State Technol. 24*(3), 65–71 (1981).
64. E. S. Anolick and G. R. Nelson, Low field time dependent dielectric integrity, in: Proceedings of the IEEE Reliability Physics Symposium (1979), pp. 8–12.
65. A. K. M. Zakzouk, General model for defect formation in silicon dioxide, *IEE Proc. 127*, 230–234 (1980).
66. S. P. Li, S. Prussin, and J. Maserjian, Model for MOS field-time-dependent breakdown, in: Proceedings of the IEEE Reliability Physics Symposium (1978), pp. 132–136.
67. P. Solomon, Breakdown in silicon oxide—A review, *J. Vac. Sci. Technol. 14*, 1122–1130 (1977).
68. T. H. DiStefano, Barrier inhomogeneities on a $Si-SiO_2$ interface by scanning internal photoemission, *Appl. Phys. Lett. 19*, 280–282 (1971).
69. T. H. DiStefano, Dielectric breakdown induced by sodium in MOS structures, *J. Appl. Phys. 44*, 527–528 (1973).
70. R. Williams and M. H. Woods, Laser-scanning photoemission measurements of the silicon-silicon dioxide interface, *J. Appl. Phys. 43*, 4142–4147 (1972).
71. J. M. Keen, Nondestructive optical technique for electrically testing insulated-gate integrated circuits, *Electron. Lett. 7*, 432–433 (1971).
72. A. G. Revesz and H. A. Shaeffer, The mechanism of oxygen diffusion in vitreous SiO_2, *J. Electrochem. Soc. 129*, 357–361 (1982).
73. J. P. Stagg and M. R. Boudry, Sodium passivation in $Al-SiO_2-Si$ structures containing chlorine, *J. Appl. Phys. 52*, 885–899 (1981).
74. P. F. Schmidt and C. W. Pearce, A neutron activation analysis study of the sources of transition group metal contamination in the silicon device manufacturing process, *J. Electrochem. Soc. 128*, 630–637 (1981).
75. J. R. Davis, A. Rohatgi, P. Rai-Choudhury, P. Blais, and R. H. Hopkins, Characterization of the effects of metallic impurities on silicon solar cell performance, in: Proceedings of the IEEE Photovoltaic Specialists Conference (1978), pp. 490–495.

Effect of Surface Contamination on the Performance of HVDC Insulators

DAVID C. JOLLY

1. Introduction

Contamination flashover, sometimes called pollution flashover, is a form of electrical insulation failure caused by conducting deposits on an insulator surface. If the contamination is sufficiently heavy, failure takes the form of a discharge growing along the surface until the terminals are bridged, at which time an electrical arc forms, and the insulation no longer supports its nominal voltage.

This type of insulation failure is most frequently encountered on outdoor electrical power transmission lines since the insulators on such lines are exposed to soluble airborne pollutants and moisture. Some of the more common outdoor high-voltage insulator types are shown in Figure 1. Contamination flashover is not confined to electric power transmission. For example, dirty spark plugs may arc along the ceramic insulation if the surface breakdown voltage falls below that of the air gap. Contamination flashover can be a design constraint in high-voltage fuel cells in which electrolyte mist deposits on electrically insulating structures. Similar processes lead to surface breakdown in all these cases, but contamination flashover has been studied most extensively in the context of outdoor power transmission.

The contaminants responsible for outdoor insulation failure are widely varied, as revealed by a survey of utilities done in 1968 by the Institute of Electrical and Electronic Engineers (IEEE).[1] The most commonly found contaminants

---

DAVID C. JOLLY • Consultant, PO Box 931, Brookline, Massachusetts 02146.

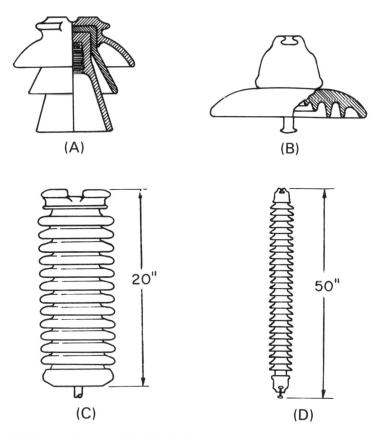

Figure 1. Major outdoor transmission line insulator types currently in use. (A) Pin-type insulator designed for 35-kV mounts on a tower cross-arm with the conductor nesting in the groove on top. (B) Modern suspension insulator. A metal cap and pin are separated by a porcelain or glass shell about 25 cm in diameter. These are hooked in series to suspend a conductor beneath a tower cross-arm, with about 12 kV appearing across each insulator. (C) Post-type insulator for 66 kV, usually used the same way as the pin type. (D) Long rod suspension insulator. Metal end-fittings are separated by a porcelain or polymer body, ribbed to increase creepage distance and shed rain water.

were sea salt, cement dust, fertilizer, fly ash, and road salt flung up by vehicle tires. Insulation failure was also reported from dried milk (no details are provided), bird droppings, and smelter emissions. Particularly severe operating conditions occur in desert areas, where windblown dust accumulates on insulator surfaces. Radiational cooling during the night leads to dew formation and flashover.[2]

Surface deposits are harmless as long as they remain electrically nonconducting. Trouble normally occurs only when the insulator surface becomes wet, and the salts dissolve to form a conducting electrolyte. Thus the IEEE survey found that about 41% of failures occurred during rain, drizzle, and mist, about 26% occurred in fog, and 16% in dew. Snow triggered a few failures, while a mysterious 0.7% were reported to occur in fair weather.

Contamination flashover problems can be quite unpredictable, as when the vast amount of ash emitted by the eruption of Mount St. Helens caused degradation of insulator performance on high-voltage equipment and transmission lines near the volcano.[3,4] The immense clouds of dust stirred up during a thermonuclear attack will similarly cause insulation problems if attempts are made to restore electric power service.

A number of DC transmission lines have been constructed or are being planned.[5-8] A major advantage of DC transmission is that a DC line is asynchronous; while, for given load conditions, an AC line must be operated carefully to maintain a certain frequency and phase relation between its two end points. By clever control of the frequency conversion equipment at the terminals of a DC line, utilities are able to improve the stability of their power grids and lessen the probability of blackouts. The high cost of the equipment at the ends of the line which convert the AC to DC and then back to AC imposes an economic penalty on DC transmission, making longer lines more economical than short ones. Conversion equipment is necessary since almost all power is generated and consumed as alternating current. An unexpected drawback of DC lines is that they are more susceptible to contamination flashover for reasons that will be discussed later.

There are several sources of information on contamination flashover, including a few bibliographies.[9-11] The IEEE bibliography[11] contains over 900 references up to 1976. The December 1972 issue of the Journal of the Franklin Institute was devoted to contamination flashover. An application guide has been prepared for installing insulators in contaminated areas.[12] Much of the literature deals with practical aspects, such as how frequently to wash deposits off outdoor insulators. Much also deals with AC transmission, although the theoretical papers tend to consider DC because modeling is simpler. The references at the end of this chapter constitute an up-to-date bibliography of papers most directly pertaining to the DC flashover problem.

2. General Overview of the Flashover Process

It has been known almost from the earliest days of power transmission that electrical insulators lose much of their insulating ability when their surfaces become contaminated. Anfossi in 1907 noted that a 25-kV line near the sea was adversely affected by salt spray.[13] In 1908 the insulators for the electrification of the New Haven Railroad were found to be inadequate when steam and soot from locomotives deposited on them, and extra insulators were installed.[14]

Early investigators blamed these failures on localized electric field concentrations. The conducting contamination layer was thought to disturb the normal field distribution in such a manner that an air discharge occurred from portions of the insulator to the metal support or the conductor. In 1911 Austin, an American engineer, proposed a criterion to optimize insulators against this type of failure.[15] As more was learned about electrical breakdown physics, this theory was seen to be erroneous. An ordinary suspension insulator having 30 cm of creepage distance can fail in service at about 7-kV peak voltage, or only about 200 V/cm. This is far below the normal air breakdown field, which is typically about 10,000 V/cm.

The modern picture of what happens during the breakdown of a contaminated insulator surface was developed over several decades of research in Germany beginning around 1930. Using photography, Obenaus found that breakdown was not a sudden process, but occurred gradually over several seconds.[16,17] Arcs were seen to bridge dry sections of the insulator as shown in Figure 2. As further drying occurred, the dry-zone edges receded and the discharges lengthened. Electrostatic attraction acting on the arc plasma was believed to aid in the final stages of arc growth. Productive investigation resumed in the 1950s as Frischmann showed with high-speed motion pictures that the root of the arc burning to the surface was mobile.[18-21] At this time Obenaus and his colleagues derived a simple quantitative criterion for flashover based on arc extinction phenomena.[22-26]

The currently accepted sequence of events leading to failure is as follows. When an energized insulator with dry, soluble contaminants is exposed to moisture deposition, an increased conduction current begins to flow along the surface. Moisture continues to accumulate on the surface until the power dissipated by the surface current is comparable to the power needed to significantly affect the deposition rate. Because of nonuniformities in contamination, deposition of water, or geometry, some portions of the surface will dissipate more power than others, and these regions will resist further deposition of water. The power dissipation will further concentrate in these dryer regions since electrical dissipation varies as the surface resistivity times the square of the surface current density, the latter varying inversely as the insulator radius and hence remaining

Effect of Contamination on HVDC Insulators

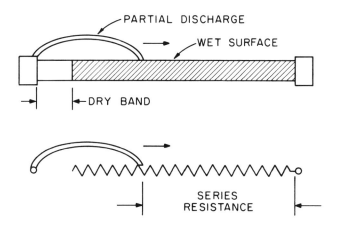

Figure 2. Schematic depiction of a dry band and a partial arc on a rod-type insulator. The lower figure shows a possible electrical model. The arrow is intended to show the direction of motion of the tip of the arc when the final stage of flashover occurs.

relatively constant over the insulator surface. The final equilibrium state under these conditions is an insulator wet over most of its surface, with one or more regions called "dry bands" or "dry zones" which interrupt the current. The power dissipation is almost entirely concentrated in these dry bands.

Under conditions usually encountered, there is intermittent discharge activity bridging the dry bands. These discharges, which bridge only part of the insulator, are called "partial discharges." These partial discharges are not normally harmful since the resistance of the wet film limits the current to several milliamperes, or, at most, about 1 A. However if the voltage is high enough, or if the contamination is severe enough, the discharge can elongate as shown in Figure 2 to bridge the insulator and trigger flashover. The long-term presence of partial discharges can be tolerated on ceramic and glass insulators, but glass-filament-reinforced polymer insulators are being used in service on a trial basis, and continued discharge activity may cause a polymer insulator to deteriorate and eventually fail.[27–34]

For purposes of discussion, it is convenient to think of contamination flashover as occurring in several stages:

1. Airborne particles of soluble salts deposit on the surface of the insulator.

2. Moisture, usually in the form of rain, fog, or dew, deposits on the insulator, dissolving the salt particles and forming a conducting film.

3. Thermal processes lead to an uneven surface conductivity, and a corresponding nonuniform distribution of the electric field.

4. When the localized electric field becomes high enough, small scintillations begin to occur sporadically. These scintillations do not bridge the entire insulator.

5. If the contamination is high enough relative to the applied voltage, one of these small scintillations may suddenly elongate, bridging the terminals of the insulator, forming an arc. The current at this point is limited largely by the impedance of the circuit, and protective devices such as circuit breakers or fuses must operate to interrupt the current.

Unfortunately for DC transmission, this sequence of events proceeds more readily than for AC. Stage 1, contaminant deposition, is much more rapid for DC. This is thought to result from an action similar to electrostatic precipitators. Stage 5, discharge growth, also seems to occur more readily for DC, perhaps because AC arcs must reignite and reform at each current-zero.[35,36] The five steps in the above sequence will be discussed in more detail below, with emphasis on stages 1 and 5.

3. Deposition of Particles

3.1. Introduction

DC insulators on outdoor electrical equipment have been observed to become contaminated more rapidly than comparably installed AC insulators. For example, in a clean air region of Germany, contamination occurred on DC insulators where AC insulators experienced no difficulty.[37] It is believed that electrostatic forces acting on airborne particles account for the difference.

Charged particles in the air surrounding an electric power line experience a force due to the electric fields present. Near AC lines, the coulomb force on a particle alternates rapidly with respect to the motion caused by gravity and wind, and there is little effect on particle trajectories. If the AC field is nonuniform, there will be a net force on the induced dipole, but this force is usually negligible. The coulomb force is steady for DC, and for typical conditions encountered in practice, the particle trajectories near insulator surfaces may be altered significantly, producing a large increase in particle deposition.

The amount of charge on an aerosol particle may vary as it nears an insulator. This is because high-voltage transmission lines and station equipment produce electrical corona, a localized form of electrical breakdown often visible at night as a faint glow. The corona injects positive or negative ions into the surrounding air where they can become attached to airborne particles. Thus a slightly charged particle approaching a power line could be strongly charged by corona and then be drawn by the electric field to the surface of an insulator.

Effect of Contamination on HVDC Insulators

This section will consider the electrical environment surrounding a DC transmission line, particle-charging mechanisms, and finally the factors governing particle deposition on insulators.

3.2. The Electrical Environment

To determine how particles behave near high-voltage equipment, it is first necessary to calculate the electric field and ion density as a function of position. In the presence of corona, the fields can be calculated using Poisson's equation (in SI-Giorgi units):

$$\nabla^2 \phi = \frac{\rho}{\epsilon_0} \quad (1)$$

where ϕ is the electrostatic potential (V), ρ is the charge density (C/m^3), and ϵ_0 is the permittivity of free space (F/m). Without corona, the charge density can be ignored. Appropriate boundary conditions must be applied, and ion motion caused by wind and the electric field must be considered. Good examples of how the field and ion density distributions can be calculated are two papers by Sarma and Janischewskyj.[38,39] Complementing such theoretical calculations, electric field and ion current measurements have been made in the vicinity of high-voltage DC lines.[40-44] Similar measurements can also be made on scale model transmission lines.[45] Although scale models permit line geometry and voltage to be easily varied, it is difficult to scale simultaneously all quantities properly.

3.3. Charging Mechanisms

An airborne particle surrounded by ions will acquire a charge. The simplest case is a spherical particle surrounded by a gas containing ions of uniform charge and mobility. Ions will diffuse to the surface of the particle. Taking positive ions as an example, as charge accumulates on the particle, its potential will increase, tending to repel further ions. Only ions sufficiently energetic to overcome this potential will reach the particle. Assuming that the ions are maxwellian, a simple expression can be obtained for the charging rate:

$$\frac{dQ}{dt} = \pi a^2 n q c \exp(-qQ/4\pi\epsilon_0 akT) \quad (2)$$

where Q = charge on the particle (C), t = time (s), n = ion density (m^{-3}), e = electron charge (C), a = particle radius (m), k = Boltzmann's constant (J/K), T = temperature of the ions (K), and c is the mean thermal velocity (m/s) of the ions:

$$c = \left(\frac{8kT}{\pi m}\right)^{1/2} \tag{3}$$

where m = the ion mass (kg). Note that the particle never stops accumulating charge, although the rate at which it accumulates charge decreases with time.

A further charging mechanism exists when an external electric field is present, as is true for particles near a high-tension power line. In this case ions swept along by the field strike the particle. As charge accumulates on the particle, its potential changes, tending to repel further particles. In the absence of diffusion, the rate of charging is given by:

$$\frac{dQ}{dt} = 3\pi a^2 nqbE\left(1 - \frac{Q}{12\pi\epsilon_0 a^2 E}\right) \tag{4}$$

where E is the applied field (V/m) and b is the ion mobility (m²/s · V). It is obvious that there is a limit to the charge on the particle. This is called the saturation charge, Q_s, given by:

$$Q_s = 12\pi\epsilon_0 a^2 E \tag{5}$$

In many cases, both diffusion and field charging must be considered simultaneously. Unfortunately, the resulting equations are unwieldy. The literature on particle charging cannot be reviewed here, but a good introductory article is the combined diffusion and field charging analysis of Liu and Kapadia.[46] It is well to remember that despite seeming theoretical precision, charging theories contain a number of assumptions and approximations. For example, particles in actuality are not spherical, and their electrical properties are not always known. These factors limit at the outset the accuracy that can be expected of any theory of particle deposition on insulators which involves particle charging.

3.4. Particle Deposition Rates

Many studies of particle deposition on high-voltage DC insulators have emphasized experimental methods. An insulator would be energized, a stream of dust blown past it, and the insulator would then be examined to determine the distribution of dust on the surface.[47-50] The DC results can be compared with AC results where the steady coulomb force is absent.[51] It is also possible to make measurements on the amount of particles deposited on insulators energized outdoors.[52-54] Such measurements provide valuable data for correlating with laboratory experiments and theory.

If the ion density, electric field, and wind velocity around a transmission line are known, it is theoretically possible to calculate the trajectories of particles approaching the line. In principle, the particle deposition rate on transmission line insulators can then be calculated. The force on a spherical particle, $\mathbf{F}(N)$, is given by[47,48,55-58]:

$$\mathbf{F} = M\mathbf{g} + Q\mathbf{E} + 6\pi\mu a(\mathbf{v} - \mathbf{v}_a) \tag{6}$$

where M = particle mass (kg), $\mathbf{g}$ = acceleration due to gravity (m/s^2), μ = viscosity of air (N · s/m^2), $\mathbf{v}$ = particle velocity (m/s), and $\mathbf{v}_a$ = air velocity (m/s). Stokes drag has been assumed. The equation of motion of the particle can be written as:

$$M\frac{d\mathbf{v}}{dt} = \mathbf{F} \tag{7}$$

The particle charge can be calculated if the charging rate is known:

$$Q(t) = Q(0) + \int_{u=0}^{t} \frac{dQ(u)}{du} du \tag{8}$$

where u is a dummy variable. Using the above equations and assuming no charging, Olsen et al. have calculated the trajectories for particles approaching a cylindrical rod-type insulator.[56] They found a qualitative agreement with the experimentally observed contamination distribution. The calculations predicted that if hemispherical electrodes were placed at the ends of the rod to make the field lines more parallel to the insulator surface, contamination should be more uniformly distributed.[56,58] This was verified by experiment. Others have also investigated the effects of external field control electrodes.[50,59]

Significant theoretical advances in understanding the deposition mechanism were made by Horenstein and Melcher.[60,61] They noted that when the divergence of both fluid velocity and electric field can be neglected, the particle density, N (m^{-3}), along any particle trajectory is constant. In practical cases, the density upwind is made uniform by turbulent diffusion, and the particle density adjacent to any point on the insulator surface connected to the upwind side by a particle trajectory is simply the upwind particle density. The deposition rate at such points is:

$$\frac{dw}{dt} = Nb|\mathbf{n} \cdot \mathbf{E}| \tag{9}$$

where w is the surface mass density of contaminant (kg/m^2) and $\mathbf{n}$ is the vector normal to the insulator surface. To find the deposition rate for the entire insulator, equation (9) can be integrated over the surface where trajectory lines terminate. Where a range of mobilities exists, equation (9) must also be integrated over mobility.

This concept was tested using 4-μm charged particles impinging on a one-fifth scale model disk insulator string. Theory and experiment agreed to within a factor of two, which can be considered good because of a number of experimental and theoretical uncertainties.[60,61] Excellent agreement was obtained between the observed particle distribution over individual insulators and a computer simulation of trajectories. A number of initially puzzling features, including distinct clean bands, could be explained with this model. Equations were also derived to characterize the situation where field charging of the particles could not be neglected. In extending such model results to practical situations, careful attention must be paid to scaling laws.

One problem in calculating the electric field near an insulator is the difficulty in modeling the potential on the insulator surface. Insulator surfaces under DC stress do not behave simply.[62] Horenstein and Melcher assume a uniform surface conductivity.[60,61] Others have assumed a linear voltage drop along the insulator surface.[59,63] Furthermore, field measurements have shown that when insulators are connected in series, the voltage drops across the individual insulators are not equal.[64]

Radun and Melcher have attempted to include the effects of turbulent transport of aerosol particles.[65,66] Their work showed that particles charged by passage near a transmission line will have an increased distribution of mobilities at each point downwind as a result of turbulent diffusion.

One area not fully studied yet is the effect of aerodynamics on deposition.[55,67] Perhaps by varying the shape of an insulator, the airflow could be modified to reduce deposition. The effects of rain washing have also not been adequately studied. More information is needed on the properties of aerosols, i.e., size distributions and natural charge. The sticking properties of particles on insulator surfaces have been largely ignored, the usual assumption being that if a particle impacts the surface, it will stay there.

It is not apparent that a technically useful model for deposition on outdoor insulators can ever be constructed. Each geographical location has its own set of conditions. The turbulence spectrum, wind velocity distribution, particle size distribution, meteorological conditions, shapes of insulators, electrode geometry, etc. can vary from one case to the next. A complete deposition model requires both detailed site information and a computer program to solve the governing equations. Neither of these two elements exist. At the present state of knowledge, the model of Horenstein and Melcher[60,61] comes closest to providing useful engineering information.

4. Deposition of Moisture

As indicated earlier, the surface of an outdoor high-voltage insulator must become wet for the surface salts to dissolve and form a conductive coating. Usual weather conditions causing this are rain, drizzle, mist, fog, dew, and the like. Heavy rain often has a beneficial effect, washing away impurities from the exposed top surfaces of insulators before the bottom surfaces are wetted. Except for possible effects of electrostatic forces on droplets, the wetting processes for DC are the same as for AC, and have been discussed in detail by Karady[68] and by others.[69-71]

5. Thermal Processes

As moisture deposits on the insulator surface, and as the deposited salts dissolve, current flows along the surface, and heat is generated. The situation is inherently unstable. If a portion of the surface is relatively dry, its resistance will also be relatively higher, leading to increased heating and further drying. The end point of this process is an insulator whose surface is completely wet except for one or more dry bands or dry zones. The high resistance of these dry zones keeps the current in the interval between partial discharges low, usually below 1 mA. There is nothing theoretically difficult about modeling the thermal processes on an insulator, but the equations in practical situations are complex and must be solved numerically. Thermal processes are similar for AC and DC insulators, and have been the subject of several investigations.[69, 72-77] This topic has already been covered by Karady[68] and will not be considered further here.

6. Localized Electrical Breakdown

A high-voltage insulator with a moist, conducting surface coating reaches an equilibrium state in which most of the surface is wet, while one or more dry bands exist. Because of the high electrical resistance of these bands, most of the voltage applied to the insulator appears across them. The electric field is consequently higher than normal at the dry bands, and if the field becomes high enough, localized electrical breakdown can occur. This localized breakdown usually consists of a low current (approximately 0.001–1 A) arc of brief (approximately 0.01–1 s) duration burning from one edge of the dry band to the other. If several dry bands exist in series, simultaneous discharges appear across all dry bands. The energy dissipated by the arc plasma and by resistive heating causes the dry band to widen, and the arc extinguishes. As more water deposits on the surface, the dry band narrows until breakdown again occurs, and the process repeats. On high-voltage insulators, this discharging is clearly visible. This repetitive discharging is not normally harmful to porcelain or glass insulators. However these discharges may erode polymer insulators.[27-34]

Little work has been directed toward the problem of dry-band discharge initiation. The prevalent belief seems to be that such discharges will always occur, and hence little would be gained by studying the factors involved in the breakdown process. With present insulator designs, this view seems to be correct. On insulators installed outdoors, there is sufficient spatial nonuniformity in deposit distribution and wetting properties to always assure a sufficiently high local concentration of electric stress to cause breakdown. A surface coating with negative resistance properties might smooth the electric field, but this has not been achieved in practice.

Should some way be found to inhibit electrical breakdown across the dry bands, flashover could be prevented. Therefore, a comprehensive study of this type of breakdown process might lead to a solution of the contamination flashover problem.

7. Discharge Growth

7.1. Introduction

Even though electrical breakdown across the dry band occurs, there may not be flashover of the total insulation length. In fact, the usual result of dry-band breakdown is for an arc to burn momentarily from one edge of the dry band to the other.[78] After a brief interval, usually less than one second, the arc extinguishes, perhaps as evaporation caused by the increased current flow widens the dry band. However, under the right conditions, the short dry-band arc may elongate, growing across the wet portion of the insulator in perhaps a few milliseconds. The physical mechanisms producing the elongation are still a subject of debate. However, by making simplifying assumptions, it is possible to make approximate calculations of the voltage at which flashover will occur. How this can be done and what factors influence the flashover voltage will be discussed below.

7.2. Extinction Theories

Our present understanding of contamination flashover rests largely on the model developed by Obenaus in the 1950s.[22–26] Figure 2 shows how he modeled an arc propagating across a contaminated insulator surface. The model consists of an arc in series with a resistor. Although simple in concept, there are practical difficulties in using the model. In particular, the arc and resistance are both nonlinear elements whose behavior is very complex. For now, these difficulties will be ignored in order to simplify the explanation.

For illustrative purposes we will try to model the flashover of a long rod-type insulator (see Figure 1). In the current range of interest (0.01–1 A), unenclosed electric arcs usually have a negative voltage–current characteristic.

Effect of Contamination on HVDC Insulators

Obenaus suggested that the characteristic can be fitted over the region of interest with an expression of the form:

$$E = AI^{-n} \tag{10}$$

where E is the longitudinal electric field in the arc (V/cm), I is the arc current (A), and A and n are constants. This expression implies that the arc properties do not vary longitudinally. Thus, an arc 2 cm in length will have twice the voltage drop of a 1-cm arc. For an insulator sufficiently long, the resistance of the portion in series with the arc can be represented by the expression:

$$R = r(L - X) \tag{11}$$

where R is the total resistance (ohms), r is the linear resistivity of the conducting layer (ohm/cm), L is the total insulator length, and X is the arc length. Since the arc and resistance are in series, their total voltage must equal the source voltage:

$$V = AXI^{-n} + Ir(L - X) \tag{12}$$

Figure 3 shows the voltage–current characteristic for a given arc length for the arc, the resistance, and the sum of the two. For a given supply voltage, the current is given by the points where the curve representing the sum equals the supply voltage. Note that there are two solutions. It can be shown that only the higher-current solution is stable to fluctuations, and the lower-current solution can be neglected. As the supply voltage is lowered, eventually a point is reached where there is only one solution, and below this voltage no solutions exist. This voltage is called the extinction voltage.

Obenaus noted that for flashover to occur the supply voltage must exceed the extinction voltage, V_{ext}, for all possible arc lengths occurring during flashover. The extinction voltage can be calculated by setting the voltage derivative of equation (12) equal to zero. The result is:

$$V_{ext} = (n + 1)(AX)^{1/(n+1)} \left[\frac{r(L + X)}{n} \right]^{n/(n+1)} \tag{13}$$

For flashover to occur, the applied voltage must exceed the maximum with respect to X of the above expression. Setting the derivative equal to zero to find the value of X where the maximum occurs, and substituting into equation (12), the flashover voltage, V_{fo}, is found to be:

$$V_{fo} = L(Ar^n)^{1/(n+1)} \tag{14}$$

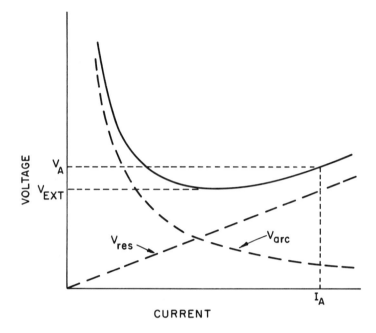

Figure 3. Voltage–current characteristic of an arc in series with a resistor. The solid curve is the sum of the arc voltage, V_{arc}, and the resistive voltage drop, V_{res}. It represents the voltage–current characteristic seen at the terminals of the insulator in Figure 2. For a given voltage, V_A, the current flow will be the intersection of the horizontal dotted line with the positively sloping portion of the solid curve, I_A. The solid curve has a minimum, V_{ext}, below which there are no solutions. Thus, below that voltage, no arc could exist and flashover would not be possible.

If current is assumed to flow in the parallel path under the arc, it is easy to show that the flashover voltage increases slightly:

$$V_{fo} = L(n+1)\left[A\left(\frac{r}{n}\right)^n\right]^{1/(n+1)} \qquad (15)$$

There are numerous variations along these basic lines.[79-108] The above expressions are based on many assumptions and approximations. However, they are a useful basis for discussion since they can be used to predict qualitatively the effects of various factors such as ambient gas composition and pressure on the flashover voltage. Before discussing these factors, attempts to remove some of the approximations inherent in equation (14) will be discussed.

First, equation (11) implies that the insulator can be regarded as a linear resistor, and neglects the contribution to the resistance caused by the constriction of current flow lines in the surface film near the arc root. It is possible to correct for this. Wilkins[87] has considered the case of flashover on rectangular strips of finite width. Flashover of a concentric electrode structure, as shown in Figure 4, has also been treated.[92,109-111] Such calculations cannot be done unless the radius of the arc root is known. The resistance in series with the arc increases as the arc root radius decreases. Although the dependence of resistance on radius is roughly logarithmic, and hence weak, some value of radius must be used. One common assumption is that the arc root has a constant current density. Thus, if the current is known, the radius can be calculated. Optical measurements have been made of the arc root radius. Wilkins[112] found a current density of about 1.45 A/cm^2, and Nacke[82,113] found a density of about 1.27 A/cm^2. Other measurements exist.[114] Systematic swept probe and photographic experiments by King[115] over a range of currents and electrolyte conductivities showed no simple relationship between radius and current. In fact, King identified a number of modes, the most significant of which involved the transition to a cathode spot structure for high electrolyte conductivity. Cohen[116] extended this work with spectroscopic observations. She also investigated arc roots on electrolytes in atmospheres of nitrogen and argon.

Equation (12) neglects the electrode voltage falls of the discharge. These are typically in the vicinity of 500–1000 V,[87,113,115,116] and simply increase the flashover voltage in an additive manner.

A further problem in calculating the series resistance is the transient thermal behavior of the water film. Between room temperature and the boiling point,

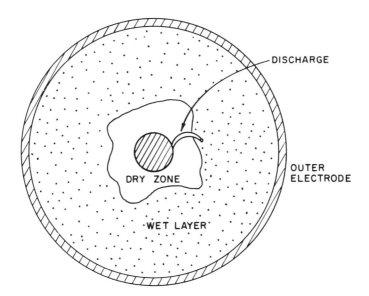

Figure 4. One type of simplified geometry used to investigate flashover, showing how the dry zone forms at the region of highest electrical stress and how a discharge forms across the dry zone.

the resistance of typical electrolytes drops by about a factor of three. Since much of the resistance is associated with the high-power-dissipation region near the arc root, a significant resistance decrease can occur as the film warms. Some attempts have been made to deal with this problem,[87] but solving the transient heating caused by a moving arc is difficult. It should be borne in mind that heating changes the surface resistance distribution, changing the current flow pattern, in turn changing the arc growth velocity, in turn changing the heating. No one has successfully closed this loop, largely because no one knows what factors govern arc growth velocity.

The voltage–current characteristic of the discharge also poses problems. The simple expression of equation (10) is not really adequate. Theoretical studies have shown that the characteristic can vary with time as the arc moves away from the surface by convective motions.[117,118] When the arc is near the surface, cooling by conduction to the surface increases the dissipation, and hence increases the voltage gradient needed to maintain an arc of a given current. In other words, the characteristic of the arc depends on the past history of the arc. This has not been incorporated into flashover models.

The accuracy of equation (15) has been tested under the simplest possible conditions using strips of tin-oxide-coated glass.[119] This material has a very low temperature coefficient of resistance. Using theoretical calculations of the discharge characteristics, a range of flashover voltages could be calculated which bounded the measured value. However, the accuracy to be expected from this approach is in the vicinity of 30–50%. For aqueous surfaces where heating and evaporation are occurring, the accuracy would be worse. Claims of more accurate theoretical values are often based on *ad hoc* assumptions which may be made after the fact to bring theory and experiment into agreement.

7.3. Effect of Polarity

There seems to be a definite effect of polarity on contamination flashover. By convention, polarity is defined as the polarity of the high-voltage end of the insulator. Thus, for a suspension-type insulator (Figure 1B), the polarity is that of the lower terminal since the lower side of the insulator is closer to the high-voltage conductor, and thus at higher potential. For laboratory specimens, polarity usually refers to that of the smaller electrode, for example, the center electrode in Figure 4. Thus, positive polarity would mean that the center electrode is positive with respect to the outer ring.

Most investigators have found that flashover occurs at a lower voltage for negative polarity. Kimoto *et al.* report that for suspension insulators, the negative polarity flashover voltage is 10–20% lower than the positive polarity voltage,[120] while Kawamura *et al.* report negative to be 15–20% lower.[121] Jolly and Poole, using a concentric electrode arrangement as in Figure 4, found that the negative flashover voltage was about 17–22% lower,[122] although for the

Effect of Contamination on HVDC Insulators

same geometry, others report negative to be higher.[52] Some typical experimental data[111] are shown in Figure 5. The reason for the polarity difference is not known, but there are several possibilities. The arc root is larger for negative polarity, and thus the series resistance is lower, lowering the flashover voltage according to the extinction model. Other possibilities are the effect of electrochemical processes at the arc root[116] and ion migration in the electrolyte film.[122] These electrochemical effects have not been thoroughly investigated yet.

Several investigators have compared the flashover voltages for AC and DC. Kimoto et al. report the DC flashover voltage to be about 10–20% lower than AC for suspension insulators,[120] while Nakajima et al.[123] report that DC can be up to 25% lower. Jolly and Poole report that DC positive was about 10% higher than AC and that DC negative was about 10% lower for the geometry of Figure 4.[122] Kim[70] found that DC could be 60% lower for this geometry, but the AC tests were done with a high-impedance transformer, which tends to result in higher flashover voltages. The slightly higher voltage often observed for AC is believed to result from reignition effects. For AC, the arc

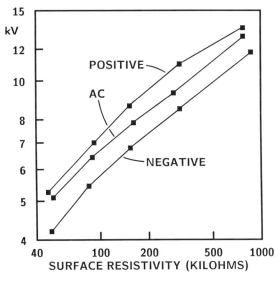

Figure 5. Typical test data obtained for the geometry of Figure 4 with the inner diameter of the outer electrode 12.7 cm and with a sodium chloride electrolyte on the surface. Voltage was applied suddenly after the surface had been wetted. For higher initial surface resistivity, the flashover voltage is seen to be higher, in accordance with the model discussed in the text. The surface resistivity is the resistance between opposite sides of a square portion of surface. Greater surface resistivity corresponds to lesser contamination. For this case, negative-polarity flashover occurs at a lower voltage than positive flashover for the same degree of contamination.

current passes through zero twice each cycle, and the arc may cool sufficiently to extinguish.[35,36]

7.4. Effect of Voltage Waveform

Even when the voltage exceeds the value given by equation (14), flashover will not occur unless the voltage is applied long enough to allow the arc to grow across the entire insulator. Unfortunately, little is known about the factors controlling arc elongation rates. It is known that thermal heating of the water layer can play a role. Apparently, even if the voltage is below the value of equation (14), ohmic heating of the electrolyte can lower the resistivity, in turn lowering the flashover voltage below the applied voltage. This shows up as a time lag of several hundred milliseconds or longer between voltage application and flashover.[124] Experiments with tin oxide films also show a relatively long time lag.[119] Some attempts have been made to model the time evolution of flashover, but they are based on many approximations and assumptions.[124-128] Too little is known about the mechanisms causing the arc to elongate. Some suggested mechanisms are reviewed elsewhere.[99,129] A number of arc elongation-rate measurements have been carried out.[95,111,126,130-132] Flashover of insulators subjected to a unipolar voltage pulse or switching surges has been the subject of a number of studies,[133-136] while pulsating DC has also been studied.[96]

7.5. Effect of the Composition of the Conducting Layer

When the conducting layer is an aqueous electrolyte, it is not unreasonable to suppose that some of the dissolved ions enter the electrical discharge. Partial discharges on NaCl-contaminated insulators are usually yellow at higher currents, suggesting a presence of ions in the discharge. Cohen[116] has observed the spectrum of discharges burning to NaCl electrolytes, and has noted the presence of sodium spectral lines as well as molecular lines of the ambient gas. Since the ionization potentials of metals are far lower than those of the constituents of air, these ions will increase the electrical conductivity of the discharge. This, in turn, will lower the electric field for a given current,[118] equivalent to lowering the value of A in equation (14), in turn lowering the flashover voltage. An electrolyte containing easily ionized cesium, for example, might produce flashover at a lower voltage for a given surface conductivity than an electrolyte of lithium ions, for example. Such an effect has been reported,[122] but the effect on voltage is only 10 or 20%, and further work needs to be done to establish the reality of the effect.

Labadie has carried out experiments using linear channels filled with NaCl dissolved in water, glycerine, and propylene carbonate.[137] The glycerine layer flashed over about 20% lower than an aqueous layer of equivalent resistivity, while the propylene carbonate differed little from water. Jolly and Chu investigated flashover on tin oxide surfaces to try to eliminate any effects of conducting film composition.[119] Free hydroxyl radicals have been detected in arcs burning to water, and their presence may increase the flashover voltage by increasing the heat conductivity of the arc plasma.[116]

It is believed that there is no direct effect of inert binders such as clay on the flashover voltage when the data are compared on the basis of surface conductivity.[138] Since corrosion of the electrodes is known to occur on DC insulators,[139] it is likely that the insulator surface will contain corrosion products which may significantly alter layer composition over long periods of time. An unusual interfacial instability also occurs when an electrolyte layer is subjected to high fields,[140] and this may play a role in altering the contaminant distribution.

7.6. Effect of Ambient Pressure

In general, increased pressure results in increased breakdown voltage. This was found to be the case for AC insulators[141, 142] and, more recently, for DC and pulsed situations.[121, 137] There is no simple dependence, and the flashover voltage for a given situation must be measured experimentally, or an attempt must be made to calculate the voltage from the high-temperature transport properties of the ambient gas.[118, 119]

7.7. Effect of Ambient Temperature

Since raising the temperature of an electrolyte decreases its resistivity, the simple discharge model presented earlier suggests that raising the temperature would decrease the flashover voltage. Limited data available so far indicate that this is the case.[143]

7.8. Effect of Ambient Gas

Little work has been done in atmospheres other than air. Labadie[137] has investigated flashover along water channels in atmospheres of various gases. Of the gases tested, sulfur hexafluoride has the highest flashover voltage, followed by nitrogen, air, and argon. For a given linear resistivity, sulfur hexafluoride had about double the breakdown field of air, while argon had about one-half that of air. Additional experiments in various gases would be very helpful in distinguishing among various theoretical models.

8. Test Methods to Evaluate Insulator Performance

In order to evaluate insulator designs, outdoor conditions are simulated in the laboratory. The goal is usually to simulate as many flashover steps as possible of those listed in Section 2. Sometimes insulators are subjected to long-term outdoor testing, but the random nature of weather conditions makes the results difficult to interpret.

While the many test procedures in use will not be discussed here, most are variations on several procedures.[144] One, the salt fog test, subjects clean, energized insulators to a salt mist. The failure voltage can be determined as a function of the spray salinity, higher salinity causing flashover at lower voltages. Another method, the clean fog test, subjects precontaminated insulators to a mist of distilled or tap water. The flashover voltage is determined as a function of the surface salt density (mg/cm^2) before fog is applied. The salt fog method is believed to best simulate seacoast conditions, while the clean fog test most closely simulates industrial regions. Another test procedure is to dip the insulator into a viscous slurry of salt water and clay, and immediately apply voltage. Insulators removed from service are also occasionally tested in laboratory fog chambers.

It is important in DC testing that the voltage source be capable of supplying the required current without significant voltage drop.[37, 145-149] A good design target is a one-percent voltage drop for a one-ampere load of one-second duration. In most cases this requires solid-state regulation. Special requirements for testing at voltages of up to 800 kV are discussed by Lloyd.[150] Numerous articles have been published giving results of contamination flashover testing of DC insulators.[49, 120, 123, 151-164]

9. Prevention of Flashover

In principle, flashover can be prevented if any of the five phases of flashover listed in Section 2 can be inhibited. How this might be done for each phase is considered below.

9.1. Deposition of Contamination

If soluble salts can be prevented from depositing on the insulator surface, a conducting film cannot form and contamination flashover will not occur. One way of reducing contamination buildup is to make the insulator easily washable by rain. This is accomplished by exposing as much of the surface as possible to the rinsing action of rain.[165] Another approach, useful for DC insulators, is to control the electric field around the insulator string to minimize electrostatic precipitation of particles on the insulator. This can be done in the laboratory by using field control electrodes.[50, 56, 58, 59, 166] This concept has not been verified under service conditions.

9.2. Deposition of Moisture

The surface of the insulator can be kept dry and nonconducting in several ways. One way is to apply to the insulator surface a slightly conducting coating which will dissipate a small amount of power continuously, on the order of 10 W for a suspension insulator.[167, 168] Although the surface power density is small, it is sufficient to ward off condensation or other light forms of precipitation. If heavy rain occurs, the top of the insulator tends to be cleaned of soluble salts before the bottom becomes wet, and flashover does not occur. Another way to prevent wetting is to design the insulator with a semi-enclosed area protected from the weather. This type of design is sometimes called a fog-bowl.[165]

The effects of surface moisture can be nullified if the surface is nonwetting. This causes the water to break up into small beads or droplets, preventing a conducting path from forming and supplying current to maintain surface discharges. Silicone or petroleum jellies and silicone rubber or poly(tetrafluoroethylene) coatings have been tried.[169-172] These methods work, but often the coating must be replaced at intervals. Beaded water films are believed to fail electrically when electrostatic forces disrupt individual droplets to form a filament bridging the electrodes.[173] It is possible to create a pool of oil on the insulator surface which interrupts the current flow, but the oil must be replaced periodically.[129]

9.3. Dry-Band Formation and Electric Field Concentration

Under practical conditions, it is impossible for a uniform water film to form. The power dissipated would lead to rapid evaporation. The surface state which usually evolves is an almost completely wet surface with one or more dry bands. The high electric field at the dry bands can lead to localized electrical breakdown across the dry region. Little effort has been directed at preventing high-electric-field regions from developing. One possible way to do this would be to use a coating with a negative voltage coefficient of resistance. In theory, this could limit the electric field at any point on the surface to a value below that required for breakdown. At present, such negative-coefficient materials exist, but they have not yet been incorporated into a practical porcelain glaze. It may be possible to control the field distribution with surface charge,[174] but the surface conductivities are probably too high for this to be practical.

9.4. Localized Electrical Breakdown across the Dry Band

Once dry banding has occurred, the electric field is spatially nonuniform. Electrical breakdown across a dry band will occur at somewhat less voltage than air breakdown for the same distance. There seems to be no way to prevent such breakdown except by limiting the electric field as discussed above.

9.5. Discharge Growth

Equation (14) indicates several ways in which the flashover voltage can be raised. The distance the arc must grow to bridge the insulator can be increased by lengthening the insulator or by making the surface more convoluted. Surface convolutions also help by increasing the series resistance. Unfortunately, there are practical limits to the length of insulators, since it becomes prohibitively expensive to build transmission towers large enough to accommodate longer insulators. Convoluting the surface can be limited by porcelain fabrication problems. Also, beyond a point, further convolutions become ineffective, since the arc simply jumps over surface indentations rather than following the surface. Other schemes have been proposed, such as using metal electrodes to intercept the arc[175, 176] or using permanent magnets to deflect or extinguish arcs on DC insulators.[177]

10. Conclusions and Future Prospects

In general, the contamination flashover process is understood qualitatively, but more work needs to be done on quantitative aspects. In particular, the properties of aerosols likely to be found near power lines should be studied for size distribution and natural charge. Quantities needed for modeling discharge propagation should be investigated, and mathematical models to incorporate these quantities need further development. It is impossible to predict how useful mathematical models of this sort will ultimately be. The conditions occurring outdoors are so variable that it may be inappropriate to try for too much precision in constructing theoretical models. However, even approximate models such as that of Obenaus have proven useful in interpreting data, and with any sort of luck, improved mathematical models will lead to better insulator design and more reliable power delivery.

References

1. E. Nasser, A survey of the problem of insulator contamination in the United States and Canada, IEEE Conference Paper, Paper no. 70 CP 240-PWR (1970).
2. A. El-Sulaiman and M. I. Qureshi, Effect of contamination on the leakage current of inland desert insulators, *IEEE Trans. Electrical Insulation EI-19*, 332–339 (August, 1984).

3. C. F. Sarkinen and J. T. Wiitala, Investigation of volcanic ash on transmission facilities in the Pacific Northwest, *IEEE Trans. Power Apparatus & Systems PAS-100*, 2278–2286 (May, 1981).
4. D. A. Greimsmann, Mount St. Helens eruptions increase knowledge of volcanic ash effects on system reliability, *Hi-Tension News (Ohio Brass Co., Mansfield, Ohio) 51*, 4–11 (July–August, 1982).
5. S. S. Low and G. R. Elder, Experience dictates future HVDC insulator requirements, *IEEE Trans. Electrical Insulation EI-16*, 263–266 (June, 1981).
6. L. O. Barthold, HVDC—now an important option, *Transmission & Distribution*, 28–32 (April, 1985).
7. HVDC inaugurated in Brazil, *Transmission & Distribution*, 8 (December, 1984).
8. J. T. Tyner, Heavyweight know-how builds intermountain HVDC line, *Transmission & Distribution*, 26–32 (June, 1985).
9. A. E. Vlastos and T. Sjokvist, Selected papers on insulator pollution and related topics, Kungliga Tekniska Hogskolan, Institutionen for Elektrisk Anlaggningsteknik, Stockholm (February, 1972).
10. E. Nasser, An annotated bibliography on the problem of insulator contamination of the electric energy system, Engineering Research Institute, Iowa State University, Special Report ISU-ERI-AMES-73220 (October, 1973).
11. IEEE Working Group on Insulator Contamination, Bibliography on high voltage insulator contamination, presented at IEEE Power Engineering Society Summer Meeting (July 11–22, 1977), Paper no. 77BL0100-8-PWR.
12. IEEE Committee Report, Application guide for insulators in a contaminated environment, paper presented at IEEE Summer Power Meeting (July 17–22, 1977), Paper no. F 77 639-8.
13. G. Anfossi, Behavior of insulators in the vicinity of the sea, *Atti della Assoc. Electrotecn. Ital. 11*, 326–334 (1907).
14. W. S. Murray, The log of the New Haven electrification, *AIEE Trans. 27*, 1615–1664 (December, 1908).
15. A. Austin, The high efficiency suspension insulator, *AIEE Proc. 30*, 1319–1344 (1911).
16. F. Obenaus, The influence of surface coating (dew, fog, salt, and dirt) on the flashover voltage of insulators, *Hescho Mitteilungen (Hermsdorf-Schomburg-Isolatoren-Gesellschaft) 70*, 1–37 (1933).
17. F. Obenaus, The flashover of contaminated insulators, *ETZ 56*, 369–370 (March, 1935).
18. W. Frischmann, Contamination flashover and arc root motion, *Dtsch. Elektrotechnik 11*, 290–295 (1957).
19. W. Frischmann, The importance of contamination factors for insulation flashover, *Dtsch. Elektrotechnik 12*, 166–171 (1958).
20. W. Frischmann, The importance of short circuit current capability for contamination flashover testing, *Dtsch. Elektrotechnik 12*, 28–31 (1958).
21. W. Frischmann, The influence of level and duration of voltage on contamination flashover, *Dtsch. Elektrotechnik 12*, 52–55, (1958).
22. F. Obenaus, Contamination flashover and creepage length, *Dtsch. Elektrotechnik 12*, 135–137 (1958).
23. G. Neumarker, Contamination state and creepage path, *Deutsche Akad. Berlin 1*, 352–359 (1959).
24. F. Obenaus, Creepage flashover of insulators with contamination layers, *Elektrizitatswirtschaft 59*, 878–882 (December 20, 1960).
25. H. Boehme and F. Obenaus, Pollution flashover tests on insulators in the laboratory and in systems and the model concept of creepage-path flashover, Conférence Internationale des Grands Réseaux Electriques à Haute Tension (CIGRE), Paris (June 8–18, 1966), Paper no. 407.
26. F. Obenaus and H. Bohme, Laboratory and service tests with contaminated suspension insulators and the model concept of creepage flashover, *Elektrie 20*, 417–422 (1966).

27. R. Mailfert, L. Pargamin, and D. Riviere, Electrical reliability of DC line insulators, *IEEE Trans. Electrical Insulation EI-16*, 267–276 (June, 1981).
28. J. E. Schroeder, Y. Zlotin, T. C. Cheng, C. T. Wu, and J. H. Dunlap, Design of a polysil DC insulator, *IEEE Trans. Electrical Insulation EI-16*, 235–241 (June, 1981).
29. D. C. Jolly, A test method for determining the outdoor lifetime of polymer transmission line insulators, Conference Record of 1982 IEEE International Symposium on Electrical Insulation (June 7–9), 1982, Paper no. 82CH1780-6-EI, pp. 248–251.
30. D. C. Jolly, A quantitative method for determining the resistance of polymers to surface discharges, *IEEE Trans. Electrical Insulation EI-17*, 293–299 (August, 1982).
31. E. A. Cherney and D. J. Stonkus, Non-ceramic insulators for contaminated environments, *IEEE Trans. Power Apparatus & Systems PAS-100*, 131–142 (January, 1981).
32. IEEE Working Group on Non-Ceramic and Composite Insulators for Transmission Lines, Minimum test requirements for non-ceramic insulators, *IEEE Trans. Power Apparatus & Systems PAS-100*, 882–890 (February, 1981).
33. K. Stimper and W. H. Middendorf, Mechanisms of deterioration of electrical insulation surfaces, *IEEE Trans. Electrical Insulation EI-19*, 314–320 (August, 1984).
34. R. Schifani, Surface discharge effects on dielectric properties of epoxy resin, *IEEE Trans. Electrical Insulation EI-18*, 504–512 (September, 1983).
35. F. A. M. Rizk, Analysis of dielectric recovery with reference to dry-zone arcs on polluted insulators, presented at IEEE Winter Power Meeting (January 31–February 5, 1971), Paper no. 71 CP 134-PWR.
36. A. Kaga, M. Sato, and H. Akagami, Reignition voltage and arc voltage on contaminated insulator surfaces, *Jap. J. Appl. Phys., Part I 23*, 1094–1100 (August, 1984).
37. F. Hirsch, H. Rheinbaben, and R. Sorms, Flashover of insulators under natural pollution and HVDC, *IEEE Trans. Power Apparatus & Systems PAS-94*, 45–50 (January, 1975).
38. M. P. Sarma and W. Janischewskyj, Analysis of corona losses on dc transmission lines: Part ii—Bipolar lines, *IEEE Trans. Power Apparatus & Systems PAS-88*, 1476–1491 (October, 1969).
39. M. P. Sarma and W. Janischewskyj, Analysis of corona losses on DC transmission lines: i—Unipolar lines, *IEEE Trans. Power Apparatus & Systems PAS-88*, 718–731 (May, 1969).
40. H. Witt, Insulation levels and corona phenomena on HVDC transmission lines, Ph.D. Thesis, Chalmers Tekniska Hogskola, Sweden (October, 1960).
41. Y. Sunaga, Y. Amano, and T. Sugimoto, Electric field and ion current at the ground and voltage of charged objects under HVDC lines, *IEEE Trans. Power Apparatus & Systems PAS-100*, 2082–2092 (April, 1981).
42. M. G. Comber and G. B. Johnson, HVDC field and ion effects research at Project UHV: Results of electric field and ion current measurements, *IEEE Trans. Power Apparatus & Systems PAS-101*, 1998–2006 (July, 1982).
43. P. S. Maruvada, R. D. Dallaire, O. C. Elye, C. V. Thio, and J. S. Goodman, Environmental effects of the Nelson River HVDC transmission lines—RI, AN, electric field, induced voltage, and ion current distribution tests, *IEEE Trans. Power Apparatus & Systems PAS-101*, 951–959 (April, 1982).
44. P. S. Maruvada, R. D. Dallaire, J. H. Bednarek, and W. H. Jones, Long-term statistical study of the corona electric field and ion current performance of a $+/-$ 900 kV bipolar HVDC transmission line configuration, *IEEE Trans. Power Apparatus & Systems PAS-103*, 76–83 (January, 1984).
45. S. A. Sebo, R. Caldecott, and D. G. Kasten, Model study of HVDC electric field effects, *IEEE Trans. Power Apparatus & Systems PAS-101*, 1743–1756 (June, 1982).
46. B. Y. H. Liu and A. Kapadia, Combined field and diffusion charging of aerosol particles in the continuum region, *J. Aerosol Sci. 9*, 227–242 (1978).

47. A. K. Gertsik, A. V. Korsuntser, and N. K. Nikolskii, The effect of fouling on insulators for HVDC overhead lines, *Direct Current 3*, 219–226 (December, 1957).
48. H. Witt, D.C. insulators, a comparison with A.C., Conférence Internationale des Grands Réseaux Electriques à Haute Tension (CIGRE), Paris (June 15–25, 1960), Paper no. 403.
49. A. Annestrand and A. Schei, A test procedure for artificial pollution tests on direct voltage, *Direct Current 12*, 1–8 (February, 1967).
50. H. Haerer, Insulators for high voltage direct current under contamination conditions, Ph.D. Thesis, University of Stuttgart, Federal Republic of Germany (July, 1971).
51. J. F. Hall and T. P. Mauldin, Wind tunnel studies of the insulator contamination process, *IEEE Trans. Electrical Insulation EI-16*, 180–188 (June, 1981).
52. T. C. Cheng and C. T. Wu, Performance of HVDC insulators under the contaminated conditions, *IEEE Trans. Electrical Insulation EI-15*, 270–286 (June, 1980).
53. T. C. Cheng, C. T. Wu, F. Zedan, G. R. Elder, S. S. Low, J. N. Rippey, and G. D. Rodriguez, EPRI-HVDC insulator studies: Part i, field test at the Sylmar HVDC converter station, *IEEE Trans. Power Apparatus & Systems PAS-100*, 902–909 (February, 1981).
54. T. C. Cheng, C. T. Wu, J. N. Rippey, and F. M. Zedan, Pollution performance of DC insulators under operating conditions, *IEEE Trans. Electrical Insulation EI-16*, 154–164 (June, 1981).
55. W. G. Thompson, The mechanism of the contamination of porcelain insulators, *IEE Journal, Part II 91*, 317–327 (1944).
56. R. G. Olsen, B. C. Furumasu, and D. P. Hartmann, Contamination mechanisms for HVDC insulators, paper presented at IEEE Winter Power Meeting (January 30–February 4, 1977), Paper no. A 77 035-9.
57. R. G. Olsen and J. Daffe, The effect of electric field modification and wind on the HVDC insulator contamination process, paper presented at IEEE Winter Power Meeting (January 29–February 3, 1978), Paper no. A 78 120-8.
58. R. G. Olsen, J. Daffe, and C. F. Sarkinen, On the origin, significance, and minimization of nonuniform contamination along HVDC insulator strings, *IEEE Trans. Power Apparatus & Systems PAS-100*, 971–980 (March, 1981).
59. M. Alem and J. B. Laghari, A prediction of deposition of contaminants on insulator surface with and without grading rings, Conference Record of 1982 IEEE International Symposium on Electrical Insulation (June 7–9, 1982), Paper no. 82CH1780-6-EI, pp. 192–196.
60. M. N. Horenstein, Particle contamination of high voltage DC insulators, Ph.D. Thesis, Massachusetts Institute of Technology, Cambridge, Massachusetts (May, 1978).
61. M. N. Horenstein and J. R. Melcher, Particle contamination of high voltage DC insulators below corona threshold, *IEEE Trans. Electrical Insulation EI-14*, 297–305 (December, 1979).
62. E. C. Salthouse, The effects of direct voltages on insulator surfaces, Ph.D. Thesis, Queen's University, Belfast, Northern Ireland (November, 1960).
63. J. R. Laghari, private communication (July 12, 1982).
64. Y. Aoshima, T. Harada, and K. Kishi, DC voltage distribution characteristics on polluted insulator string, *IEEE Trans. Power Apparatus & Systems PAS-100*, 948–955 (March, 1981).
65. A. Radun, Particle charging in a turbulent air stream, Ph.D. Thesis, Massachusetts Institute of Technology, Cambridge, Massachusetts (February, 1981).
66. A. V. Radun and J. R. Melcher, DC power line charging of macroscopic particles and associated electrical precipitation on insulators, *IEEE Trans. Electrical Insulation EI-16*, 165–179 (June, 1981).
67. H. Bohme and H. Zeh, Contamination deposition by wind on insulators, *Elektrie 21*, 339–240 (July, 1967).
68. G. Karady, Effect of surface contamination on high voltage insulator performance, in: *Surface Contamination: Genesis, Detection and Control* (K. L. Mittal, ed.), pp. 945–965, Plenum Press, New York (1979).

69. H. H. Woodson and A. J. McElroy, Insulators with contaminated surfaces, part iii: modelling of dry zone formation, *IEEE Trans. Power Apparatus & Systems PAS-89*, 1868-1876 (November/December, 1970).
70. J. H. Kim, Characterization of contaminated insulator flashover parameters and analysis of insulator wetting mechanisms, Ph.D. Thesis, University of Southern California, Los Angeles, California (June, 1975).
71. M. Leclerc, R. Bouchard, Y. Gervais, and D. Mukhedkar, Wetting processes on a contaminated insulator surface, *IEEE Trans. Power Apparatus & Systems PAS-101*, 1005-1011 (May, 1982).
72. E. C. Salthouse, Initiation of dry bands on polluted insulation, *Proc. IEE 115*, 1707-1712 (November, 1968).
73. D. O. Lavelle, Thermal considerations in the surface behavior of insulators, Ph.D. Thesis, Queen's University, Belfast, Northern Ireland (October, 1970).
74. J. O. Loberg and E. C. Salthouse, Dry-band growth on polluted insulation, *IEEE Trans. Electrical Insulation EI-6*, 136-141 (September, 1971).
75. Yu. N. Shumilov and V. A. Aksenov, Surface layer electrophysical processes on flashover of fouled insulators, *Electric Technology U.S.S.R.* 8-18 (1983).
76. A. Saad and R. Tobazeon, Surface conduction and losses of an insulator wetted by a liquid dielectric, *IEEE Trans. Electrical Insulation EI-19*, 193-199 (June, 1984).
77. M. Nishida, N. Yoshimura, and F. Noto, Process of dry belt formation on surface of organic insulation materials in tracking breakdown, *Elect. Eng. Japan (USA) 103*, 26-37 (1983).
78. E. Nasser, Some physical properties of electrical discharges on contaminated surfaces, *IEEE Trans. Power Apparatus & Systems PAS-87*, 957-963 (April, 1968).
79. L. Alston and S. Zoledziowski, Growth of discharges on polluted insulation, *Proc. IEE 110*, 1260-1266 (July, 1963).
80. C. H. W. Clark, P. Dey, W. A. McNeill, J. S. Forrest, K. W. Huddart, D. M. Cherry, S. Zoledziowski, L. I. Alston, B. F. Hampton, C. H. A. Ely, and P. J. Lambeth, Discussion of flashover of polluted insulation, *Proc. IEE 111*, 1589-1592 (September, 1964).
81. B. Hampton, Flashover mechanism of polluted insulation, *Proc. IEE 111*, 985-990 (May, 1964).
82. H. Nacke, Stability of contamination layer discharges and flashover theory, *ETZ-A 87*, 577-585 (August 5, 1966).
83. S. Hesketh, General criterion for the prediction of pollution flashover, *Proc. IEE 114*, 531-532 (April, 1967).
84. S. Hesketh, The propagation of arcs over a water surface, Proceedings of the 8th International Conference on Ionization Phenomena in Gases (1967), Paper no. 3.2.11.6, p. 255.
85. A. Rumeli, The mechanism of flashover of polluted insulation, Ph.D. Thesis, Univ. of Strathclyde, Glasgow, Scotland (1967).
86. S. Zoledziowski, Flashover of polluted insulation, Proceedings of the 8th International Conference on Ionization Phenomena in Gases (1967), Paper no. 3.2.11.5, p. 254.
87. R. Wilkins, Flashover voltage of high voltage insulators with uniform surface pollution films, *Proc. IEE 116*, 457-465 (March, 1969).
88. A. Baghdadi, The mechanism of flashover of polluted insulation, Ph.D. Thesis, The Victoria University of Manchester, Manchester Institute of Technology, England (May, 1970).
89. F. A. M. Rizk, Application of dimensional analysis to flashover characteristics of polluted insulators, *Proc. IEE 117*, 2257-2260 (December, 1970).
90. H. Woodson and A. J. McElroy, Insulators with contaminated surfaces, part ii: modelling of discharge mechanisms, *IEEE Trans. Power Apparatus & Systems PAS-89*, 1858-1867 (November/December, 1970).
91. P. Claverie, Predetermination of the behavior of polluted insulators, *IEEE Trans. Power Apparatus & Systems PAS-90*, 1902-1908 (July/August, 1971).

92. D. C. Jolly, Physical processes in the flashover of insulators with contaminated surfaces, Ph.D. Thesis, Massachusetts Institute of Technology, Cambridge, Massachusetts (May, 1971).
93. Katsuo Isaka, Basic research on the breakdown phenomena upon the polluted surface of the insulator, Ph.D. Thesis, University of Tokyo, Japan (March, 1971).
94. F. A. M. Rizk, A criterion for AC flashover of polluted insulators, presented at IEEE Winter Power Meeting (January 31–February 5, 1971), Paper no. 71 CP 135-PWR.
95. R. Wilkins and A. Baghdadi, Arc propagation along an electrolyte surface, *Proc. IEE 118*, 1886–1892 (December, 1971).
96. W. Bundschuh, The insulation strength of contaminated insulators stressed with pulsed direct current, *Tech. Mitt. AEG-Telefunken 62*, 334–337 (1972).
97. D. Goulsbra, The behaviour of electrical discharges on polluted insulation, Ph.D. Thesis, The Victoria University of Manchester, Manchester Institute of Technology, England (April, 1972).
98. B. F. Hampton, Arc propagation along an electrolyte surface, *Proc. IEE 119*, 1228 (August, 1972).
99. D. C. Jolly, Contamination flashover, part i: Theoretical aspects, *IEEE Trans. Power Apparatus & Systems PAS-91*, 2437–2442 (November/December, 1972).
100. E. Nasser, Contamination flashover of outdoor insulation, *ETZ-A 93*, 321–325 (1972).
101. C. Huraux and A. Rahal, Analysis of the instability of a discharge on the surface of an insulator starting with the one-dimensional model of Obenaus, *C.R. Acad. Sci. Paris (Series B) 278*, 823–826 (1974).
102. T. C. Cheng and C. T. Wu, EPRI-HVDC insulator studies: Part iii, theories on flashover processes, paper presented at IEEE Summer Power Meeting (July 15–20, 1979), Paper no. A 79 537-2.
103. L. Higginbottom, S. Zoledziowski, and J. H. Calderwood, Flashover along a conducting surface, 3rd International Conference on Dielectric Materials, Measurements and Applications, University of Aston, Birmingham, England (September 10–13, 1979), pp. 291–293.
104. A. M. Rahal and C. Huraux, Flashover mechanism of high voltage insulators, *IEEE Trans. Power Apparatus & Systems PAS-98*, 2223–2231 (November/December, 1979).
105. J. Gers, S. Zoledziowski, and J. H. Calderwood, Criteria for discharge elongation along a conductive surface, Proceedings International Symposium on Pollution Performance of Insulators and Surge Diverters, Indian Institute of Technology, Madras, India (February 26–27, 1981), Paper no. 1.05.
106. A. M. Hizal and Y. Demir Rumeli, Analytical estimation of flashover performances of polluted insulators, Proceedings International Symposium on Pollution Performance of Insulators and Surge Diverters, Indian Institute of Technology, Madras, India (February 26–27, 1981), Paper no. 1.02.
107. M. Tantawy, and M. El-Maghraby Y. Abed, Digital computation of flashover voltage over polluted insulators under several constraints, Proceedings International Symposium on Pollution Performance of Insulators and Surge Diverters, Indian Institute of Technology, Madras, India (February 26–27, 1981), Paper no. 1.03.
108. T. C. Cheng, C. Y. Wu, and H. Nour, DC interfacial breakdown on contaminated electrolytic surfaces, *IEEE Trans. Electrical Insulation EI-19*, 536–542 (December, 1984).
109. A. J. McElroy, Flashover mechanisms of insulators with contaminated surfaces, Ph.D. Thesis, Massachusetts Institute of Technology, Cambridge, Massachusetts (June, 1969).
110. T. C. Cheng, Mechanism of flashover of contaminated insulators, Ph.D. Thesis, Massachusetts Institute of Technology, Cambridge, Massachusetts (May, 1974).
111. J. Melcher and D. C. Jolly, Contamination Flashover Mechanisms of DC Transmission Line Insulators, U.S. Dept. of Energy, Final Report, Contract E(49-18)-2068 (January, 1978).
112. R. Wilkins, Mechanisms of failure of high-voltage insulation with surface contamination,

Ph.D. Thesis, The Victoria University of Manchester, Manchester Institute of Technology, England (January, 1968).
113. H. Nacke, Arc resistance and leakage current resistance of insulating materials, Ph.D. Thesis, Technical University of Berlin, Federal Republic of Germany (May, 1962).
114. H. P. Mercure and M. G. Drouet, Dynamic measurements of the current distribution in the foot of an arc propagating along the surface of an electrolyte, *IEEE Trans. Power Apparatus & Systems PAS-101*, 725–736 (March, 1982).
115. D. J. King, Measurements of the properties of arcs near electrolyte surfaces, B.S. Thesis, Massachusetts Institute of Technology, Cambridge, Massachusetts (June, 1975).
116. V. Cohen, Anode and cathode phenomena of arc burning to electrolytes, M.S. Thesis, Massachusetts Institute of Technology, Cambridge, Massachusetts (September, 1979).
117. E. J. Los, Time constants of low current arcs near flat surfaces, M.S. Thesis, Massachusetts Institute of Technology, Cambridge, Massachusetts (February, 1974).
118. E. J. Los and D. C. Jolly, Static and dynamic properties of arcs near plane surfaces, *Z. Physik B 20*, 3–11 (1975).
119. D. C. Jolly and S. T. Chu, Surface electrical breakdown of tin oxide coated glass, *J. Appl. Phys. 50*, 6196–6199 (October, 1979).
120. I. Kimoto, J. Fujumura, and K. Naito, Performance of insulators for direct current transmission line under polluted conditions, *IEEE Trans. Power Apparatus & Systems PAS-92*, 943–949 (May/June, 1973).
121. T. Kawamura, M. Ishii, M. Akbar, and K. Nagai, Pressure dependence of DC breakdown of contaminated insulators, *IEEE Trans. Electrical Insulation EI-17*, 39–45 (February, 1982).
122. D. C. Jolly and C. D. Poole, Flashover of contaminated insulators with cylindrical symmetry under DC conditions. *IEEE Trans. Electrical Insulation EI-14*, 77–84 (April, 1979).
123. Y. Nakajima, T. Seta, K. Nagai, H. Horie, and K. Naito, Performance of contaminated insulators energized by DC voltage, Conférence Internationale des Grands Réseaux Electriques à Haute Tension (CIGRE), Paris (August 21–29, 1974), Paper no. 33-07.
124. S. Zoledziowski, Time to flashover characteristics of polluted insulation, *IEEE Trans. Power Apparatus & Systems PAS-87*, 1397–1404 (June, 1968).
125. M. Rea, Arc dynamics on insulating surfaces partly covered with conducting deposits, *L'Energia Elettrica 44*, 145–154 (1967).
126. D. C. Jolly, T. C. Cheng, and D. M. Otten, Dynamic theory of discharge growth over contaminated insulator surfaces, paper presented at IEEE Winter Power Metting (January 27–February 1, 1974), Paper no. C 74 068-3.
127. A. El-Arabaty, A. Nosseir, E. Nasser, A. El-Sarky, and S. El-Debeiky, Measurement and analysis of dynamic discharge propagation on h.v. polluted insulators, 3rd International Symposium on High Voltage Engineering, Milan, Italy (August 28–31, 1979), Paper no. 54.01.
128. L. Higginbotton, S. Zoledziowski, and J. H. Calderwood, The dynamic model of flashover along a conductive surface, 3rd International Symposium on High Voltage Engineering, Milan, Italy (August 28–31, 1979), Paper no. 54.03.
129. D. C. Jolly, Contamination flashover and insulator design, *J. Franklin Institute 294*, 483–500 (December, 1972).
130. F. D. A. Boylett, Electric discharges on water surfaces, *Electronics Letters 5*, 47–48 (February 6, 1969).
131. F. D. A. Boylett and I. G. Maclean, The propagation of electric discharge across the surface of an electrolyte, *Proc. Roy. Soc. A324*, 469–489 (1971).
132. T. Matsumoto, M. Ishii, and T. Kawamura, Optoelectronic measurement of partial arcs on a contaminated surface, *IEEE Trans. Electrical Insulation EI-19*, 543–549 (December, 1984).
133. C. H. A. Ely and W. J. Roberts, Flashover of polluted h.v. insulators under switching surges, *Proc. IEE 115*, 443 (March, 1968).
134. H. Matsuo, Y. Yunoki, T. Oshige, and N. Mita, Impulse discharge on contaminated surface, *Elect. Eng. Japan 89*, 26–34 (1969).

135. W. Mosch and E. Lemke, Switching surge flashover of insulators under polluted conditions, Proceedings International Symposium on Pollution Performance of Insulators and Surge Diverters, Indian Institute of Technology, Madras, India (February 26-27, 1981), Paper no. 1.06.
136. A. H. Qureshi, A. Rumeli, and M. Hizal, Flashover along a water column under impulse voltages, Proceedings International Symposium on Pollution Performance of Insulators and Surge Diverters, Indian Institute of Technology, Madras, India (February 26-27, 1981), Paper no. 1.01.
137. J. Labadie, Study of the validity of the electrical model of flashover of contaminated high voltage insulators, Ph.D. Thesis, l'Université Paul Sabatier, Toulouse, France (May, 1977).
138. D. C. Jolly, Contamination flashover, part ii: Flat plate model tests, *IEEE Trans. Power Apparatus & Systems PAS-91*, 2443-2451 (November/December, 1972).
139. I. M. Crabtree, K. J. Mackey, K. Kito, N. Naito, A. Wanatabe, and T. Irie, Studies on electrolytic corrosion of hardware of DC line insulators, *IEEE Trans. Power Apparatus & Systems PAS-104*, 645-654 (March, 1985).
140. D. C. Jolly and D. L. Murray, Spatially periodic instability occurring in moving boundary electrophoresis experiments, *J. Electroanal. Chem. 160*, 103-116 (1984).
141. G. N. Alexandrov and R. S. Burchanov, Flashover voltage of lightly contaminated suspension insulator strings at reduced air density, *Elektrie 21*, 370-371 (1967).
142. V. I. Bergman and O. I. Kolobova, Some results of an investigation of the dielectric strength of contaminated line insulation in reduced air-pressure conditions, *Soviet Elect. Eng. (USA) 54*, 54-56 (1983).
143. M. Ishii, M. Akbar, and T. Kawamura, Effect of ambient temperature on the performance of contaminated DC insulators, *IEEE Trans. Electrical Insulation EI-19*, 129-134 (April, 1984).
144. L. Pargamin and S. Tartier, A comparison of contamination test methods for DC line insulators, *IEEE Trans. Electrical Insulation EI-16*, 224-229 (June, 1981).
145. H. Rasokat, The influence of load characteristics of test equipment for the direct current withstand voltage characteristics of contaminated insulators, *ETZ-A 90*, 691-692 (1969).
146. H. Rasokat, Loading characteristics of test voltage sources for AC and DC, and their influence on the withstand voltage characteristics of contaminated insulators, Ph.D. Thesis, Technical University of Berlin, Federal Republic of Germany (December, 1970).
147. T. M. Ishii, M. Akbar, and K. Nagai Kawamura, Stabilized DC source for testing of polluted insulators, 3rd International Symposium on High Voltage Engineering, Milan, Italy (August 38-31, 1979), Paper no. 43-09.
148. K. J. Lloyd and M. G. Comber, HVDC contaminated insulator tests—Leakage currents and their influence on the power supply, paper presented at IEEE Summer Power Meeting (July 15-20, 1979), Paper no. A 79 530-7.
149. Y. Beausejour and F. A. M. Rizk, Feedback-controlled cascade rectifier source for HV testing of contaminated DC insulators, *IEEE Trans. Power Apparatus & Systems PAS-100*, 3525-3534 (July, 1981).
150. K. J. Lloyd, Testing contaminated insulators at Project UHV for voltage levels of the future, *IEEE Trans. Electrical Insulation EI-16*, 220-223 (June, 1981).
151. R. S. Geus and R. F. Stevens, High voltage DC test program of Bonneville Power Administration, *IEEE Trans. Power Apparatus & Systems PAS-82*, 1054-1061 (December, 1963).
152. M. G. Poland, W. M. Scarborough, H. L. Hill, and P. E. Renner, BPA's extra high voltage DC tests: i—Contaminated insulators, *IEEE Trans. Power Apparatus & Systems PAS-86*, 1146-1152 (October, 1967).
153. B. Macchiaroli and M. Rea, Flashover voltage of artificially contaminated surfaces, *Proc. IEE 118*, 271-274 (January, 1971).
154. T. Seta, N. Arai, and T. Udo, Natural pollution test of insulators with DC high voltage, *IEEE Trans. Power Apparatus & Systems PAS-93*, 878-883 (May/June, 1974).

155. R. Sorms, The flashover behaviour of naturally contaminated DC transmission line insulators, Ph.D. Thesis, Technischen Universität Berlin (1974).
156. H. L. Hill, A. Capon, O. Ratz, P. Renner, and W. D. Schmidt, Transmission Line Reference Book HVDC to $+/-$ 600 kV, Electric Power Research Institute, Palo Alto, California (1976).
157. T. C. Cheng and C. T. Wu, The performance of insulators with different surface pollutants under HVDC conditions, 3rd International Conference on Dielectric Materials, Measurements and Applications, University of Aston, Birmingham, England (September 10–13, 1979).
158. C. T. Wu, Flashover mechanisms of contaminated HVDC insulation, Ph.D. Thesis, University of Southern California, Los Angeles, California (June, 1979).
159. T. C. Cheng, C. T. Wu, Y. B. Kim, and S. Yokayama, EPRI-HVDC insulator studies: Part ii, laboratory simulation studies, *IEEE Trans. Power Apparatus & Systems PAS-100*, 910–920 (February, 1981).
160. M. G. Comber and R. J. Nigbor, Performance of contaminated insulators tested from 200 to 1000 kV DC, *IEEE Trans. Electrical Insulation EI-16*, 230–234 (June, 1981).
161. T. Fujimura, K. Naito, and Y. Suzuki, DC flashover voltage characteristics of contaminated insulators, *IEEE Trans. Electrical Insulation EI-16*, 189–198 (June, 1981).
162. T. Seta, K. Nagai, K. Naito, and Y. Hasegawa, Studies on the performance of contaminated insulators energized with DC voltage, *IEEE Trans. Power Apparatus & Systems PAS-100*, 518–527 (February, 1981).
163. G. Peyregne, A. M. Rahal, and C. Huraux, Flashover of a liquid conducting film, part 1: Flashover voltage, *IEEE Trans. Electrical Insulation EI-17*, 10–14 (February, 1982).
164. G. Peyregne, A. M. Rahal, and C. Huraux, Flashover of a liquid conducting film, part 2: Time to flashover-mechanisms, *IEEE Trans. Electrical Insulation EI-17*, 15–19 (February, 1982).
165. J. J. Taylor, Insulators to withstand airborne deposits, *AIEE Trans. 67*, 1436–1441 (1948).
166. H. Bocker, Proposal for a DC insulator, *ETZ-A 90*, 690 (1969).
167. R. W. Sanders and D. R. Holmes, Conducting coatings for high-voltage insulator stabilization, *Nature 195*, 170–171 (1962).
168. M. J. Billings and R. Wilkins, Considerations of the suppression of insulator flashover by resistive surface films, *Proc. IEE 113*, 1649–1653 (October, 1966).
169. J. E. Conner and A. D. Lantz, The insulator contamination problem as influenced by silicone surface coatings, *AIEE Trans. Part III 77*, 1101–1112 (December, 1958).
170. J. E. Toms and A. B. Suttie, Insulator surface treatments, *Electrical Review*, 412–415 (September 17, 1965).
171. P. J. Lambeth, J. S. T. Looms, A. Stalewski, and W. G. Todd, Surface coatings for h.v. insulators in polluted areas, *Proc. IEE 113*, 861–869 (May, 1966).
172. R. M. Radwan and G. M. El-Salam, Effect of silicon grease on the DC electrical characteristic of polluted insulators, 3rd International Symposium on High Voltage Engineering, Milan, Italy (August 28–31, 1979), Paper no. 54.04.
173. T. C. Cheng, D. C. Jolly, and D. J. King, Surface flashover of water repellant insulators under moist conditions, *IEEE Trans. Electrical Insulation EI-12*, 208–213 (June, 1977).
174. Y. Yamano, S. Kobayashi, and T. Takahashi, Reduction of surface charge-induced electric field enhancement and increase in AC flashover voltage, *IEEE Trans. Electrical Insulation EI-20*, 529–536 (June, 1985).
175. T. C. Cheng, G. Wilson, and D. C. Jolly, High-Voltage Electrical Insulator Adapted to Prevent Flashover, U.S. Patent 3,963,858 (June 15, 1976).
176. D. A. Swift, Flashover across the surface of an electrolyte: Some methods of arresting arc propagation, 3rd International Symposium on High Voltage Engineering, Milan, Italy (August 28–31, 1979), Paper no. 54.05.
177. D. C. Jolly and D. J. King, High-Voltage Electrical Insulator Having Magnetic Elements to Prevent Flashover, U.S. Patent 4,010,316 (March 1, 1977).

8

Effect of Surface Contamination on Electric Contact Performance

MORTON ANTLER

1. Introduction

An electrical contact is a junction between two or more current-carrying members which provides electrical continuity at their interfaces. Components having contacts include connectors, terminals, bus bars, circuit breakers, switches, relays, and slip rings. Most electrical contacts are degraded by contamination. To understand why this is so, it is necessary to consider the nature of solid surfaces and the effect of foreign materials on current flow.

Contact surfaces are irregular on a microscopic scale. Even nominally plane surfaces have a waviness on which a roughness is superimposed with peak-to-valley dimensions typically from tenths of micrometers to several micrometers. For example, a typical gold-plated contact is illustrated in Figure 1. When two contacts are brought together at a light load, they touch at only a few asperities. As the load is increased, more asperities come into contact and the surfaces move together. The true area of contact depends, therefore, on normal load and on the hardness of the metal. This area is only a small fraction of the apparent contact, except at very high loads where the surfaces can be severely deformed.

When the surface is covered by a nonconductive layer such as an oxide film, the area of metallic contact will be zero provided the film is unbroken. If the film is discontinuous, or is punctured on making closure, the load is borne by both film and metal. Figure 2 schematically illustrates this condition where the apparent area of contact, the metallic regions, and the places with an insulating layer are differentiated. The lines of current flow converge at the regions of metallic contact, called "a" spots, as illustrated schematically in Figure 3. Constriction resistance is the increase of resistance beyond that of a continuous

---

MORTON ANTLER • AT&T Bell Laboratories, 6200 East Broad Street, Columbus, Ohio 43213.

Figure 1. Surface of electroplated gold contact. Scale bar: 0.01 mm.

solid, i.e., not having an interface, and originates in this current line convergence. If the film on the surface is very thin, up to a few atom layers, some current can pass through it by a phenomenon termed tunnel conduction. When the voltage is large, insulating films can be electrically punctured (1 V will puncture approximately 100 Å of insulating film). For this reason, in most cases, contact resistance is measured at small potentials where the maximum open-circuit voltage is too low to electrically puncture surface films.

Curves of contact resistance versus force for typical contact metals appear in Figure 4. Contact resistance decreases with increasing load. The softer and more conductive the metal, the lower will be its contact resistance at a given force.

Separable electronic connectors typically have many contacts which carry small currents; connectors used in utility-power distribution systems carry high current in high-voltage circuits, and are made by severe plastic deformation of the contact surface. They are generally permanent, and not intended to be re-

Effect of Contamination on Electric Contact Performance

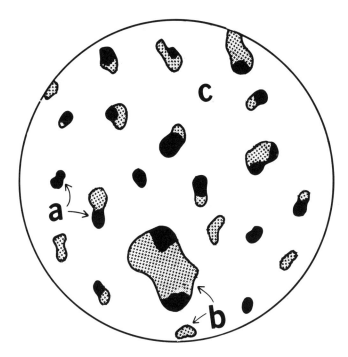

Figure 2. Schematic illustration of contact surface. The regions of metallic contact, a (solid areas), in most cases are only a tiny fraction of the apparent area indicated by the circle. Contact at b (shaded) is with insulating contaminant. Region c does not touch.

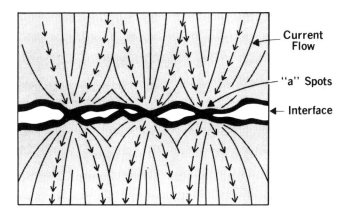

Figure 3. Microscopic view of contact interface (schematic); constriction resistance originates in the constriction of current flow at the touching metallic junctions (''a'' spots) of the mating surfaces.

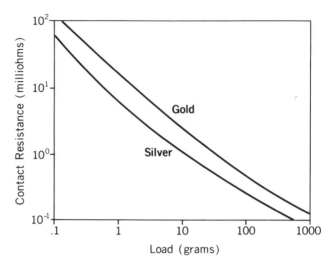

Figure 4. Contact resistance–load characteristic curves for solid gold and silver mated to themselves.

placed once having been made. Typical electronic and power connectors are shown in Figure 5. Although solid base metals such as aluminum are used in power applications, it is generally necessary to employ noble metals for electronic contacts.

A contact contaminant can be defined as a foreign substance whose presence on the surface is undesirable because it prevents or reduces metallic contact and thereby degrades contact resistance. Some contaminants, while initially innocuous, corrode surfaces or their surrounding structures and eventually cause contact failure. A contaminant for one contact material in a given application may be innocuous in another application. Thin oxide films can prevent electrical contact from being made when the contact force is low, yet have no effect under conditions where surfaces are joined at large forces which cause high deformations. For example, a 100-Å oxide film on a polished copper flat will prevent metallic contact from being made when mated to a 0.3-mm diameter hemispherically-ended gold rod at 10 g force, yet be severely disrupted when the normal force is increased to 1000 g. Some foreign materials, such as certain lubricants which reduce friction and wear, may be desirable, and therefore are not considered to be contaminants.

2. Sources of Contamination

Any classification of contact contaminants is arbitrary. The various forms of contamination and their causes reflect as much the use of particular materials

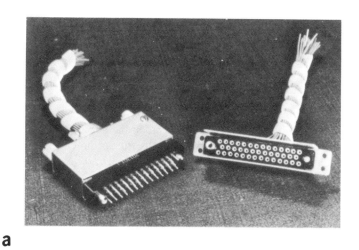

a

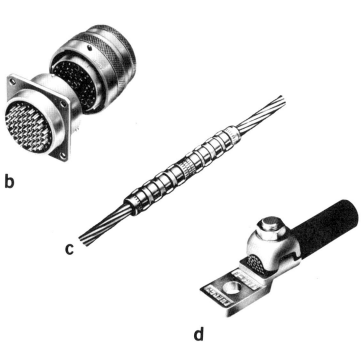

b

c

d

Figure 5. Typical electronic (a, b) and power (c, d) connectors. A: Courtesy of AMP, Inc.; b–d: Courtesy of Burndy Corporation.

and designs of contact-containing components as they do external contributing factors. However, a division of contamination into the following categories has been useful.

2.1. Oxidation and Corrosion

Base metals develop coatings of oxide and corrosion products on exposure to air. Except for metals which become passive, these layers thicken with time, particularly when air pollutants such as H_2S, SO_2, and HCl are present and the relative humidity is high. If water condenses on the surface, corrosive attack may be particularly aggressive. Tarnish and corrosion products are contaminants that can adversely affect electrical performance, which is why lightly loaded contacts are made of noble metals.

An example[1] of the effect of films on the contact resistance of a base metal is shown in Figure 6. Coupons of nickel were exposed for various lengths of time in telephone central office buildings, and contact resistance was determined by probing, as described below (Section 4). Contact resistance rose more rapidly in a non-air-conditioned industrial indoor location (site F) compared to

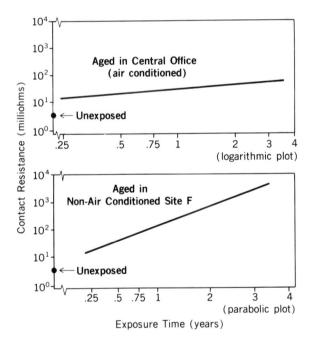

Figure 6. Contact resistance of nickel after exposure indoors in typical air-conditioned telephone central offices and in the non-air-conditioned basement (site F) of one of the central office buildings. Probed at 100 g with a solid gold rod.

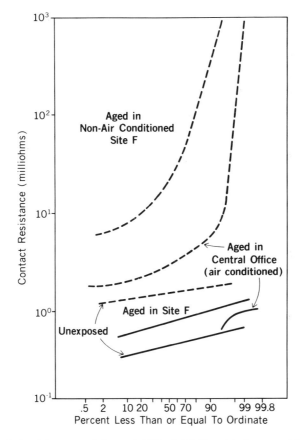

Figure 7. Contact resistance cumulative probability distribution plots for pure gold electrodeposits (solid lines) and clad palladium (dashed lines). Exposures were for 3 to 4 years in the same locations as described for Figure 6. Probed at 100 g with a solid gold rod.

other sites where humidity was controlled. The contact resistance of nickel was still higher, exceeding 1000 Ω, after one year of exposure out-of-doors in a sheltered chamber at site F.

Even metals that are generally considered to be noble may be attacked, and an example[2] is palladium that had been exposed in the same non-air-conditioned location as nickel. In Figure 7 the contact resistance of palladium at site F rose to large values within three years, which was found to be due to attack by traces of unidentified chlorine-containing compounds in the environment when relative humidity exceeded approximately 55%. These pollutants were present[2] at an average level of 0.63 parts per billion, expressed as Cl_2. Under similar conditions, gold was virtually unaffected, which makes it and high-gold alloys the preferred contact metals for aggressive environments.

When a noble metal is used as a layer on a base substrate (to conserve material because of its high cost), the coating may be thin enough to be porous. If the noble metal is applied to only part of the contact structure, tarnish and corrosion products from adjacent base surfaces may encroach on the noble metal. This is dramatically shown in Figure 8 with gold-plated silver where, after exposure to a sulfiding environment, insulating silver sulfide was found to have migrated for a considerable distance across the gold surface as well as spreading radially from pores in the deposit. Spreading films can also occur with gold-plated copper, and this has been a problem with connectors and printed circuit boards that were stored in contact with kraft paper, which releases corrosive gases in humid environments.[3] Figure 9 illustrates voluminous but nonspreading corrosion products at pores in gold plate on nickel that was exposed to a SO_2-containing atmosphere.

2.2. Particulates

Particulate contamination is of two types, airborne and that generated by the contacts themselves or the structures in which they are used or attached.

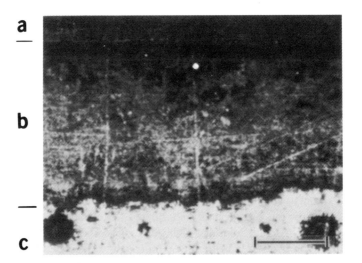

Figure 8. Gold-plated silver after exposure to hydrogen sulfide. Unplated area, *a*, is severely tarnished, and silver sulfide film, *b*, has advanced over the interface between the silver and gold. Note sulfide creep from pores in region *c*. Scale bar: 1 mm.

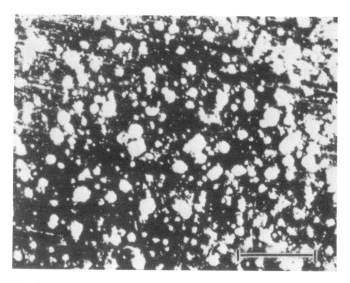

Figure 9. Corrosion products on porous gold-plated nickel after exposure to sulfur dioxide at elevated relative humidity. Scale bar: 1 mm.

The origin and structure of airborne dusts have been treated extensively in the literature.[4] Locally formed particulates include wear debris from separable contacts produced during engagement and such diverse causes as printed circuit board wear from sliding in metal apparatus guides. Air cooling is common in electronic systems, and the impingement of airborne particulates on electronic components may cause dust to accumulate on contacts.

Figure 10 illustrates contaminants that have been observed on contact surfaces.[2] Specimens were exposed in telephone central office buildings, and a particularly bad case of contamination from an air-conditioned equipment room is shown in Figure 10a. After three years of exposure, the surface was found to be covered with fine dust particles, below 0.5 μm in size, having a population density of about $10^8/\text{cm}^2$. The particles consisted of both a solid and an oily substance, and are believed to have originated from the building air humidification system. A sample from a non-air-conditioned area in the same building which did not contain operating equipment is shown in Figure 10b. Particles on the surface ranged in size from submicron to 50 micrometers and, when examined by energy-dispersive X-ray analysis, were shown to contain calcium, iron, sodium, chlorine, and sulfur. This material is believed to have come in part from snow-melting compounds used on nearby streets, having entered the room through open windows.

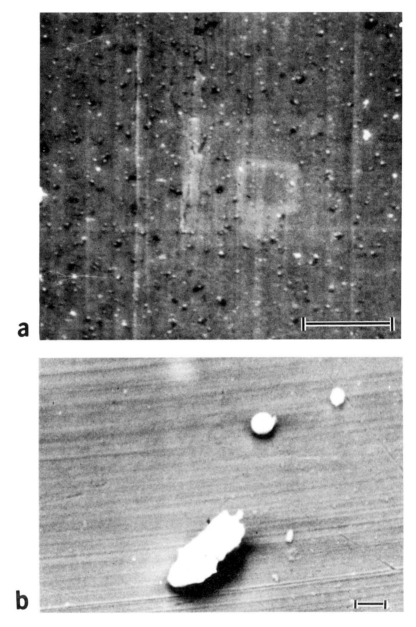

Figure 10. Particulate contaminants on contacts exposed indoors: (a) from air-conditioned telephone central office (scale bar: 10 μm); (b) from a non-air-conditioned site (scale bar: 10 μm).

2.3. Thermal Diffusion

Since noble contact metals are used mainly as thin layers on base substrates, thermal diffusion of substrate metal through the contact material to its surface can occur.[5] The base metal is transformed at the surface to oxides and other insulating compounds which may, in time, build up to levels sufficient to degrade contact resistance. Some noble contact metals contain small amounts of base elements for hardening and these also can diffuse to the surface.

These processes are illustrated in Figure 11 for copper contacts plated with either pure gold, both with and without a nickel underplate, or gold containing 0.26% cobalt. During aging in air at 150°C, contact resistance at 100 g was determined periodically with a gold probe. Surface analysis showed contaminating compounds of base metals to be present after aging. These were primarily cobalt compounds from the cobalt–gold plating (Figure 11, curve a) and copper compounds or copper plus nickel compounds from the substrate which had diffused through the thin pure gold plating (Figure 11, curves b and c). If

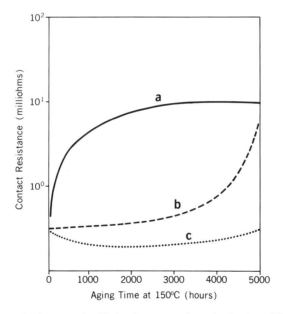

Figure 11. Contact resistances of gold-plated contacts after aging in air at 150°C: (a) 3.3-μm-thick 0.26% cobalt–gold on copper; (b) 1-μm-thick pure gold on copper; (c) same as (b), but having 2.6-μm-thick nickel between the gold layer and copper substrate. Probed at 100 g with a solid gold rod.

the probe had been made of the same thermally aged materials as the contacts rather than pure gold, the increases in contact resistance would have been much greater.

2.4. Fretting

Fretting is a small-amplitude oscillatory motion between two touching solid surfaces.[6] Contact wear occurs and if the debris that is formed oxidizes or corrodes, as will ordinarily be the case with base metals, insulating particles build up and increase contact resistance. When the contacts are made of platinum group metals, such as palladium, rhodium, ruthenium, or their alloys, polymeric solid contaminants which originate in adsorbed organic air pollutants will form. Fretting is inherent in the operation of some electrical components, such as wire spring relays that have contacts which wipe a distance of tens of micrometers during operation. Separable connectors, although normally at rest during service, may be subjected to vibrations and mechanical shock from nearby equipment and during transport. Thermal excursions can cause fretting of as much as 100 micrometers due to differential thermal expansions and contractions of the structures to which the contacts are joined.

An example of frictional polymer contamination is shown in Figure 12 for palladium contacts, and fretting corrosion products from tin–lead-solder-plated

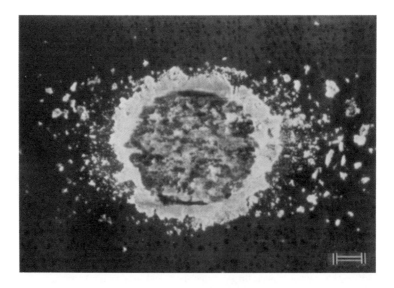

Figure 12. Surface of palladium contact after fretting in air; viewed with scanning electron microscope at 20 kV. The frictional polymer is the charged (light-colored) material on and surrounding the wear spot. Scale bar : 0.1 mm.

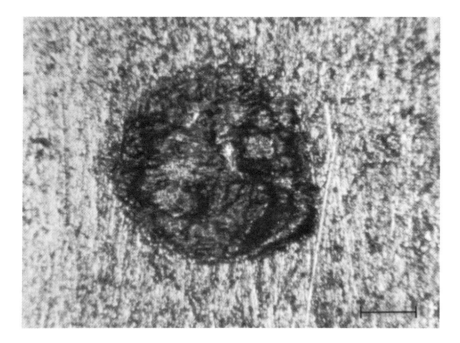

Figure 13. Worn 60Sn40Pb-plated contact after fretting. Note dark oxide on surface (optical microscope). Scale bar: 0.1 mm.

contacts in Figure 13. Figure 14 shows contact resistance degradation which can occur during fretting in laboratory tests of various metals mated to themselves. After a few thousand cycles of fretting on a 20-μm-long track with the contacts mated at 50-g force, both the solder and palladium systems developed very high contact resistances. On the other hand, contacts plated with 0.6-μm-thick cobalt-hardened gold had nearly constant contact resistance; although wear debris formed, the gold particles were chemically inert and electrically conductive. However, if the contacts had been made with very thin (below 0.25 μm) gold, the plating might have worn out during the experiment to expose underlying metals which, in turn, would have produced contaminating oxides or frictional polymers.

2.5. Manufacturing Processes

Electrical contacts are subjected to numerous manufacturing steps which can result in contaminated surfaces. Among the more common sources of contamination are:

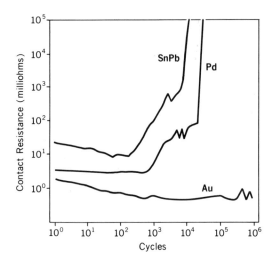

Figure 14. Contact resistance of various materials after fretting in air.

1. Contact plating salts that are incompletely washed in the rinsing cycle.
2. Rosin flux residues due to incomplete removal following soldering. Some flux cleaning procedures have been found[7] to dissolve traces of solder from the component, which can redeposit on adjacent contacts.
3. Cover-coat residues on printed circuit board contacts due to improper masking during application; such lacquer-like coatings are used to protect printed circuit conductors from abrasion and environmental attack.
4. Metal can transfer to contact surfaces during the fabrication of inlays. Minute amounts of copper from the substrate move to the rolls during reduction and subsequently can be back-transferred to the noble metal inlay. Coining is used to form contact weldments on springs and to make rivet relay contacts; this has sometimes been found to cause contamination from metal transfer.[7]
5. Handling can cause fingerprint contamination which may be a serious problem with contacts which are mated below about 100 g.
6. Excessive amounts of solid lubricants, such as microcrystalline waxes, on connector and printed circuit board contacts. Thick coatings of solid lubricants, unlike fluid lubricants, are not easily penetrated to make metallic contact required for low contact resistance.
7. Traces of silicone fluids from dielectric materials and potting compounds which migrate to relay contacts. Subsequent operation can cause solid insulating products to form.[8]

An example[9] of the effect of manufacturing process contaminants on gold-plated contacts of printed circuit boards is given in Figure 15. The cumulative plot of contact resistance determined by probing at 60 g with a gold rod showed a well-behaved distribution to about the 75th percentile. The incidence of elevated contact resistance then rose sharply so that 1% of the contacts had a contact resistance of over 20 mΩ. The contact resistance of clean contacts was 2 mΩ. Solder flux and some of the other contaminants described earlier are believed to have contributed to the elevated contact resistances.

2.6. Outgassing and Condensation on Contact Surfaces of Volatiles from Noncontact Materials

Connector housings, printed circuit boards, wire insulation, adhesives, elastomers, and other materials used in the construction of electronic components can decompose thermally and evolve volatiles that condense on contacts in their vicinity. In some cases these products are also corrosive. An example[10] is shown in Figure 16 which gives the cumulative probability distributions of contact resistance of contact metals before and after aging 30 days at 120°C in sealed vessels with organic materials. Two vessels were used; one (called "connector") contained inlay contacts of gold and DG R-156 (an alloy of 60 Pd40Ag having a gold-rich surface) with glass-filled nylon connector bodies, and the second (called "board") contained inlay gold and DG R-156 with glass epoxy boards. The contact resistance increases of the contacts were large and depended on both the contact metal and the organic materials with which they

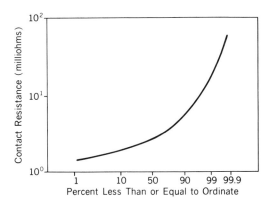

Figure 15. Effect of manufacturing process contaminants on the contact resistance of gold-plated contacts. Probed with solid gold rod.

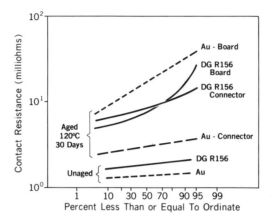

Figure 16. Effect of organic contaminants from heat-aged connector and printed circuit board materials on the contact resistance of gold and DG R-156 (a Au–Pd–Ag alloy).

were enclosed. Although these aging conditions were rather drastic, much electronic equipment operates at elevated temperature, and overheating occasionally occurs. It is obvious also that accelerated thermal life testing of contact-containing components which have organic materials must be conducted under conditions which do not cause unrealistic degradation.

3. Effects of Contamination

Surface contamination affects electric contacts by interfering with the attainment of low contact resistance when the contacts are joined. There are, in addition, many indirect contaminant effects which degrade contact reliability.

3.1. Direct

When the connection is intended to carry large currents and contact resistance is higher than that associated with clean surfaces, significant heat is generated at the interface. This promotes oxidation of the contact material, further reducing the area of metallic contact. Heat can destroy wire insulation and other materials in the vicinity. Fluctuating contact resistance may be an intermediate stage in the failure of the contact.[11]

Figure 17 illustrates typical contact resistance–load characteristic behaviors. Figure 17a is that of a clean noble metal surface such as gold where contact resistance decreases monotonically with increasing load. On unloading, contact resistance changes little until the parts are separated. Because asperity deformation is plastic in nature, the area of metallic contact remains practically unchanged as the load is decreased.

Figure 17b is from a dirty contact where there was abrupt fracture of the surface contaminant during loading, which caused contact resistance to fall rapidly. Depending on the type and thickness of contaminant, contact resistance may be similar to that of a clean contact at sufficiently high load. Surface coverage is rarely uniform and contact resistances from replicate determinations may be highly variable. With dirty surfaces, irregularities often appear in the curve on both loading and unloading. For example, contact resistance can increase over small ranges of force as load is increased, and on unloading "crossover" behavior (higher contact resistance than on loading) may occur.

Figure 17c is typical of contacts where film breakup is not abrupt or where conduction through the film is primarily by electron tunneling.[12] Smooth hard metals covered with uniform thin films generally display this behavior.

Another form of contaminant is particulate matter, including heterogeneous corrosion products, as shown in Figures 9 and 10. Contact geometry, particle size, and surface roughness have an effect on the contact.[13] Figure 18 schematically illustrates this condition where contact has occurred between a

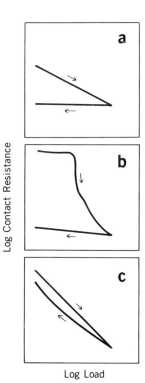

Figure 17. Typical contact resistance–load characteristic behaviors: (a) clean gold; (b) gold coated with relatively thick insulating solid; (c) contact coated with thin tough insulating solid where conduction is primarily by electron tunneling.

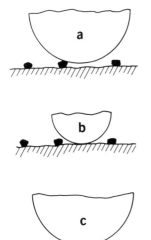

Figure 18. Idealized contact between a hemisphere and a flat covered with insulating particles: (a) contact members fail to touch; (b) small-diameter contact is less susceptible to interference by the particles; (c) particles are less likely to prevent contact when the flat is rough than when it is smooth.

dome-shaped member and a flat covered with insulating particles. In Figure 18a the contact members fail to touch. Figure 18b shows that a contact having a smaller radius of curvature is less affected by particulates. In Figure 18c many of the particles have disappeared in the roughness of the contacts and do not interfere. From this model, it is clear that particle size, particle density, and the physical dimensions of the contacts affect contact resistance.

Finally, contacts which are designed to slide a short distance on engagement before coming to rest may displace loose contaminants as well as insulating films. However, below some force that is characteristic of the system, contact resistance may rise as contaminations accumulate ahead of the contact members.

3.2. Indirect

Indirect effects of contamination on contact performance are many, and some of the better-known examples are described. Particulate contaminants can promote abrasive wear of sliding and wiping contacts if they are hard, sharp, and coarse. The force involved in engaging a contaminated connector may rise. For this reason, care is exercised to exclude external contaminants from a contact system. However, wear debris generated from sliding and wiping contacts is also a contaminant because it promotes further metal loss which subsequently oxidizes.

Gaseous pollutants may cause contact surfaces to tarnish or corrode, as described earlier. Solid contaminants can have a similar effect. When such solids are hygroscopic, the water that they absorb may cause them to dissolve and ionize. Droplets of electrolyte on base metals promote corrosion. Even nonhygroscopic or nonionizable solid contaminants may induce corrosion by forming differential aeration cells.

Hygroscopic, ionizable solid contaminants on dielectric materials can cause electrical leakage between adjacent contacts and other conductors at high relative humidity. This can also promote further corrosion of the metals involved.

A special problem encountered some years ago[14] involved airborne nitrate contaminating particles that originated in the high levels of nitrogen oxide pollutants from combustion sources. The accumulation of such particles caused stress corrosion of nickel–brass contact springs with resulting fracture.

It is necessary to be able to solder some contact materials or the structures to which they are attached. Contaminated surfaces are difficult to solder and may require rigorous cleaning treatments in order to be used.[15] In some cases, the expense of surface protection with noble contact platings for solderability protection is borne in order to minimize the effects of contamination which might be particularly severe were base metals used.

4. Contact Resistance Probes for the Detection and Characterization of Contamination

The most useful, and often the simplest, method for detecting contamination on an electric contact surface is to determine its contact resistance with an instrument called a contact resistance probe. Although probing generally cannot identify the composition of a contaminant, practical information may be obtained concerning its probable influence on the behavior of a component. When it is necessary to determine the composition of the contaminant, traditional surface analytical methods such as chemical spot tests, energy-dispersive X-ray analysis, Auger, and electron spectroscopy for chemical analysis (ESCA) methods are employed in conjunction with probing.[16]

Probes have been used for many years in contact contamination studies, and recently an ASTM "Standard Practice for Construction and Use of a Probe for Measuring Electrical Contact Resistance" was adopted.[17] "Procedures for Measuring the Contact Resistance of Electrical Connectors (Static Contacts)"[18] is a guide to the required electrical instrumentation for probing.

4.1. Description of Probes

A probe generally consists of the following:

1. Fixtures for holding specimens and for attaching electrical leads to them.
2. A mechanism for applying a mechanical load to the specimen and for measuring its value.

3. A reference surface (the probe), usually made of a noble metal, that is pressed against the specimen.

4. A current source with meters for determining current and voltage drop through the contact.

5. Additional electrical circuitry may be included to permit related measurements, such as the voltage breakdown characteristics of film-covered surfaces, to be carried out.

Figure 19 illustrates a simple probe. In this instrument, an electronic force transducer is located between the micrometer and the probe contact. The micrometer screw is turned to advance the probe against the specimen manually or with a motor drive. Contact resistance and force are read out simultaneously on a two-channel oscillographic recorder or an XY-plotter. Transducers having load limits of 1, 10, and 100 g may be interchanged in the probe head.

4.2. Determination of Contact Resistance

Contact resistance is interfacial resistance, and if a simple two-wire ohmmeter is used to determine it, the measured resistance will include an additional resistance from the bulk metal of the contact members and the lead wires. To reduce these as much as possible, the four-wire method of measurement is used.[18] This technique makes use of four leads from an appropriate instrument, two for current and two for voltage. The current leads are attached to the probe and specimen, and voltage leads are connected as close as possible to the point of contact. Crossed rod specimen geometry (Figure 20) can eliminate bulk resistance entirely, but ordinarily it is not practical to use this configuration because of limitations in the shape and size of the specimens.

Self-contained four-wire AC instruments having a low open-circuit potential are commercially available. Their circuitry is designed so that contact resistance can be read directly on a meter.

The probe tip preferred by the ASTM method is a smooth hemispherically-ended solid gold rod having a radius of 1.6 mm. Since the probe tip is slightly deformed plastically, after mounting in the instrument it is conditioned before use by pressing against a hard smooth surface several times so as to provide an equilibrium shape.

Repeated contact of a probe with a contaminated surface can cause such materials to transfer to it. It is therefore necessary to verify the cleanliness of the probe tip by determining the contact resistance of reference metals, such as freshly cleaned gold or a specimen having a uniform insulating film with known contact properties. Electroplated SnNi on a metal substrate has been used for this purpose since it develops a thin passive oxide on air aging which is rugged and chemically stable. If the probe tip becomes contaminated, it can usually be

Figure 19. Contact resistance probe.

Figure 20. Four-wire technique for measuring contact resistance. The only circuitry common to the current path and the voltage probes is across the interface between the rods.

cleaned by wiping with lens tissue which has been moistened with 1,1,1-trichloroethane or other volatile solvent. Another, more complex, design of probe involves a mechanism for rotating an inclined hemispherical probe tip between measurements, or indexing a gold-plated wheel whose rounded edge is the probe surface, so that a fresh region is brought into contact with the specimen for each determination.

4.3. Modes of Operation

There are several different ways that probes can be used, depending on the kind of information desired. For example, probes may be operated at

1. Low (20 mV maximum) open-circuit voltage with current limited to 100 mA, or so-called "dry circuit" conditions, where:

   a. At small forces the presence and distribution of films can be assessed, or
   b. At progressively higher forces film fracture tendencies can be determined. This has practical use in showing whether the films can be deleterious in components containing contacts.

2. Fixed force with tangential wipe to simulate the motion of contacts in some devices.

3. Low force with progressive increase of voltage while current is measured. Electrical puncture occurs at a particular voltage, which can then be related to film thickness; semiconducting films leak current, and may display different resistances according to probe polarity when DC circuits are used.

An interesting variation of probing techniques, although having limited applicability, was proposed[19] where the current–voltage relationship of two touching surfaces is determined at sufficiently high force to cause limited metallic contact. If nonohmic behavior is exhibited because of the presence of surface films, the contaminant's softening or melting point can be determined. In effect, the probe is a "hot stage" which operates on the contaminant *in situ*.

Another probe has been described[20] which traverses the surface automatically with a fine wire in continuous contact loaded at about 7 mg. A capacitor-type integrating circuit measures the product of contact resistance and time to give an estimate of surface contamination.

An example[21] of probing to determine the spatial distribution of contaminants is shown in Figure 21. Contact resistance was determined along a specimen with indexing by a small increment between determinations. The presence of migrated sulfide contaminations and the comparative mechanical characteristics of films from silver and two copper alloys are given. The same approach may be taken to determine the distribution of contamination in the x,y-plane of an extended surface.

Automatic probes for detecting contamination which lend themselves to factory use have been described. Figure 22 is an example[22] of an instrument employed to monitor printed circuit boards. A microcomputer directs the positioning of test probes, processes the test measurements, makes pass/fail decisions, and prints out test data. The system can be programmed for a variety of circuit-pack profiles and performs tests many times faster than is possible with a manual instrument.

Automatic contact resistance probes for laboratory use[23] have also been described which are capable of obtaining large numbers of measurements with little operator intervention. It is desirable to make many contact resistance measurements on a given sample because of variability in the distribution and the properties of surface contaminants. Statistical treatments of contact resistance are useful, such as those illustrated in Figures 7, 15, and 16.

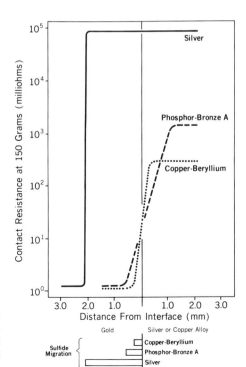

Figure 21. Contact resistance of sulfided samples. Data obtained with probe by stepping a line of points from clean gold, across migrated sulfide, over interface, and onto basis metal.

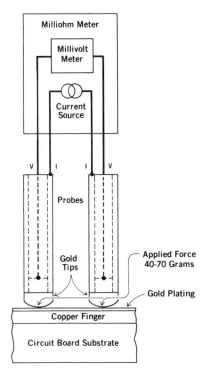

Figure 22. Contact resistance probe for determining presence of manufacturing process contaminants on printed circuit boards.

5. Summary

Surface contamination is one of the most serious causes of failure of connectors, relays, switches, slip rings, and other devices that have electrical contacts. The most common types of contamination are oxide and corrosion products, particulates, films formed by thermal diffusion, debris produced by mechanical wear and fretting, outgassing and condensation on contact surfaces of volatiles from noncontact materials, and those which originate in manufacturing processes. This review gives examples of contaminations, their effects on contact resistance and device reliability, and the principles of contact resistance probing for the detection and characterization of contaminants.

References

1. M. Antler, Field studies of contact materials: Contact resistance behavior of some base and noble metals, *IEEE Trans.*, *CHMT* 5(3), 301–307 (1982).
2. M. Antler, M. H. Drozdowicz, and C. A. Haque, Connector contact materials: Effect of

environment on clad palladium, palladium-silver alloys, and gold electrodeposits, *IEEE Trans., CHMT* 4(4), 482–492 (1981).
3. N. R. Stalica, Corrosion studies of Bell System connectors in central office applications, in: Proceedings of the Eleventh Annual Connector Symposium, Electronic Connector Study Group, Inc., Cherry Hill, N.J. (Oct. 25–26, 1978), pp. 376–392.
4. S. Friedlander, *Smoke, Dust and Haze*, J. Wiley and Sons, New York (1977).
5. M. Antler, Gold plated contacts: Effect of heating on reliability, *Plating* 57, 615–618 (1970).
6. M. Antler, Fretting of electrical contacts, in: *Materials Evaluation under Fretting Conditions*, Special Technical Publication, STP 780 (S. R. Brown, ed.), American Society for Testing and Materials, Philadelphia, Pa. (1982), pp. 68–85.
7. C. A. Haque, M. H. Drozdowicz, R. A. Frank, and J. T. Hanlon, Extraneous metal deposits from production processes on contact materials, *IEEE Trans., CHMT* 6(1), 55–60 (1983).
8. G. J. Witter and R. A. Leeper, A comparison for the effects of various forms of silicon contamination on contact performance, in: Proceedings of the Ninth International Conference on Electric Contact Phenomena, Illinois Institute of Technology, Chicago, Ill. (1978), pp. 371–376.
9. D. E. Tompsett and G. C. Emo, Development of a contact resistance probe for detecting contamination on connector contacts, in: Proceedings of the Electronic Components Conference, IEEE CHMT Society and EIA, Cherry Hill, N.J. (May 14–16, 1979), pp. 243–246.
10. S. P. Sharma and S. DasGupta, Reaction of contact materials with vapors emanating from connector products, in: Proceedings of the Electronic Components Conference, IEEE CHMT Society and EIA, Orlando, Fla. (May 16–18, 1983), pp. 418–424.
11. J. Aronstein and W. E. Campbell, Failure and overheating of aluminum-wired twist-on connections, *IEEE Trans., CHMT* 5(1), 42–50 (1982).
12. R. Holm, *Electric Contacts Handbook*, 4th Ed., Springer-Verlag, New York (1967).
13. J. B. P. Williamson, J. A. Greenwood, and J. Harris, The influence of dust particles on the contact of solids, *Proc. Roy. Soc. A237*, 560–573 (1956).
14. H. W. Hermance, C. A. Russell, E. J. Bauer, T. F. Egan, and H. V. Wadlow, The relation of airborne nitrate to telephone equipment damage, *Environ. Sci. Technol.* 5, 781–781 (1971).
15. C. J. Thwaites, Soft soldering, in: *Gold Plating Technology* (F. H. Reid and W. Goldie, eds.), pp. 225–245, Electrochemical Publications, Ltd., Ayr, Scotland (1974).
16. C. A. Haque, Temperature–time effects on film growth and contact resistance of a plated copper-tin-zinc alloy used as a surface finish on electronic components, in: Proceedings of the Twelfth International Conference on Electric Contact Phenomena, Illinois Institute of Technology, Chicago, Ill. (1984), pp. 385–389.
17. Standard Practice for Construction and Use of a Probe for Measuring Electrical Contact Resistance, B667-80, American Society for Testing and Materials, Philadelphia, Pa.
18. Procedures for Measuring the Contact Resistance of Electrical Connections (Static Contacts), B539-80, American Society for Testing and Materials, Philadelphia, Pa.
19. R. E. Cuthrell and L. K. Jones, Surface contaminant characterization using potential–current curves, in: Proceedings of the Holm Conference on Electrical Contacts, Illinois Institute of Technology, Chicago, Ill. (1977), pp. 157–162.
20. S. W. Chaikin, J. R. Anderson, and G. J. Santos, Jr., Improved probe apparatus for measuring contact resistance, *Rev. Sci. Instrum.* 32, 1294–1296 (1961).
21. M. Antler, New developments in the surface science of electric contacts, *Plating* 53(12), 1431–1439 (1966).
22. M. P. Asar, F. G. Sheeler, and H. L. Maddox, Measuring contact contamination automatically, *The Western Electric Engineer* 23(3), 32–38 (1979).
23. M. Antler, L. V. Auletta, and J. Conley, Automated contact resistance probe, *Rev. Sci. Instrum.* 34, 1317–1322 (1963).

9

The Role of Surface Contaminants in the Solid-State Welding of Metals

J. L. JELLISON

1. Introduction

Although no unified theory of solid state welding has been developed, the generally agreed upon requisite for producing a solid state weld is that metallic surfaces must be brought sufficiently close together that short-range interatomic attractive forces operate. In general, this involves both (1) intimate mating of surfaces and (2) removal of surface barriers to atomic bonding.

One form of solid state welding, called deformation welding, is accomplished by subjecting the surfaces to be welded to extensive deformation. The other form, called diffusion welding, is accomplished by bringing the surfaces to be welded together under moderate pressure and elevated temperature in a controlled atmosphere so that a coalescence of the interfaces or faying surfaces can occur. In general, diffusion welding involves little bulk deformation and is conducted at temperatures greater than one-half the absolute melting point ($>1/2\ T_m$). Conversely, deformation welding typically is conducted at temperatures ranging from room temperature to $<1/2\ T_m$. These distinctions are not very precise, since deformation welding often is performed at elevated temperatures and, conversely, significant deformation sometimes accompanies diffusion welding. For example, some deformation processes such as forge welding, roll welding, and extrusion welding may be performed at high temperatures. Another distinction is that deformation welding is usually accomplished within seconds, whereas diffusion welding typically requires weld durations ranging from minutes to hours. Another generalization is that surface films are thought to be disrupted mechanically in the case of deformation welding, whereas for

---

J. L. JELLISON • Process Metallurgy, Division 1833, Sandia National Laboratories, P.O. Box 5800, Albuquerque, New Mexico 87185.

diffusion welding, thermal mechanisms such as the diffusion of interstitials dominate.

2. Role of Contaminants in Preventing Solid-State Welds

Intimate mating of surfaces is generally easily achieved by both deformation welding and diffusion welding. Therefore, we conclude that barriers such as oxide films or other surface contaminants are responsible in most cases when extensive metallic bonding does not result. Milner and Rowe have reviewed the literature regarding deformation welding and have found that, with the exception of localized defects, mating of faying surfaces always occurs at levels of deformation significantly below that required for deformation welding.[1] Some threshold deformations for cold welding of common metals are summarized in Table 1. These data illustrate that very high deformations are required to initiate welding, particularly in the case of indentation welding. Yet faying surfaces are brought into intimate contact at deformations of the order of 10%.[2] The fragmentation of surface oxides was hypothesized by Tylecote to control welding and to determine the threshold deformation for welding.[3] It is interesting to note, however, that gold exhibits a threshold deformation for indentation welding of 24 to 30%, even though it is free of a surface oxide. Tylecote later concluded, on the basis of experimental data, that although fragmentation of surface oxides may be important in deformation welding, other factors must be considered in determining the threshold deformation for cold welding.[4]

The thermodynamic driving force for solid state welding is a lowering of surface energy; two free surfaces are replaced by one solid–solid interface and, in fact, this interface may, in turn, be destroyed by some high-temperature mechanisms such as recrystallization and grain growth. In the case of solid state welding, the weld interface is energetically similar to a large-angle grain bound-

Table 1. Threshold Deformations for Cold Welding

Metal	Threshold deformation, %	
	Indentation welding	Roll bonding
Aluminum	40	25
Copper	73	45
Gold	24–30	—
Iron	81	—
Lead	10	10
Silver	80	—

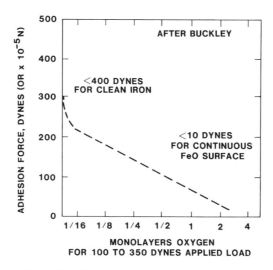

Figure 1. Effect of oxygen on the adhesion of iron.

ary. The interfacial free energy of such an interface is about one-third that of a free surface. Therefore, substitution of a weld interface for two free surfaces results in a reduction of surface energy to about one-sixth of the original level. The goal of solid state welding processes is to bring metallic surfaces sufficiently close to one another so that bonding is not only thermodynamically favorable but also kinetically favorable. For strong adhesion to occur, opposing surfaces must be brought within the range that results in mutual short-range interatomic attractive forces. This distance is no more than about 1 nm for most metals. We can now begin to understand the role of surface contaminants. If the thickness of a surface contaminant exceeds 1 nm, the attractive forces between surfaces are too low to result in adhesion. Actually, the effect of surface contamination on solid state welding is worse than implied by viewing contaminants simply as physical barriers to adhesion. The interatomic attractive forces are due to the fact that surface atoms on a clean surface desire to share electrons with additional atoms. When a surface becomes contaminated, surface metallic atoms become satisfied by forming bonds with contaminant atoms. Thus, attractive forces at the surface are reduced. The marked effect that a thin contaminant layer can have on metal adhesion is illustrated by the work of Buckley on the role of oxygen in the adhesion of iron[5] (see Figure 1). He found that the adhesion of clean iron was drastically reduced by as little as a one-half monolayer of oxygen. The arrangement of oxygen atoms on iron for this condition is schematically shown in Figure 2. Note that for as little as a one-half monolayer of oxygen, each iron atom has a near-neighbor oxygen atom. Conse-

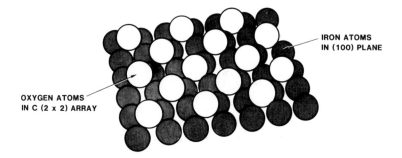

Figure 2. Schematic representation of one-half monolayer of oxygen on iron.

quently, all of the iron atoms are partially satisfied with regard to electron sharing. Thus the thermodynamic driving force for solid state welding is significantly reduced.

We conclude that unless surface contaminants are removed or disrupted, they will prevent metal–metal welding. Before proceeding, we must consider the possibility of the opposite effect, i.e., metal–contaminant bonds might form so that contaminant films or inclusions might actually contribute to the strength of a solid state weld. Tylecote[2] and Mohamed and Washburn[6] discount the contribution of metal oxide bonding to overall weld strength. However, Ludemann[7] extensively studied metal oxide welds and concluded that such welds can form under diffusion welding conditions and contribute significantly to joint strength. Ludemann summarized the requirements for a large metal oxide contribution to joint strength as: (1) both metals must bond strongly to the oxide layer, (2) the oxide must be strong, and (3) the oxide layer must be thin enough so that its brittle properties do not influence joint strength. Good diffusion welds were obtained between aluminum and anodized beryllium, aluminum and oxidized iron, copper and oxidized iron, and oxidized copper and iron. However, the strength of joints involving iron oxide was limited to the strength of the oxide. Ludemann concludes that residual oxide films in diffusion welds are not necessarily detrimental and, in fact, in the case of some dissimilar metal welds, may be beneficial by reducing intermetallic compound formation. Metal oxide bonds are less likely to be formed under conditions of deformation welding than during high-temperature diffusion welding, although examples of the former type have been reported by some workers.[8,9]

Whereas oxide films and possibly other inorganic films might contribute some strength to a solid state weld, most organic contaminants are either too weak or too weakly bonded to metals to make much contribution to joint strength. In summary, most contaminants prevent metal–metal bonding and, because they are intrinsically weak, comprise defects if not eliminated from the weld interface. We conclude that, in general, solid state welding must be based on a good understanding of the contaminants present and ways of eliminating their influence on solid state welding.

3. Classification of Surface Contaminants

3.1. Inorganic Films

Metal oxides are by far the most common of inorganic films. Fortunately, the growth of oxide films on most metals occurs at a logarithmic rate at room temperature; this results in limiting film thicknesses of 2 to 10 nm. Small amounts of other contaminants such as water, carbon dioxide, and sulfur dioxide can significantly increase the limiting oxide film thickness.[2] Also, oxide films formed by reaction with chemical solutions or electrochemical reactions may be thicker or of different structure. This possibility must be kept in mind during surface preparation of metals for solid state welding. If sufficient oxygen is available, oxide films formed at elevated temperatures are often much thicker than those formed at room temperature. Consequently, the use of vacuum, inert, or reducing atmospheres must be considered during solid state welding processes conducted at elevated temperatures.

Other inorganic films such as chlorides,[7] nitrides,[10] and carbides[10] can hinder welding. Nitrides form on some metals (such as titanium) by reaction with air.[2] Chlorides and phosphates can form by reaction with acids. Even degreasing solvents can react to form some metal halides.[7] Therefore, caution must be taken during treatments used to remove organic contaminants and oxide films because these same treatments may result in the formation of other undesirable inorganic films.

Although adsorbed films are generally thought of as organic contaminants, inorganic gases can also be adsorbed. Water vapor is a commonly adsorbed inorganic film. Other adsorbed inorganic gases are carbon monoxide and carbon dioxide. Buckley observed that carbon dioxide dissociated on iron surfaces resulting in the formation of iron oxide and carbon monoxide.[11]

3.2. Organic Contaminants

Organic contaminants are generally found to at least monolayer thicknesses on all metals unless very special precautions have been taken to remove organic surface films. Organic contaminants can be divided into two categories: those that are adsorbed from the atmosphere and those that are incurred by contact, such as oils, greases, and waxes.

Generally, organic atmospheric contaminants consist of relatively small, volatile molecules. Atmospheric contaminant films tend to be limited to monolayers because once a monolayer of the organic contaminant is adsorbed, active sites are no longer available. The resulting barrier films usually do not exceed 4 nm in thickness and can be much thinner, as has been determined by Auger electron spectroscopy.[12] Exceptions are situations where a second species, perhaps even an inorganic like water, attaches itself to the adsorbed monolayer of the first species. However, even though the resulting barrier film is relatively thin, it generally is difficult to remove and constitutes a significant barrier to welding.

Organic contaminants resulting from direct contact often result in surface barriers of considerable thickness. Thicknesses of nonatmospheric contaminants can easily exceed 5 μm. Aside from being thicker, nonatmospheric organic contaminants tend to differ from atmospheric contaminants by consisting of larger, more strongly bonded molecules and may exhibit cross-linked structures. It should be noted that contaminants incurred by contact do not need to be thick to hinder welding. An example of a thin contact contaminant is the common fingerprint, which typically includes sodium chloride, potassium chloride, calcium chloride, lactic acid, stearic acid, sebacic acid, and urea.

3.3. Particulate Contaminants

Film-type contaminants are generally the most injurious to solid state welding. However, particulate contaminants must also be considered, particularly in the case of microminiature welds employed in electronic circuitry. In this latter case, common particulate contaminants can be significant in size compared to the size of the desired welds. It is common for microminiature solid state welds to be made in clean-room environments. Common particulate contaminants consist of fibers from clothing, hair, and, in some areas, sand. Control of these contaminants is based on both a knowledge of their sources and the maintenance of good clean-room practices, which generally requires air filtration.

4. Role of Contaminant Properties

For both inorganic and organic contaminants, the effect of a surface film as a barrier to welding depends on the properties of the film as well as its thickness. Compared with the metals themselves, inorganic films tend to be more brittle, whereas organic contaminants are more mobile and tend to smear. However, variations within each category exist and both types of contaminants are often encountered together. A consideration of the physical and mechanical properties of these films is particularly important in the case of deformation welding, which depends extensively on mechanical disruption of surface films.

4.1. Inorganic Films

The influence of oxide films on deformation welding primarily depends on how the oxide films fragment and distribute themselves along the weld interface and the ability of the metal to extrude through breaks in the oxide film. Although several workers[3,4,13-15] have attempted to elucidate the role of oxide films in deformation welding, a consistent correlation between weldability and the properties of metal oxides has not been determined. Experiments in the field of friction by Whitehead[16] and Wilson[17] suggest that the coefficient of friction correlates with the hardness of the oxide relative to that of the substrate metal. A hard oxide on a soft metal was found to break more easily than soft oxides or hard oxides on hard metals. Tylecote *et al.* attempted to apply this concept to deformation welding, but found little correlation between the ratio of oxide hardness to metal hardness and weldability.[4] The correlation observed for friction experiments probably breaks down for deformation welding because most oxides fragment fairly easily at the high deformations associated with welding. Surprisingly, even relatively thick oxide films such as produced by anodization do not prevent welding, provided deformation is sufficiently high. Vaidyanath and Milner[14] and Donelan[18] found that anodized layers of 2 to 6 μm permitted aluminum welds to be made that were nearly as good as those with thin oxides, if the parts were first baked to remove adsorbed contaminants and deformation was high enough to fragment and distribute the oxide films. There is some evidence that thick, soft oxides such as those formed on copper are particularly detrimental to welding because they tend to smear along the weld interface. Vaidyanath and Milner[14] observed that whereas two opposed thick oxide films on aluminum generally fragment together during roll welding so as to provide a high fraction of welded interface, copper oxide is sheared into small fragments so that the total accumulative length of oxide fragments becomes more than twice the original length of the specimen. They found, in fact, that only 18% of the interface was made available for welding at 60% deformation. Tylecote suggests that the poor weldability of zinc is also due to the softness of its oxide.[2]

Gilbreath[19] showed that the effect of gas adsorption on metal adhesion correlated with the heat of adsorption for the particular metal/gas reaction. Little loss of adhesion occurred when gases were physisorbed on metals (heats of adsorption less than 5 kcal/mol) as for the adsorption of argon, hydrogen, and nitrogen on aluminum or copper. Loss of adhesion in those metal/gas systems having heats of adsorption above 20 kcal/mol (chemisorbed) occurred at exposures of 10^{-4} to 10^{-2} torr/s. In the case of gold, which does not form compounds with most gases, chemisorbed species such as water, carbon dioxide, carbon monoxide, and sulfur dioxide must be eliminated before adhesion can be obtained at room temperature.[19] Conversely, oxygen, nitrogen, and noble

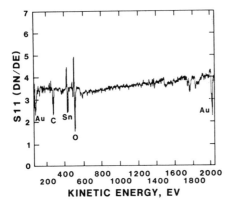

Figure 3. Auger electron spectrum for gold surface that exhibited poor weldability during thermocompression welding (after Vaughn and Raut[21]).

gases, which are only physisorbed on gold, have little effect on adhesion. Bowden and Throssell found that water hindered gold welding far worse in the form of condensed water films than when restricted to one or two chemisorbed monolayers.[20]

Metallic contaminants have been known to influence solid-state welding processes either because of the presence of a metallic contaminant at the welding interface or oxidation of the metallic contaminant. An example of a metallic contaminant influencing a deformation welding process was reported by Vaughn and Raut.[21] They concluded that problems encountered in the thermocompression welding of gold ribbons to gold metallization were due to tin deposits stemming from hydrogen peroxide that had been stabilized with sodium stannate. Figures 3 and 4 show Auger electron spectra for gold surfaces that exhibit poor and good weldability, respectively. Tin contamination was found on the surface with poor weldability. The high oxygen peak seen in Figure 3 suggests that tin oxide contributed to the welding problems.

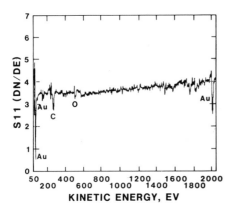

Figure 4. Auger electron spectrum for gold surface that exhibited good weldability during thermocompression welding (after Vaughn and Raut[21]).

4.2. Organic Films

Much of the best work conducted to determine the effect of the chemical and structural properties of organics on metal–metal welding has involved adhesion and friction experiments, where the extent of plastic deformation is negligible compared to most industrial deformation welding processes. While high plastic deformation is known to reduce the effect of organic surface contaminants, the qualitative observations made during adhesion experiments generally would be expected to apply to deformation welding processes.

Chemical activity is a dominant property of organic contaminants with regard to their influence on adhesion. As indicated earlier, Gilbreath observed that for adsorbed contaminants, the effect of environment on metal adhesion is related to the heat of adsorption.[19] Organic gases are generally chemisorbed by metal surfaces and consequently nearly always degrade adhesion. The extent of this degradation depends on the chemical and mechanical properties of the adsorbed species. Buckley has extensively studied the effect of adsorbed gases on metal adhesion, particularly with regard to chemical activity.[5,11,22,23] The dependence of the adhesion of iron on the chemical structure of hydrocarbons, as determined by Buckley, is summarized in Table 2. The simple one- and two-carbon-atom saturated hydrocarbons, methane and ethane, had very modest effects on adhesion. The more pronounced effect on iron adhesion due to the adsorption of ethylene and acetylene is explained on the basis of the progressive reduction in bond saturation for these hydrocarbons.[23] Although the rigidity of the double and triple carbon atom bonds, as in the case of ethylene and acetylene, respectively, as compared with single bonds and saturated hydrocarbons probably contributed to the difficulty in disrupting unsaturated hydrocarbons, the primary effect is thought to be their greater chemical activity. An even greater loss of adhesion was observed by the addition of chemically active atoms such as chlorine (halogenation) or oxygen to a hydrocarbon, as in the cases of vinyl chloride and ethylene oxide, respectively.[22] Buckley also conducted

Table 2. Effect of Various Hydrocarbons on Adhesion of Clean Iron, (011) Planes

Hydrocarbon	Projected molecular formula	Adhesive force[a] (newtons $\times 10^4$)
Methane	CH_4	40
Ethane	H_3C-CH_3	28
Ethylene	H_2C-CH_2	17
Acetylene	$HC-CH$	8
Vinyl chloride	$H_2C-CHCl$	3
Ethylene oxide	$H_2C-O-CH_2$	1

[a]Contact load, 2×10^{-4} newtons.

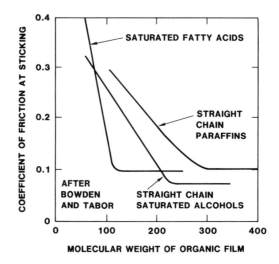

Figure 5. Coefficient of friction (on steel) as a function of molecular weight.

studies on the influence of adsorbed saturated hydrocarbons on the coefficient of friction of iron.[11] Ethane (two carbon atoms) reduced the coefficient of friction by more than a factor of two. As the number of carbon atoms in the hydrocarbon chain was increased up to eight, the coefficient of friction tended to decrease slightly.

Studies on boundary friction by Bowden and Tabor disclosed that friction on steel is inversely dependent on the molecular chain length of lubricants (for up to 14 carbon atoms), which consisted of homologous series of paraffins, alcohols, and fatty acids[24] (see Figures 5 and 6). Also borrowing from the field of lubrication, the effect of an organic contaminant on welding would be

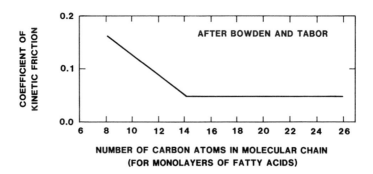

Figure 6. Effect of molecular chain length on kinetic friction (stainless steel sphere on glass surface).

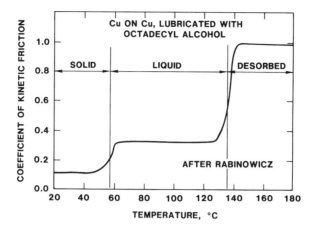

Figure 7. Effect of state of organic film on sliding friction.

expected to be dependent on the state in which the contaminant exists on the surface. For example, Rabinowicz found that the coefficient of friction for copper lubricated with octadecyl alcohol underwent transitions at the melting point and desorption temperature of the lubricant[25] (see Figure 7). Similarly, a friction transition temperature was observed by Bowden and Tabor for a steel surface lubricated with fatty acids.[24] However, in this case the transition temperature occurred at a higher temperature than the melting point of the fatty acid (see Figure 8). Bowden and Tabor have shown that the friction transition tem-

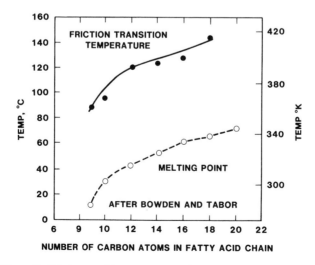

Figure 8. Friction transition temperature of fatty acids on steel surfaces.

perature for fatty acids occurs at approximately the melting point of the metallic soap that would be expected to form by reaction of the metal and the fatty acid. The high chemical activity of the fatty acids accounts, in part, for the greater dependence on molecular chain length than for the other homologous series. Bowden and Tabor indicate that the high melting points and the strong lateral attraction between molecular chains also contribute to the excellent lubricating properties of metallic soaps.[24] Recall that fingerprints include fatty acids capable of forming metallic soaps. These fatty acids account for the effectiveness of fingerprints in hindering deformation welding. By means of experiments involving adhesion tests between indium and platinum, Cuthrell observed that the coefficient of adhesion is inversely proportional to the length of the hydrocarbon chain.[26] After conducting adhesion tests on a large number of organic contaminants as diverse as alcohols, halogenated hydrocarbons, and salts of organic acids, he further concluded that adhesion is very sensitive to the contaminant layer thickness, as is graphically illustrated in Figure 9.

A strong dependence in the deformation welding of gold on contaminant layer thickness was shown in the studies by Holloway and Bushmire[27] and Jellison.[12] The effect of photoresist contamination as observed by Holloway and Bushmire is shown in Figure 10. In Figure 11, from Jellison's studies, the dependence of weld strength on welding temperature is seen to increase with increasing contaminant layer thickness.

The effect of the chemical activity of contaminants on the high-deforma-

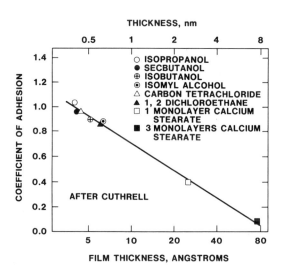

Figure 9. Effect of organic film thickness on adhesion.

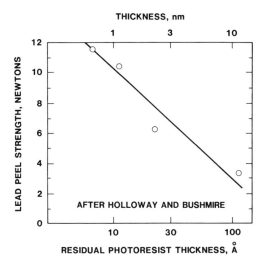

Figure 10. Effect of photoresist contamination on the thermocompression bonding of gold.

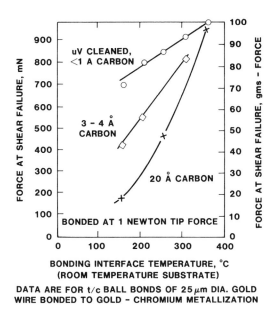

Figure 11. Effect of atmospheric (lab air) contamination on the thermocompression bonding of gold (after Jellison[12]).

tion cold welding of aluminum, copper, nickel, and chromium was studied by Baranov[28] and Semenov.[29] Polar contaminants (such as water, ethanol, oleic acid) were found to impede welding much more than nonpolar materials (e.g., heptane and petrolatum). Baranov also found that polar contaminants adhere more tightly to chromium and copper than to nickel.

The effect of mechanical properties of organic films on solid state welding has not been systematically studied and, in individual experiments, is difficult to separate from chemical effects. Experiments by Jellison involving thermocompression gold ball bonding to thin-film gold have shown that a strongly bonded photoresist impedes solid state welding more than comparable or even greater thicknesses of contaminants adsorbed from the atmosphere.[30] Wetzel indicates that strongly cross-linked organic films exhibit a threshold effect on welding.[31] These hard layers prevent welding at lighter loads, but at higher loads these fragment somewhat like inorganic film layers so that welding becomes satisfactory. Wetzel's observation implies that at high deformations highly cross-linked organics might actually be less detrimental than weaker, more mobile organics.

Whether synergistic effects involving oxide film and organic contaminants exist has not been determined. Excluding charge effects, organic materials are more weakly adsorbed on oxides than on metals. This weak adhesion can facilitate removal of contaminants. However, some investigators[7, 14, 32] suspect that an organic that is more weakly adsorbed on the metal oxide than on the metal itself might be particularly troublesome because it would be expected to have high mobility on the oxide until it became readsorbed on the higher-energy nascent metal surface created by deformation. Vaidyanath and Milner[14] suggest that high threshold deformations are required for welding because adsorbed contaminants quickly transfer to expose nascent metal at lower deformations. Consequently, the small areas of freshly contaminated metal become nearly as difficult to weld as those remaining covered with oxide.

4.3. Particulate Contaminants

Normally, contaminants are in the form of surface films, as has just been discussed. However, particulate contaminants also must be considered as they have been known to prevent welding. A particle on a surface can prevent intimate contact over an area much larger than its own cross section. The effect of particulate contaminants depends on their hardness, the hardness of the metallic surface, the amount of deformation, and the number of particles. Tylecote et al.[4] studied the effect of alumina powder on the deformation welding of tin. They found that alumina was far less detrimental to welding than grease and that the effect of alumina powder was small at high deformations where the particles became embedded in the metals (see Figure 12). Similar observations were made by Williamson et al. in their study of the effect of carborundum

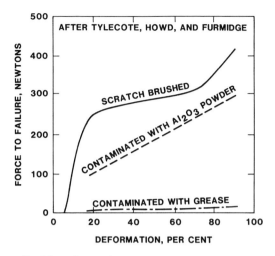

Figure 12. Effect of contaminants on the deformation welding of tin.

powders and polyethylene powders on the welding of gold.[33] They observed that the soft polyethylene particles were compressed, thus permitting greater metal contact than that for carborundum powders.

5. Mechanisms for Elimination of Surface Barriers during Solid-State Welding

As discussed earlier, welding is attained when surfaces are brought into intimate contact with no barriers (nonmetals) between faying surfaces. Theoretically and in the laboratory these barriers can be removed prior to welding. Although precleaning of surfaces is common in industrial welding processes, surfaces are seldom clean in the atomic sense and oxide films are generally unavoidable. Therefore, contaminant barriers must be eliminated during the welding process itself.

How detrimental contaminants are depends on the process and metals involved, because these factors govern which mechanisms are available for elimination of surface films. Deformation welding is generally impaired by organic films, whereas inorganic films constitute the major barrier to diffusion welding. Depending on the amount of surface extension, the level of surface shear stresses, the temperature, and the ability of the metal to assimilate surface films by diffusion, precleaning may only involve wiping the surfaces with a dry cloth[28] or, at the other extreme, require procedures capable of reducing films to less than a monolayer in thickness.[12] At very low levels of plastic deformation, as in the case of adhesion tests, a monolayer of oxygen may be sufficient to prevent welding.[5] However, the limiting oxide film thicknesses on most metals (typically 2 to 10 nm) generally do not prevent welding at the high levels of deformation employed in most deformation welding processes. Of course, the required deformation can be significantly reduced if oxide layers are not present. In conclusion, when considering the mechanisms which overcome surface barriers to welding, we are mostly concerned with the effect of organic contaminants on deformation welding processes and of oxide films on diffusion welding processes. Generally, surface films are thought to be disrupted mechanically in the case of deformation welding, whereas for diffusion welding, thermal mechanisms disperse the oxide film. For convenience, the following sections will discuss the various mechanisms for elimination of surface barriers for diffusion welding and deformation welding. It should be noted that these distinctions are not very precise, and mechanisms which are discussed under deformation welding may also play a role in diffusion welding and, conversely, mechanisms which are discussed with regard to diffusion welding may also play a role in deformation welding.

5.1. Thermal Mechanisms Occurring during Diffusion Welding

Thermal mechanisms are involved in diffusion welding in both the mating of surfaces and the elimination of contaminants. The mating of surfaces involves diffusion-controlled creep, diffusion-controlled sintering, and evaporation and condensation. It is obvious that these mechanisms, whose principal role results in the mating of surfaces, also may be involved in the elimination of contaminants. For example, during evaporation the contaminant films may also be eliminated.

If surfaces are atomically clean to begin with, the mechanisms just mentioned are sufficient to result in complete welding. Of course, in the real world, contaminants are invariably present and, depending on the materials being welded, mechanisms must exist for the dispersion of contaminants away from the faying surfaces. Metals that have a high solubility for interstitial contaminants such as oxygen can easily accommodate removal of these contaminants from the faying surfaces by assimilation into the base metal by volume diffusion. Thus, the surface is decontaminated during welding by diffusion and short-range interatomic forces can then operate to produce welds across the interface. Metals such as titanium, copper, iron, zirconium, niobium, and tantalum fall into this class and are the easiest to diffusion weld.

Other thermal mechanisms are involved in both diffusion welding and deformation welding. These mechanisms may operate during precleaning steps or may be important during the actual heating of parts in the process of welding.

As already mentioned, the principal mechanisms for destruction of oxide films during diffusion welding is assimilation of the films into the bulk metal by diffusion. Bryant has reviewed available data on hot isostatic welding (gas pressure welding) and has shown that the weldability can be predicted on the basis of oxygen solubility, free energy of formation of the most stable oxide, and the mechanical properties.[34] Bryant's graphical summary of hot isostatic welding data is given in Figure 13. He has refined the common rule of thumb that diffusion welding should be performed at a homologous temperature of one-half or higher. For metals such as copper, silver, and molybdenum, whose oxides are only moderately stable (free energy of formation, 100 kcal/mol or greater) or metals with high solubility for oxygen (greater than 1 atomic percent) a homologous temperature of approximately one-half is adequate. He calls these Group I metals. At the other extreme, aluminum, which has a very stable oxide and negligible solubility for oxygen, requires a homologous temperature greater than 0.9 for diffusion welding (Group 3 metals). Bryant classifies metals with intermediate oxide stabilities (e.g. Co, Fe, Ni, and Mn) as Group 2 metals.

As has been shown by Kinzel,[35] Fine *et al.*,[36] and Ham,[37] the time required for dissolution of the oxides can be readily calculated, if the diffusion constants are available. The perhaps surprising result of such a calculation is that if the temperature is increased significantly above a homologous temperature of one-half for metals such as iron, copper, and titanium, the required diffusion time becomes only a few minutes. This results, of course, from the

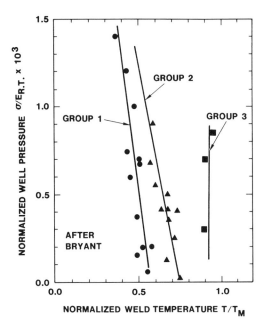

Figure 13. Minimum pressure–temperature conditions for hot isostatic pressure welding.

fact that diffusion rates are highly temperature dependent. The practical result is that much or all of the oxide is actually eliminated by diffusion during processes that might normally be thought of as hot deformation welding.

Another way in which oxide films are eliminated is by reaction with carbon from the substrate metal. Ham has calculated reaction times for reduction of molybdenum and iron oxides.[37] In 1 min, 1 μm, 7 μm, and 100 μm of oxide can be removed from molybdenum at 673 K, alpha iron at 773 K, and gamma iron at 1183 K, respectively.

Finally, the possibility of the reduction of the oxide by a reducing atmosphere should not be ignored. For metals with moderately stable oxides such as iron and copper, addition of a reducing atmosphere to the process is entirely feasible, and often more effective than a vacuum or inert atmosphere. A pitfall of this approach is that an atmosphere which is reducing at welding temperatures may actually be oxidizing the same metal during heating. Therefore, the heating rate and time at welding temperatures must accommodate the kinetics of the oxidation and reduction processes.

5.2. Deformation Welding

5.2.1. Mechanical Mechanisms

On the basis of both theoretical considerations and experimental results, Ahmed and Svitak[38] summarized the fundamental parameters that influence deformation welding as follows: (1) a centrally located nonplastic zone may exist where welding is difficult, (2) the magnitude of the normal stress above that required for intimate contact is unimportant, (3) mutual extension of two normally clean surfaces in intimate contact beyond some threshold deformation is a sufficient condition for welding, (4) the required surface extension is greatly reduced by simultaneous application of an interfacial shear stress, and (5) neither interfacial sliding nor high interfacial shear stress result in welding unless surface extension occurs. Their experimental results were for gold thermocompression welding where a surface oxide is not encountered. Oxidized surfaces are generally expected to require greater surface extensions. Although confirming studies in other systems must be completed before Ahmed and Svitak's conclusions can be generalized, some of the more important conclusions seem to have wide applicability. In particular, the roles of both friction and geometric restraint in controlling state of stress are fundamental to deformation of metals and, therefore, can be expected to generally control deformation welding.

As noted by Ahmed and Svitak, geometric and frictional restraint may give rise to two conditions that influence solid state welding: (1) a nonuniform distribution of normal stresses (pressure hill) and (2) a nonplastic zone where, although isostatic compressive stresses may be high, the difference in principal stresses is insufficient to cause plastic flow. A phenomenon is often observed in roll welding where, due to the pressure hill effect, insufficient pressure is applied near the edges of the weld. Often a change in the method of pressure application, such as from uniaxial pressure to isostatic pressure, will minimize pressure hill effects. Deformation welding requires both mating of surfaces and extensive surface deformation. It is common for deformation welds to be incomplete due to the presence of a nonplastic zone within a portion of the weld interface. This occurs, of course, at the point of greatest restraint. In the case of lap welding with indenter tools, geometric restraint often creates nonplastic zones near the edges of the indenters. Conversely, in butt welding, a nonplastic zone is often centrally located due to frictional restraint.

In general, there are five stages involved in deformation welding: (1) mating of surfaces, (2) fracture of the surface oxide and/or strain-hardened surface, (3) extension of the surface to expose areas of nascent metal, (4) extrusion of the metal through the cracks of the surface layer, and (5) establishment of welds between the nascent metal that is extruded through the surface layer. Because plastic deformation does not occur uniformly along a faying surface, various stages of the deformation welding process may occur simultaneously at different locations. Also, for some processes and some materials all five stages may not be required. For example, in the deformation welding of gold, which is free of oxide, contact between nascent metal occurs without the necessity of the fracture of the surface layer, although surface extension is still useful in overcoming the effects of organic surface contaminants. In the case of slide welding, nascent metal surfaces are exposed by a plowing action which minimizes the need for surface extension and extrusion of the metal through fractured surface layers.

As indicated in Section 1, the typical minimum deformation for cold welding is significantly greater than that required for either general mating of surfaces or fracture of brittle surface layers (such as oxides). It is generally believed that these high deformations are required for surface extension and extrusion of metal through gaps in the fractured surface layer. Several mathematical models have been developed on the basis of the amount of overlapping nascent metal produced by lap welding or roll welding. Some of the general observations that have been made and which influence the assumptions necessary in modeling are: (1) Opposing surface layers generally break up independently in the case of lap welding. (2) If surface preparation has eliminated most of the organic contaminants, opposing surface layers generally break up together in the case of roll welding. Consequently, the minimum threshold deformation for welding is often lower for roll welding than lap welding. (3) Interfacial shear displacements reduce the amount of surface extension required for lap or butt welding. Interfacial shear displacements are promoted by dissimilar metal welds or, in the case of lap welds, unequal thicknesses of weld pairs.

However, excessive differences in yield strengths in dissimilar metal welds can result in too little surface extension in the stronger alloy.

The first principal effort to model deformation welding on the basis of the extension of surfaces was that of Vaidyanath et al.[13] for the case of roll welding of anodized aluminum, where opposing oxide layers tend to adhere and, therefore, break up together. They calculated weld strength on the basis of the amount of exposed metal. More recent work by Wright et al.[39] modifies the Vaidyanath calculations on the basis of their observation that some exposed metal surfaces still do not contribute to welding even though they are free of the oxide layer. These metal surfaces appear to be surfaces which are covered with organic contaminants. They suggest that threshold deformations required for welding consist not only of the amount of deformation required to expose significant oxide-free areas but also an additional amount needed to expose areas which are free of organic contaminants. Work by Gulyaev and Skoblova suggests that the application of mathematical modeling to deformation welding must critically consider the actual process conditions, as these govern which assumptions apply.[9] That is, some experimental verification of the assumptions may be required to support the model. For example, they showed, in contrast with assumptions made by earlier workers, that opposing oxide layers can be at least partially welded to each other and contribute to weld strength if welding is done immediately following wire brushing. On the other hand, they showed that the assumptions which had been made by Vaidyanath regarding opposing oxide layers breaking up as units are invalid for the case where the oxide layers are significantly contaminated with organic materials. Mohamed and Washburn[6] have calculated the strength of aluminum lap welds on the basis of thin oxide layers fragmenting independently.

As indicated earlier, Ahmed and Svitak[38] observe that the simultaneous application of interfacial shear stresses with normal stresses reduces the required surface extension. Stapleton evaluated the role of interfacial shear in the deformation welding of gold and found that better welds could be obtained at 7% surface extension in the presence of high interfacial shear than at 25% surface extension in the absence of interfacial shear stresses.[40] Both Anderson[41] and Ahmed and Svitak[38] believe that interfacial shear results in higher localized deformation leading to mixing of contaminants.

Another factor which has been considered in deformation welding is interfacial sliding or relative movement between surfaces.[1,2,37,42–45] The general observation was that much stronger welds could be made for the same degree of bulk deformation. The effect of interfacial sliding and welding is thought to be analogous to the plowing effect observed in friction experiments, where contaminants are pushed aside.[24] Tylecote has commented that interfacial sliding would be expected to be more important for oxidized surfaces than for oxide-free surfaces.[46] The maximum benefit of sliding is generally achieved within sliding distances of 8 mm or less.

5.2.2. Thermal Mechanisms

Mechanisms which change the properties of, reduce the amount of, or eliminate organic contaminants can be very effective in improving both deformation welding and diffusion welding. As discussed earlier, melting of organic compounds significantly increases friction between metal surfaces and, therefore, would be expected to aid deformation welding.

Evaporation and dissociation of organic contaminants can markedly improve weldability. Investigators of adhesion[10, 19, 24, 47] have often emphasized the inability to obtain good adhesion on surfaces exposed to high temperature or high vacuum. These observations hold for conditions of low deformation because a monolayer of a chemisorbed contaminant can prevent adhesion. However, the significant reduction in the thickness of contaminant layers brought about by dissociation and/or evaporation greatly aids welding processes employing high deformation. This point is illustrated by the weldability of gold. Under conditions of low deformation, gold welds can be very difficult to obtain even for surfaces which have been heated to greater than 1000 K in vacuum.[2, 48, 49] Tylecote et al. showed that a one-minute heat treatment at 1173 K was required in order to cold weld gold at room temperature with 30% deformation.[4] Conversely, Jellison found that atmospherically contaminated gold surfaces could be restored to excellent weldability with regard to thermocompression welding by heating to 500 K in air.[12] Auger electron spectroscopy revealed that a residual carbon layer of approximately 0.2 nm remained which would be sufficient to inhibit welding under low-deformation conditions, but not for the high deformations encountered in thermocompression ball bonding (typically 70%). Where complete desorption on heating is possible, as probably is the case for some organics adsorbed on metal oxides, even greater improvements in weldability would be expected. Vaidyanath and Milner found that the ability to roll weld aluminum was much improved by a prebake and vacuum at 773 K followed by cooling in a desiccator.[14] Since surface analyses were not conducted, whether organic contaminants were desorbed or merely reduced to tolerable levels is not known. It is also possible to dissociate inorganic contaminants such as metal oxides and metal chlorides at temperatures encountered in deformation welding. A very common example is the dissociation of silver oxide, which can occur below 500 K.

Jellison,[12, 30] English and Hokanson,[50] and Hayasaka and Hattori[51] have observed a time-related phenomenon in the welding of gold which cannot be explained on the basis of assimilation of surface films by diffusion, because the solubilities of oxygen, carbon, and hydrogen in gold are negligibly small. These workers each observed that faying surfaces that were only partially welded by mechanical disruption of surface films would continue to grow in strength as a function of time at modest elevated temperatures (423 to 648 K), whether or not an external load was applied. In some cases weld strengths increased by

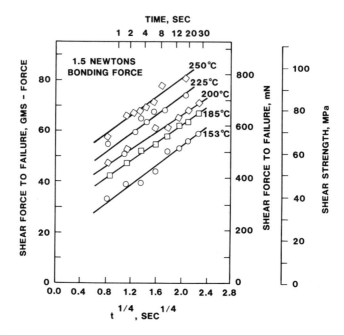

Figure 14. Time–temperature dependence of the strength of gold ball bonds made to photoresist contaminated gold.

nearly an order of magnitude following initial deformation (see Figures 14 and 15). As proposed earlier by Jellison,[12] the growth of the metal–metal weld interface during the second stage of welding or during postheating appears analogous to a sintering phenomenon. The metal–metal interfaces grow at the expense of the higher-energy metal–contaminant interfaces. In contrast with most

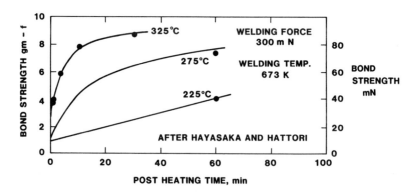

Figure 15. Increase in gold beam lead bond strength on postheating.

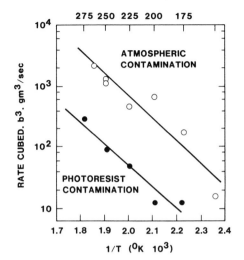

Figure 16. Arrhenius relations for the increase in strength of thermocompression gold ball bonds during postheating.

sintering processing, however, the growth of gold–gold welds appears to be far more stress dependent than temperature dependent. The rate of weld growth with an external load applied was found to be several times that without external loads. Also, within experimental accuracy, the rate of weld growth under an external load does not appear to be temperature dependent. Conversely, the growth process was found to be thermally activated in the absence of an external load (see Figures 16 and 17). In plotting Figure 17, a parabolic rate law is

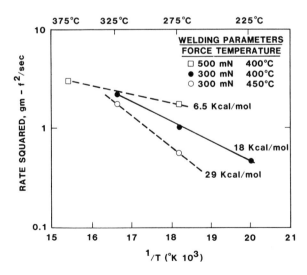

Figure 17. Arrhenius relations for the increase in strength of beam lead bonds during postheating.

assumed, which seems more consistent with Figure 15 than the linear rate law assumed by Hayasaka and Hattori. The increased rate dependence on temperature in the absence of external load is consistent with a stress-assisted thermally activated process such as low-temperature creep of gold[52] and other face-centered cubic metals,[53] where the activation energy decreases with increasing stress.

In summary, a sintering phenomenon appears responsible for the growth of welds following initial deformation. Whether this phenomenon is due to stress-assisted thermally activated plastic flow, as proposed by Jellison, bulk diffusion as assumed by Hayasaka and Hattori, or surface diffusion as suggested by Tylecote[46] requires further experimental and theoretical clarification. In any case, the process appears to be a potentially important way of improving weld strength.

6. Surface Preparation

Several good reviews have been written on surface preparation for solid-state welding.[1,2,7] The consensus of these reviews is that some form of surface preparation is generally required prior to welding. The amount and type of surface preparation required depend on the welding process to be employed. Diffusion welding often requires flatter surfaces than deformation welding but removal of contaminants may not be as important because contaminants can be assimilated by diffusion or reaction. The degree to which organic contaminants must be removed prior to deformation welding depends on the amount of surface extension, interfacial shear, and interfacial sliding.

The surface preparation methods that have been employed can generally be classified as solvent cleaning, chemical pickling, mechanical, and atmospheric techniques. Solvent cleaning based on degreasing solvents such as trichloroethylene or detergents may be sufficient in some cases. Solvent cleaning normally leaves some residual contaminants but for processes employing high deformation or high temperatures, the residual level of contamination may be tolerable. In addition to adsorption of the solvent, the solvent may react with the metal to form a compound, as in the case of chlorinated solvents with copper.

For many common forms of solid-state welding, such as diffusion welding, lap welding, butt welding, and roll welding, solvents are used in the removal of organic contaminants. Vapor degreasing and ultrasonic cleaning are commonly employed. In the case of ultrasonic cleaning, detergents are used to remove the more polar contaminants; detergent cleaning is followed by water rinses and a final alcohol rinse to promote drying.

Solvent cleaning adequately removes organic contaminants for processes involving high surface extension, such as multiple upset butt welding. However, the residual contaminants left from solvent cleaning are detrimental to many processes. For cold roll welding and lap welding, the most common procedure for removal of residual contaminants is to follow solvent cleaning by scratch brushing with rotary stainless steel wire brushes. Scratch brushing results in thermal desorption of organic contaminants. Typical wire diameters for rotary wire brushing are $\frac{1}{4}$–$\frac{1}{3}$ mm. Surface speeds of 10 to 20 m/s are common. Scratch brushing also results in a strain-hardened surface layer which helps control the fragmentation of surface oxides. Parts should be welded immediately following scratch brushing to avoid recontamination from the atmosphere, which has been observed to occur in less than 10 min. Vaidyanath and Milner[14] have shown that following degreasing by scratch brushing is far more effective than degreasing following scratch brushing (see Figure 18). Following solvent cleaning by dry machining to expose virgin metal also provides weldable surfaces but not as effectively as scratch brushing (also see Figure 18).

Aside from the ability of machining and scratch brushing to remove contaminants, the effect of the surface roughness they create must be considered. In diffusion welding of high-strength alloys, excessive surface roughness can prove detrimental by preventing complete mating. Generally, however, some roughness is desirable. If significant interfacial sliding occurs, rough surfaces aid in the plowing of contaminants. Also, since the magnitude of shear stresses are orientation dependent, a rough surface assures that part of the surface is oriented so that significant interfacial shear stresses exist.

Processes in which parts were scratch brushed or surface machined and welded in vacuum (by rolling or lap welding) are characterized as those which require very low deformations for welding (5 to 10%).[8,54,55] The drastic effect

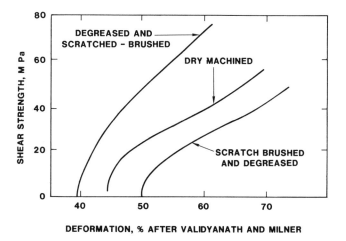

Figure 18. Effect of surface preparation on deformation welding.

of machining in vacuum is believed to be related to the avoidance of both surface oxides and recontamination by organic contaminants.

Chemical pickling used by itself is generally not very effective as a surface preparation with one possible exception: that of preparation for diffusion welding. Tylecote[2] indicates that it is apt to produce hydrated oxides which are even more detrimental to welding than the thin oxides formed following machining. However, chemical pickling may be necessary to remove heavy oxide scales from iron and copper base alloys. Chemical pickling, followed by heating (such as vacuum baking),[14,28] affords better weldability. A somewhat related process, electrochemical polishing, has occasionally been employed with a high degree of success for both deformation welding and diffusion welding. For example, Albright[56] found electropolishing to be highly effective for the diffusion welding of copper. In the same study they found that chemical pickling with hydrochloric acid was only modestly effective, resulting in only about 20% of the faying surface being welded.

In still another related process some workers have used electrodeposition to produce weldable surfaces. For example, Spurgeon et al.[57] found that by electroplating 410 stainless steel with copper they could obtain a very weldable surface because of the low stability of copper oxide at elevated temperatures. In this process, the parts were first cleaned anodically, in effect by electropolishing; then, by reversing the current in the same bath, the parts were electroplated with a thin layer of copper. During subsequent handling the copper surface layer became oxidized, but the oxide was easily reduced in hydrogen or vacuum during welding.

Other methods of desorbing or dissociating organic contaminants such as vacuum baking or hydrogen firing[2] have been used, but recontamination from air atmospheres generally is a problem unless welding immediately follows.

More recently, processes such as UV/ozone cleaning[12] and RF plasma cleaning[58] have been introduced. These processes have a potential of reducing organic contaminants to nondetectable levels,[12,27,30,59,60] but must be used with care because they can also promote oxide film growth on metals. Consequently, they have proved most suitable on gold and gold–platinum surfaces.

In general, atmospheric cleaning approaches for removal of organic contaminants, particularly simple hydrocarbons, depend on the oxidation of the organic materials. This normally yields volatile products such as carbon monoxide and carbon dioxide. A caution must be noted here, for the same processes which result in oxidation of hydrocarbons may yield insoluble, nonvolatile residues of contaminant materials, which may then be very difficult to remove by any other means. UV radiation, which is generated in RF plasma cleaning and UV/ozone cleaning, will break organic chemical bonds, which then permits atomic and ionic oxygen species to oxidize the volatile hydrocarbon by-products. Also, hydrogen plasmas may[61] be used to clean surfaces, and may be particularly desirable for those metals where an oxidizing plasma would produce an undesirable amount of metal oxide. A somewhat similar approach is to use low-energy hydrogen ion bombardment for surface cleaning. Another form of plasma cleaning is to use reactive plasma cleaning with a species such as HCl to remove both hydrocarbons and oxide layers.

7. Concluding Remarks

The amount of contamination control required for reliable solid state welding depends on both the processes and metals involved. For most diffusion welding processes, solvent cleaning is adequate. Solvent cleaning followed by vacuum baking or rotary wire brushing to promote desorption of organic contaminants has been proved to be highly effective for deformation welding processes. Processes such as friction welding, ultrasonic welding, and multiple upset butt welding which result in very high deformation at the weld interface can tolerate more contaminants, and thus require less stringent surface preparation. In fact, some production welding using these processes involves as little as simply wiping the surfaces to be welded with a dry cloth. Conversely, many small parts used in the electronics industry, for which deformation is restricted by geometry, may require processes such as UV/ozone cleaning and RF plasma cleaning to obtain adequately clean parts.

The ability to eliminate contaminant layers either by precleaning or during welding is also dependent on the metal involved. Some metals, such as aluminum and gold, possess little ability to assimilate contaminants by diffusion, whereas others, such as iron and zirconium, readily dissolve contaminants. In deformation welding, the ductility of the metal is extremely important, since ductility determines the ability to fragment oxide films, extrude the metal through the gaps in the oxide, and deform the metals sufficiently to displace other contaminants.

In conclusion, for successful solid-state welding one must consider the properties of the materials being joined in order to select the appropriate process. Then for contamination control one must consider both the selected process and the metals involved.

References

1. D. R. Milner and G. W. Rowe, Fundamentals of solid phase welding, *Metal. Rev.* 7, 433–480 (1962).
2. R. F. Tylecote, *The Solid Phase Welding of Metals*, St. Martin's Press, New York (1968).
3. R. F. Tylecote, Investigations on pressure welding, *Br. Welding J. 1*, 117–135 (1954).
4. R. F. Tylecote, D. Howd, and J. E. Furmidge, Influence of surface films on pressure welding of metals, *Br. Welding J. 5*, 21–38 (1958).
5. D. H. Buckley, LEED and Auger Studies of Effect of Oxygen on Adhesion of Clean Iron

(001) and (011) Surfaces, NASA TND 5756, Lewis Research Center, NASA, Cleveland, Ohio (1970).
6. H. A. Mohamed and J. Washburn, Mechanism of solid state pressure welding, *Welding J. 54*, 302s–309s (1975).
7. W. D. Ludemann, A Fundamental Study of the Pressure Welding of Dissimilar Metals through Oxide Layers, Lawrence Radiation Laboratory Report UCRL-50744, Livermore, Calif. (1969).
8. K. Tanuma and T. Hashimoto, Effect of surface treatment in high vacuum on solid phase weldability, *Trans. Natl. Research Institute Metals 14*, 123–133 (1972).
9. A. S. Gulyaev and L. V. Skoblova, The quality of the joint in cold roll welded clad metal, *Automatic Welding 31*, 36–39 (1978).
10. W. P. Gilbreath, in: *Adhesion or Cold Welding of Materials in Space Environments*, STP No. 431, pp. 128–148, American Society for Testing and Materials, Philadelphia, Pa.
11. D. H. Buckley, Influence of Chemisorbed Films on Adhesion and Friction of Clean Iron, NASA TND 4775, Lewis Research Center, NASA, Cleveland, Ohio (1968).
12. J. L. Jellison, Effect of surface contamination on the thermocompression of gold, *IEEE Trans. Parts, Hybrids and Packaging PHP-11*, 206–211 (1975).
13. L. R. Vaidyanath, M. G. Nicholas, and D. R. Milner, Pressure welding by rolling, *Br. Welding J. 6*, 13–28 (1959).
14. L. R. Vaidyanath and D. R. Milner, Significance of surface preparation in cold pressure welding, *Br. Welding J. 7*, 1–6 (1960).
15. S. B. Ainbinber and F. F. Klovoka, Adhesion of Metals and Their Stability under Complex Stresses, Latv. PSB Zinat Akad. Vestis (1953), No. 4, pp. 113–116.
16. J. R. Whitehead, Surface deformation and friction of metals at light loads, *Proc. Roy. Soc. A201*, 109–124 (1950).
17. R. Wilson, Influence of oxide films on metallic friction, *Proc. Roy. Soc. A212*, 450–452, 480–482 (1952).
18. J. A. Donelan, Industrial practice in cold pressure welding, paper presented at a conference organized by Univ. of Birmingham, *Br. Welding J. 6*, 5 (1959).
19. W. P. Gilbreath, The Influence of Gaseous Environments on the Self-adhesion of Metals, NASA Report, NASA TND 4868, Moffett Field, Calif. (1968).
20. F. P. Bowden and W. R. Throssell, Adsorption of water vapour on solid surfaces, *Proc. Roy. Soc. A209*, 297–308 (1951).
21. J. G. Vaughn and M. K. Raut, Tin contamination during surface cleaning for thermocompression bonding, in Proceedings 1984 International Symposium on Microelectronics, Dallas, Tex. (1984), pp. 424–427.
22. D. H. Buckley, Interaction of Methane, Ethane, Ethylene, and Acetylene with an Iron (001) Surface and Their Influence on Adhesion Studied with Leed and Auger, NASA TND 582, Lewis Research Center, NASA, Cleveland, Ohio (1970).
23. D. H. Buckley, Absorption of Ethylene Oxide and Vinyl Chloride on an Iron (001) Surface and Effect of These Films on Adhesion, NASA TND 5999, Lewis Research Center, NASA, Cleveland, Ohio (1970).
24. F. P. Bowden and D. Tabor, *The Friction and Lubrication of Solids*, Clarendon Press, London (1954).
25. E. Rabinowicz, *Friction and Wear of Materials*, John Wiley and Sons, New York (1965).
26. R. E. Cuthrell, Quantitative Detection of Molecular Layers with the Indium Adhesion Tester, Sandia Labs Report SC-DR-66-300, Albuquerque, N.M. (1966).
27. P. H. Holloway and D. W. Bushmire, Detection by Auger electron spectroscopy and removal by ozonization of photoresist residues, in: Proceedings of the Twelfth Reliability Physics Symposium Las Vegas, Nev. (1974), pp. 180–186.
28. I. B. Baranov, Cold Welding of Plastic Metals, OTS: 63-21785, Joint Publication Reserve

Service: 19103 (1963), English translation, available from Office of Technical Services, U.S. Government Printing Office.
29. A. P. Semenov, Phenomenon of seizure and its investigation, *Wear* 4(1), 1-9 (January/February, 1961).
30. J. L. Jellison, Kinetics of thermocompression bonding to organic contaminated gold surfaces, *IEEE Trans. Parts, Hybrids and Packaging PHP-13*, 132-137 (1977).
31. F. H. Wetzel, Discussion on paper by P. F. Bowden, in: *Adhesion and Cohesion* (P. R. Weiss, ed.), p. 143, Elsevier, TP968 S95 (1962).
32. J. K. Nesheim, The effects of ionic and organic contamination on wirebond reliability, in: Proceedings of the 1984 International Symposium on Microelectronics, Dallas, Tex. (1984), pp. 77-78.
33. J. P. Williamson, J. B. P. Greenwood, and J. Harris, The influence of dust particles on the contact of solids, *Proc. Roy. Soc. A237*, 560-573 (1956).
34. W. A. Bryant, Method for specifying hot isostatic pressure welding parameters, *Welding J.* 54, 433s-435s (1975).
35. A. B. Kinzel, Adam's Lecture—Solid phase welding, *Welding J.* 23, 1124-1144 (December, 1944).
36. L. Fine, C. H. Maak, and A. R. Ozanich, Fundamentals affecting the bond in pressure welds, *Welding J.* 25, 517-529 (1946).
37. J. L. Ham, Mechanisms of surface removal from metals in space, *Aerospace Eng.* 20-21, 49-54 (1961).
38. N. Ahmed and J. J. Svitak, Characterization of gold-gold thermocompression bonding, in: Proceedings of the 25th Electronic Components Conference, IEEE, Washington, D.C. (1975), pp. 52-63.
39. P. K Wright, D. A. Snow, and C. K. Tay, Interfacial conditions and bond strength in cold pressure welding by rolling, *Metals Technol.* 5(1), 24-31 (Jan. 24, 1978).
40. R. P. Stapleton, Surface elongation and interfacial sliding during gold-gold thermocompression bonding, Master's Thesis, Department of Metallurgy, Lehigh University, Bethlehem, Pa. (1974).
41. O. L. Anderson, Role of surface shear strains in adhesion of metals, *Wear* 3, 253-273 (July-August, 1960).
42. F. P. Bowden and G. W. Rowe, The adhesion of clean metals, *Proc. Roy. Soc. A233*, 429-442 (1956).
43. E. Holmes, Influence of relative interfacial movement and frictional restraint in cold pressure welding, *Br. Welding J.* 6, 29-37 (1959).
44. V. W. Cooke and A. Levy, Solid phase bonding of aluminum alloys to steel, *J. Metals 1*, 28-35 (1949).
45. J. L. Knowles, High Strength, Low-Temperature Bonding of Beryllium and Other Materials, Lawrence Radiation Laboratories Report, UCRL-50766, Livermore, Calif. (1970).
46. R. F. Tylecote, Solid phase bonding of gold to metals, *Gold Bull.* 11, 74-80 (1978).
47. K. L. Johnson and D. V. Keller, Jr., Effect of contamination on the adhesion of metallic couples in ultrahigh vacuum, *J. Appl. Phys.* 38, 1896-1904 (1967).
48. J. L. Jellison, Solid Phase Welding of Transition Members of Radioisotope Thermoelectric Generators, Sandia Laboratories Report, SAND75-0053, Albuquerque, N. Mex. (1975).
49. R. F. Tylecote, Investigations on pressure welding, *Br. Welding J. 1*, 117-135 (1954).
50. A. T. English and F. L. Hokanson, Studies of bonding mechanisms and failure modes in thermocompression bonds of gold plated Ti-Au metallized substrates, in: Proceedings of the Ninth Annual Reliability Physics Symposium, Las Vegas, Nev. (1971), pp. 178-186.
51. T. Hayasaka and S. Hattori, On the mechanism of thermocompression bonding of gold beam leads, *Review Electronic Communication Laboratory 23*, 344-352 (1975).

52. R. R. Herring and M. Meshii, Thermally activated deformation of gold single crystals, *Metal. Trans. 4*, 2109–2114 (1973).
53. K. A. Osipov, *Activation Processes in Solid Metals and Alloys*, American Elsevier Publishing Company, Inc., New York (1964).
54. R. M. Brick, Hot roll bonding of steel, *Welding J. 49*, 440s–444s (1970).
55. R. Blickensderfer, Bonding of titanium and molybdenum to iron by vacuum rolling, *Thin Solid Films 54*, 342–343 (1978).
56. C. E. Albright, Solid State Bonding, SME-AD75-853, Society of Mechanical Engineers, Dearborne, Michigan (1975).
57. W. M. Spurgeon, S. K. Rhee, and R. S. Kiwak, Diffusion bonding of metals, *Bendix Technical J.—Materials and Processing, 2*(1), 24–29, (1969).
58. J. A. Weiner, G. V. Clatterbaugh, H. K. Charles, and B. M. Romenesko, Gold ball bond strength effects of cleaning, metallization, and bonding parameters, in: Proceedings of the 1983 Electronic Components Conference, Orlando, Fla. (1983), pp. 208–220.
59. D. F. O'Kane and K. L. Mittal, Plasma cleaning of metal surfaces, *J. Vac. Sci. Technol. 11*, 567–569 (1974).
60. K. L. Mittal, Surface contamination: An overview, in: *Surface Contamination: Genesis, Detection and Control* (K. L. Mittal, ed.), pp. 3–45, Plenum Press, New York (1979).
61. D. M. Mattox, Substrate preparation for thin film deposition—A survey, *Thin Solid Films 124*, 3–10 (1985).

10

Surface Contamination and Contact Electrification

K. P. Homewood

1. Introduction

When two materials are touched together and then separated, we find, in general, that charge has transferred between them. This phenomenon is known as contact electrification or contact charging. It is most apparent when one or both of the contacting materials is an insulator, because of the ability of an insulator to retain charge. Contact charging is one of the oldest studied phenomena in physics but is still incompletely understood. Despite this, considerable use is made of these effects in industrial processes. It is the basis of xerography, electrostatic precipitators, electrostatic paint and crop spraying, and ink jet printing and is made use of in a wide variety of other applications. Contact charging can also be a considerable nuisance, causing irritating electric shocks in the home and office environment. Many semiconductor devices are highly susceptible to damage from acquired static charges. The buildup of static charges and their subsequent discharge can be extremely hazardous in flammable and explosive environments.

The present state of understanding of contact electrification (of solids) is summarized in a recent review by Lowell and Rose-Innes.[1] Further information can be found in the proceedings of the various electrostatics conferences.[2-4]

The contact charge on insulators resides either on surface states, which may be intrinsic or extrinsic, or bulk states which happen to be near the surface. Indeed, recent experiments on poly(methyl methacrylate) (PMMA) show that, at least in this material, the charging is limited to the top 0.4 nm of the poly-

---

K. P. Homewood • Joint Laboratory of Physics and Electrical Engineering, University of Manchester Institute of Science and Technology, Sackville Street, Manchester 1, United Kingdom. *Present address:* Department of Electronic and Electrical Engineering, University of Surrey, Guildford, Surrey GU2 5XH, United Kingdom.

mer.[5] It follows that contact charging is primarily a surface phenomenon, and it is for this reason that the effect of surface contamination has to be carefully considered. In the early study of contact electrification, the subject acquired a reputation for being one in which experiments gave irreproducible and non-quantifiable results; much of this may be attributable to the effects of surface contamination. Charged surfaces are particularly prone to particulate contamination (the charge creates a field which polarizes the dust particle which it then attracts). For this and other reasons it has been common practice in recent work on contact electrification to conduct experiments in vacuum (primarily to prevent discharge of the sample by sparking), and a general improvement in reproducibility seems to have occurred. I do not intend, therefore, to say any more in this chapter about the effects of such "gross" contamination. I also intend to concentrate on the more recent work in this field. Earlier work on the problem is covered in the book *Contact and Frictional Electrification* by Harper.[6]

2. Types of Contaminants

2.1. Adsorbed Molecules

The question may be asked whether contact electrification experiments carried out in the vacuums used in most experiments (i.e., pressures ~ 10^{-5} torr) have any real significance, because at this pressure, surfaces will be covered by adsorbed molecules (at a pressure of 10^{-6} torr, enough molecules hit the surface to cover it with a monolayer in one second, were they all to stick). Consequently, the contact charging may be largely governed by what molecules happen to be adsorbed on the surface. It has been pointed out[7,8] that an "ideal" insulator should not charge on contact with metals. Consequently, it might be that the contact charging of insulators is due to extrinsic surface states associated with molecules adsorbed onto the surface of the insulator. If this were true, it would have serious implications for many theories of contact electrification. The fact that it is possible to obtain consistent results on specimens prepared in different ways suggests that adsorbed molecules do not always dominate contact charging. Nevertheless, the effect, if any, of adsorbed molecules needs to be investigated, if only to answer criticism.

Lowell[9] has used a simple but ingenious technique to assess the importance of extrinsic surface states on the contact charging of polymers. The technique relies on the fact that contactors of different shape will deform a soft polymer differently when contacted. Lowell compared contacts to a soft, flat polymer surface made by a metal sphere and a metal cone (see Figure 1). He pointed out that if a large, hard sphere is pressed against a flat polymer surface, then the area of contact is approximately equal to the projected surface area πr^2, where r is the radius of the contact area. For a conical contactor, however,

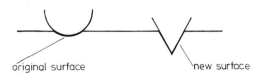

original surface   new surface

Figure 1. Diagram illustrating how a conical contactor touches new surface whereas a spherical contactor touches largely the original surface.

the area of mutual contact, between the polymer and the metal, is greater than the projected area of the indentation by a factor given by the cosecant of the cone's half-angle. Consequently, when contact is made between a metal sphere and a polymer flat, one would expect that most of the contact area will be original surface. But when contact is made between a metal cone and the polymer flat, then at least some of the contact area must be "new" polymer surface. It follows that if extrinsic surface states play a role in the contact charging of a polymer, then different contact charge densities will be obtained when that polymer is touched by a spherical and a conical metal contactor. Lowell used this technique to investigate a number of polymers. His experiments used metal spheres of 1-cm diameter and metal cones of half-angle 30°. With this geometry, at least half of the polymer surface in contact with the metal cone must be "new" surface (as cosec 30° = 2).

Some of Lowell's results for poly(tetrafluoroethylene) (PTFE) and polyethylene (PE) are shown in Figure 2. The contact charge densities obtained on PTFE are the same for spherical and conical contacts, whereas for PE the contact charge density obtained with a spherical contactor is much greater than when the conical contactor is used. The conclusion is that extrinsic surface

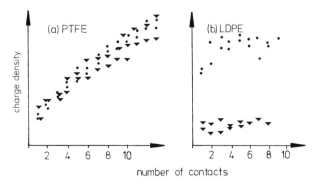

Figure 2. Contact charge densities obtained when a polymer is touched by a metal sphere (circles) and a metal cone (triangles): (a) PTFE; (b) LDPE. After Lowell (from Reference 9).

states are unimportant in the contact charging of PTFE but that they play a dominant role in the contact electrification of PE.

Homewood[10] has carried out the direct experiment of investigating the contact charging of "new" surfaces of insulators freshly formed under ultrahigh vacuum (UHV) conditions. The experiment consisted of preparing an atomically clean surface of an insulator, contacting the surface with a metal, and then measuring the charge transferred from the metal to the insulator. This charge is then compared with that obtained on the same surface after exposure to air. To prepare the atomically clean surface and maintain its cleanliness, the experiments were conducted in an ultrahigh vacuum system. The system was maintained at a pressure of better than 10^{-10} torr during the experiment. At a pressure of 10^{-10} torr, it takes about three hours for a monolayer of gas to be adsorbed, assuming a sticking probability of one. The initial clean surface of the insulator was prepared by cleaving the specimen in the ultrahigh vacuum. Cleaving rather than cutting was preferred as a means of preparing the surface of the insulator. Soft specimens could be made brittle prior to cleaving by lowering their temperature.

Experiments were conducted on the polymer PTFE, the contactor being an aluminum sphere. The newly cleaved polymer surface was found to possess a small residual surface charge density, but this was only about 1.0% of the charge density transferred to PTFE when contacted by a metal and so had a negligible effect on the experiment. The fresh surfaces of PTFE were contacted with the metal sphere in a time short compared with the time for a monolayer of gas molecules to adsorb on the surface at the system pressure used.

The results for PTFE are shown in Figure 3a in the form of histograms of the contact charge density obtained, for a number of contacts, when the polymer was contacted with aluminum; each contact was to different spots on the insulator. Figure 3b shows the results of the same experiment on the same surfaces after exposure to air for about twenty-four hours. There is no significant difference between the charge densities obtained in the two cases. The charge density obtained is typical of that achieved in more conventional apparatuses at more usual pressures.[11]

The results indicate that molecules adsorbed from the air onto the surface have no significant effect on the contact charging of PTFE by metals; this implies that in this material, charge transfer is to states "intrinsic" to the polymer.

Preliminary results on PE indicate a difference in the contact charging of fresh surfaces formed in UHV and the same surfaces after exposure to clean air. For materials in which adsorbed molecules do play a role in their contact charging, this experiment could provide a means of identifying the molecule responsible—by selected exposure of the fresh surface to the suspected contaminant followed by measurement of the contact charge.

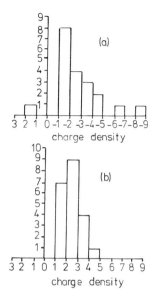

Figure 3. Histograms of contact charge densities transferred to PTFE: (a) on surfaces freshly cleaved in UHV; (b) on the same surfaces after exposure to air. After Homewood (from Reference 10).

2.2. Ionic Contamination

A form of contamination that has to be carefully considered with regard to contact charging is ionic contamination because clearly this is charged. It has been suggested by several authors[7, 12-16] that contamination of the surfaces of insulators by ions may contribute to contact charging (because transfer of ions from one surface to the other could be an explanation for contact charging). In particular, Kornfeld[15, 16] has invoked the transfer of ions as the primary mechanism for the contact electrification of insulators. Kornfeld proposes that, in general, solid insulators possess a net intrinsic internal charge, and as the result of exposure for some time to the atmosphere, this charge is compensated by adsorbed ions, the total amount of adsorbed contamination depending on the "history" of the sample. He then suggests that during contact with another material some or all of the charged contamination transfers to the contactor, leaving a net charge on the insulator.

It is particularly important to assess the contribution, if any, of adsorbed ions to the charging, as it is common practice[1] in contact charging experiments to discharge the insulator, prior to an experiment, by exposure to ions produced by a radioactive source or an electrical discharge. The mechanism of discharge has not been closely investigated and is not well understood. Indeed, it is not known whether the insulator is neutralized by being coated with ions of opposite sign to the surface charge, or whether the neutralizing ions accept electrons from the insulator and then leave the surface as neutral molecules. The difference in these two alternative mechanisms could clearly be of immense importance if surface ions play an important role in contact charging.

The experiments of Homewood[10] described in the previous section are relevant to this problem. Firstly, when the polymer was cleaved in UHV it was found to be only negligibly charged, and secondly, the contact charging experiments were made on surfaces that had not been discharged. Furthermore, the charge density obtained was the same as that found in a "normal" experiment. Clearly, in a vacuum of 10^{-10} torr, the pressure used, ionic contamination of the surfaces could not occur during the period of the experiment, and it follows that the contact charging of PTFE cannot be the result of ionic transfer.

Other authors have also investigated the effect of the adsorption of ions on the contact electrification of insulators. Robins et al.[14] have investigated the contact charging of pyroelectric insulators. Pyroelectrics are materials that have a permanent dipole moment (below their critical temperature). Under normal atmospheric conditions, the surfaces of the pyroelectric will have become covered by a dense layer of compensating ions. The pyroelectrics investigated were poly(vinylidene fluoride) (PVF_2) and triglycine sulfate (TGS). The experiment consists of comparing the contact charging of the pyroelectric insulator when polarized and exposed to ions with the contact charging of the same insulator that has been polarized but not compensated by ion exposure. For PVF_2 the contact charging is similar in both cases, and the conclusion is that ion transfer is not the mechanism for charge transfer in PVF_2. For TGS, Robins et al. found that there was a difference in the contact charging of the compensated and the uncompensated insulator and concluded that compensating ions, on the surface of this material, can influence the contact charging but not in a way consistent with the ion transfer model.

2.3. Adsorbed Water

One of the most important contaminants that has to be considered in electrification phenomena is adsorbed water. This is particularly true under "real" (normal atmospheric) conditions and in industrial environments. Water can affect the contact charging of insulators in two ways. Firstly, it can affect the contact charging (generation) process itself, and secondly, and of more general importance, it can affect the dissipation of charge on the insulator surface. It is not usually possible to prevent the occurrence of contact charging, and so static nuisances and hazards are normally controlled by ensuring, as far as possible, that the charge can be neutralized or dissipated.

Adsorbed water increases the rate of dissipation of charge from an insulator surface by decreasing the surface resistivity of the material. The term resistivity is used rather loosely, as it is strictly only applicable to ohmic materials. In "dynamic" charging situations, the resistivity of a material is probably the most important parameter in determining the maximum levels to which charge will build up on a material (the rate of generation is clearly as important but is not usually alterable).

A good and common example, of considerable importance, is web/roller contact charging in which an insulating film passes over a metal (usually grounded) roller. As the insulating film passes over the roller, contact charge exchange will occur between the roller and the film. If the film is infinitely resistive, then the contact charge density on the film will be equal to the transferred contact charge density at the roller. However, if the film possesses some finite conductivity, then some of the charge can flow back to the grounded roller (the field is such as to attract the charge) and the charge density on the film away from the roller will be less.

This situation has been thoroughly analyzed[17, 18] and verified experimentally.[19] For any particular value of conductivity, there is a critical velocity above which charging will occur but below which there is no charging. This critical velocity is greater, the higher the conductivity of the film. Clearly, if charging is to be avoided, a process that reduces the resistivity of insulators (without affecting their other properties) is desirable. It is found that the surface resistivity of many insulators is reduced as they adsorb water[20, 21]; the reduction in the resistivity is often found to vary, at least approximately, as the exponential of the relative humidity, and the drop can be dramatic. It is often sufficient, especially with natural fibers, just to increase the humidity of the environment to avoid static problems.

It should also be noted that many antistatic coatings applied to plastics are surfactants that rely on subsequent adsorption of water to lower the surface resistivity, and they can become ineffective in low-humidity conditions. For readers interested in pursuing the points raised in this section, References 22–27 provide a sample of the available literature.

3. Identification of Extrinsic Traps

When extrinsic surface states are shown to be playing a role in the contact charging of a polymer, it is clearly of interest to identify the impurity responsible. It has been frequently noted that the contact charging of low-density polyethylene (LDPE) increases when fresh surfaces of the polymer are exposed to the atmosphere.[9, 28, 29] The change in the contact charging of LDPE can be considerable; Figure 4, taken from Hays,[28] shows an order of magnitude increase in the contact charging of branched (low-density) polyethylene when exposed to laboratory air for twenty hours.

By means of a number of "elimination" techniques, Hays[28, 29] decided

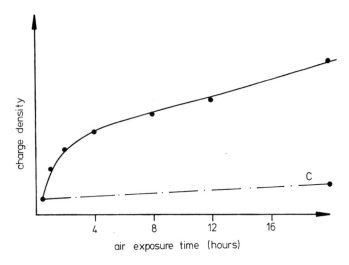

Figure 4. Contact charge densities obtained when mercury is contacted against LDPE which has been exposed to laboratory air. The data denoted by C were obtained on control samples retained in a container. After Hays (from Reference 28).

that the surface modification was caused by a trace gaseous component of the air, and he suspected ozone to be the contaminant responsible. He also found that if PE was exposed to air near a xenon lamp or corona source (both of these are known to produce ozone), the contact charge increased very rapidly, reaching in a few minutes the same value that would have taken a day in the absence of these ozone producers. However, a number of other reactive species are produced by these sources, and so Hays constructed an ozone generator. When exposed to the ozonizer, the contact charging of the LDPE increased considerably after only a few seconds. Hays used infrared (IR) spectroscopy to identify the active surface centers produced by the ozone. Additional absorption bands appeared in the infrared spectrum of ozonized PE and the bands due to unsaturated vinyl groups vanished. The new bands were associated with ozonide and carbonyl groups.

Much of this work has been confirmed by the experiments of Selders et al.[30] on corona-treated PE. If the surface state density after exposure to ozone is dependent on the number of unsaturated bonds in the original surface, then more highly saturated polyethylenes should be much less influenced by ozone exposure or by exposure to the atmosphere. This is indeed found to be the case.[29,30]

Several workers have observed changes in the contact charging of insulators exposed to corona sources and have suggested that this is good evidence for attributing the mechanism of contact charging to ion transfer. However, Bauser,[31] who reports an increase in the contact charging of corona-treated pyrene, found that the contact charging still depended on the metal's work function. A dependence of the charging on work function almost certainly means that charging is due to electron transfer.[32] It seems, therefore, that changes in the contact charging of corona-treated insulators are due to an increase in the surface state density of electron-accepting or electron-donating centers, caused by exposure of the surfaces to reactive molecular species present in the corona discharge.

It might be thought that the modern surface analysis techniques, such as X-ray photoelectron spectroscopy (XPS), ultraviolet photoelectron spectroscopy (UPS), Auger electron spectroscopy (AES), secondary ion mass spectrometry (SIMS), and others, could be used to investigate the role of surface contamination in contact electrification. Unfortunately, insulators pose problems for these techniques. In particular, secondary emission of electrons causes the insulator to charge up; this can seriously reduce the emission efficiency and badly distort the spectra in most of these techniques. XPS is least affected by surface charging, as the secondary electrons are emitted at sufficiently high energies to be little influenced by charge on the sample. It is, therefore, probably the most useful technique presently available for the study of insulator surfaces. However, the contact charge densities normally obtained on insulators only correspond to about one active site for every ten thousand surface atoms, and most of these techniques, including XPS, fall far short of this sensitivity.

Nevertheless, XPS has been used to investigate some aspects of contact electrification phenomena. In particular, Salanek *et al.*[33] have, using XPS, investigated the role of material transfer in contact electrification. They show that considerable material is transferred when a polymer is contacted by a metal or another polymer, typically 10^{-2} monolayers (of a polymer). Transfer of such large amounts of material between contacting surfaces means that its role in contact charging must be considered. Experiments by Lowell[34] show that despite the large amounts of material transfer occurring on contact, this is not the primary mechanism of contact electrification. He points out that when contacts are repeated to the same spot, with discharging between contacts, the contact charging is the same on each contact; but most material transfer occurs at the first contact and much less on subsequent contacts.

A fuller discussion of modern surface analysis techniques in the study of contact electrification is given in a review by Briggs.[35]

4. Effect of Contamination on the Metal

Most of the fundamental investigations of contact electrification have involved touching the insulator with a metal. We should, therefore, consider the effect of the contamination on the metal. The property of the metal of primary interest in contact charging is its work function or contact potential. It has been shown that the adsorption of even a fraction of a monolayer of contamination can cause large shifts in a metal's work function.[36] Published values of work function measurements on metals are made on carefully cleaned or prepared surfaces, whereas in contact electrification experiments no special precautions are usually taken to clean the metal. It is, therefore, important when a work function dependence of the contact charging is suspected that the work function of the metal is measured during the experiment and preferably *in situ*. Published values of work functions may differ substantially from that of the metal actually used in an experiment.

5. Conclusion

The serious investigation of contact electrification of insulators is still to some extent at the "data-gathering" stage. Consequently, certain experiments may have been made only once and often on a limited number of materials. As a result, it is difficult to be sure just how important and widespread the effects of contamination are on the contact charging of insulators. It is, however, clear that contamination is not the primary cause of the phenomenon as there are materials, for example, PTFE, in which extrinsic surface states play no part. Nevertheless, when extrinsic states are important, as in LDPE, their effect can be considerable. Those materials in which extrinsic surface states are significant, and in which the contaminant can be identified, are likely to be useful for the understanding of contact electrification in the more general case.

References

1. J. Lowell and A. C. Rose-Innes, Contact electrification, *Adv. Phys.* 29(6), 947–1023 (1980).
2. Proceedings of Conferences on Static Electrification, *Institute of Physics Conference Series*, No. 11 (1971), No. 27 (1975), No. 48 (1979), No. 66 (1983).
3. Proceedings of the 4th International Conference on Electrostatics, The Hague (1981), *J. Electrostat.* 10 (1981).
4. Proceedings of the 5th International Conference on Electrostatics, Uppsala (1985), *J. Electrostat.* 16 (1985).
5. K. P. Homewood, An experimental investigation of the depth of penetration of charge into insulators contacted by a metal, *J. Phys. D* 17, 1255–1263 (1984).
6. W. R. Harper, *Contact and Frictional Electrification*, Oxford University Press, Oxford (1967).
7. H. Bauser, W. Klopffer, and H. Rabenhorst, On the charging mechanism of insulating solids, *Adv. Static Elec.* 1, 2–9 (1970).
8. G. A. Cottrell, C. Reed, and A. C. Rose-Innes, Contact electrification of ideal insulators: Experiments on solid rare gases, in: *Static Electrification, Institute of Physics Conference Series*, No. 48, 249–256 (1976).
9. J. Lowell, Surface states and contact electrification of polymers, *J. Phys. D.* 10, 65–71 (1977).
10. K. P. Homewood, Do 'Dirty' surfaces matter in contact electrification experiments? *J. Electrostat.* 10, 299–304 (1981).

11. K. P. Homewood, An experimental investigation of the contact electrification of insulators by metals, Ph.D. Thesis, University of Manchester, England (1981).
12. P. S. H. Henry, Generation of static on solid insulators, *J. Text. Inst.* **48**, 5–25 (1957).
13. E. S. Robins, A. C. Rose-Innes, and J. Lowell, Are adsorbed ions involved in the contact charging between metals and insulators?, in: *Static Electrification, Institute of Physics Conference Series, No. 27*, 115–121 (1975).
14. E. S. Robins, J. Lowell, and A. C. Rose-Innes, The role of surface ions in the contact electrification of insulators, *J. Electrostat.* **8**, 153–160 (1980).
15. M. I. Kornfeld, Nature of frictional electrification, *Soviet Phys. Solid State* **11**, 1306–1310 (1969).
16. M. I. Kornfeld, Frictional electrification, *J. Phys. D* **9**, 1183–1192 (1976).
17. T. Horvath and I. Berta, The effective location of eliminators in the electric field of moving sheet materials at conducting rollers, in: Proceedings of the International Conference on Static Electricity, Grenoble (1977), p. 32(a).
18. T. Horvath and I. Berta, Mathematical simulation of electrostatic hazards, in: *Static Electrification, Institute of Physics Conference Series, No. 27*, 256–263 (1975).
19. J. F. Hughes, A. M. K. Au, and A. R. Blythe, Electrical charging and discharging between films and metal rollers, in: *Static Electrification, Institute of Physics Conference Series, No. 48*, 37–44 (1979).
20. J. S. Forrest, Methods of increasing the electrical conductivity of surfaces, *Br. J. Appl. Phys.* **4**, Suppl. 2, S37–39 (1957).
21. Y. Awakuni and J. H. Calderwood, Water vapour adsorption and surface conductivity in solids, *J. Phys. D* **5**, 1038–1045 (1972).
22. G. W. Brundrett, A review of the factors influencing electrostatic shocks in offices, *J. Electrostat.* **2**, 295–315 (1976/1977).
23. S. P. Hersh, Review of electrostatic phenomena on textile materials, *Dechema Monographs* **72**, 199–216 (1974).
24. J. E. McIntyre, Antistatic fibers, *Rep. Prog. Appl. Chem.* **59**, 99–108 (1974).
25. E. L. Zichy, Antistatics for plastics, *Dechema Monographs* **72**, 147–161 (1974).
26. A. R. Blythe, Device for controlling static charge levels on film, in: *Static Electrification, Institute of Physics Conference Series, No. 27*, 238–245 (1975).
27. J. Boyd and D. Bulgin, The reduction of static electrification by incorporating viscose rayon containing carbon, *J. Text. Inst.* **48**, 66–99 (1957).
28. D. A. Hays, The effect of oxidation and an electric field on the contact electrification of polyethylene by mercury, *Dechema Monographs* **72**, 95–103 (1974).
29. D. A. Hays, Contact electrification between mercury and polyethylene: Effect of surface oxidation, *J. Chem. Phys.* **61**, 1455–1462 (1974).
30. M. Selders, F. K. Dolezalek, O. Frenzl, and H. Rabenhorst, Contact electrification of corona treated polyethylenes, *J. Electrostat.* **10**, 315–320 (1981).
31. H. Bauser, Static electrification of organic solids, *Dechema Monographs* **72**, 11–28 (1974).
32. D. K. Davies, The generation and dissipation of static charge on dielectrics in a vacuum, in: *Static Electrification, Institute of Physics Conference Series, No. 4*, 29–36 (1967).
33. W. R. Salanek, A. Paton, and D. T. Clark, Double mass transfer during polymer–polymer contacts, *J. Appl. Phys.* **47**, 144–147 (1976).
34. J. Lowell, The role of material transfer in contact electrification, *J. Phys. D* **10**, L233–235, (1977).
35. D. Briggs, The role of modern surface analysis techniques in understanding electrification phenomena, in: *Static Electrification, Institute of Physics Conference Series, No. 48*, 201–213 (1979).
36. J. M. Chen, Mechanism of work function reduction by oxygen adsorption, *J. Appl. Phys.* **41**, 5008–5011 (1971).

11

Surface Contamination and Biomaterials

BUDDY D. RATNER

1. Introduction

Synthetic materials for medical applications (biomaterials) are widely used in the United States today. The types of materials used to fabricate medical implant devices include plastics, elastomers, fibers, metals, carbons, ceramics, glasses, and composites. The extent of the use of synthetics in medicine is emphasized by the numbers of implants itemized in Table 1. Although high success rates are generally realized with these implants, there is much room for improvement. It is reasonable to surmise that an enhanced understanding of the surface of these implants, the region directly interfacing with the body, can lead to improvements in performance of these devices.

With the development of an understanding of the surface chemistry and surface structure of implants will come an understanding of the surface contamination on them. Surface contamination, defined in this context as the fouling of a surface with an undesired or unexpected chemistry, has not received great attention from biomaterials researchers. This is surprising since there is a small but significant body of literature which would suggest that a consideration of surface contamination is important in the preparation and study of biomaterials.[1]

From the perspective of the biomaterials scientist, two aspects of surface contamination must be considered. First, exactly what are the unintended or undesirable substances on the surface of a material, and second, what importance do these have for the biological response to the material. In this article, a number of important papers dealing with biomaterials and surface contami-

---

BUDDY D. RATNER • National ESCA and Surface Analysis Center for Biomedical Problems (NESAC/BIO), Center for Bioengineering and Department of Chemical Engineering, BF-10, University of Washington, Seattle, Washington 98195.

Table 1. Biomedical Applications of Synthetic Polymers

Ophthalmologic	
Intraocular lenses	500,000[a]
Contact lenses	2,000,000[a]
Retinal surgery implants	50,000[b]
Prostheses after enucleation	5,000[b]
Cardiovascular	
Vascular grafts	250,000[a]
Heart valves	75,000[a]
Pacemakers	100,000[b]
Blood bags	30,000,000[a]
Reconstructive	
Breast prostheses	100,000[b]
Nose, chin	10,000[b]
Penile	5,000[b]
Other devices	
Ventricular shunts	21,500[b]
Catheters	200,000,000[a]
Oxygenators	500,000[a]
Renal dialyzers	16,000,000[a]

[a] Numbers indicate devices used in Western countries and Japan in 1981 (from Reference 32).
[b] Numbers represent approximate annual usage in U.S. (from Reference 33).

nation will be reviewed, and the significance of surface contamination for the performance of implants will be discussed.

2. General Principles of Surface Contamination

For the purposes of this chapter, it is worthwhile to consider some general concepts related to surface contamination. The most important general principle is that surfaces are different in composition and organization from the bulk. Differences result from adsorption and retention of components from the surrounding environment, surface molecular orientation, and surface reaction. Adsorption of material from the environment constitutes the major source of surface contamination. However, the other factors can be important in contributing to contamination and, therefore, will be discussed.

The primary driving force for surface contamination, particularly where adsorption, retention, and molecular orientation are concerned, is a reduction of interfacial energy. Perfectly clean metal, ceramic, or glass surfaces will be recontaminated with hydrocarbons, water, and other adsorbed gases extremely rapidly. Polymers have significantly lower surface free energies than metals or inorganics. Consequently, the driving force to reduce interfacial energy is lower, which may result in a slower contamination rate for these materials. However, contamination will still occur on most polymeric surfaces. For example, consider poly(tetrafluoroethylene) (PTFE), a polymer with a low surface energy of 18.5 dyn/cm. Theoretically, only a substance with a surface energy lower than that of PTFE should be able to adsorb to it to contaminate its surface. There are few substances (e.g., fluorosurfactants) which can reduce the air–PTFE interfacial energy by adsorption. Still, electron spectroscopy for chemical analysis (ESCA) studies inevitably show trace amounts of hydrocarbon contamination on presumably clean PTFE surfaces (Figure 1). Such contamination may be associated with topographic defects (e.g., crevices) on the surface or with oxidation of other non-PTFE materials at the surface. However, the general principle is illustrated. All surfaces are subject to contamination.

The presence of contamination has special implications for biomaterials testing and application. In surface physics studies and for some microelectronic preparations, ultrahigh vacuum conditions and ion etching can be employed to

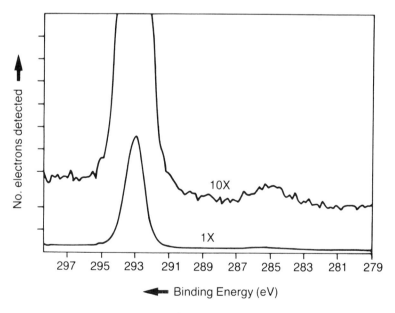

Figure 1. C1s ESCA spectrum of a clean poly(tetrafluoroethylene) film surface. A 10× magnification of the y-axis clearly shows the presence at 285.0 eV of a small amount of hydrocarbon contamination.

produce a surface almost free from contamination. However, the surgeon, physician, or biomaterials scientist is constrained to work in an atmospheric pressure environment with specimens which must have, as emphasized above, certain amounts of contamination.

Therefore, two important points must be made about surface contamination on biomaterials. First, since there will always be some level of surface contamination, we must learn to live with it. Second, the level of contamination should be minimized. Thus, in a study of the effectiveness of various techniques for cleaning glass,[2] relatively stable glass surfaces could be prepared in a laboratory environment with surface carbon/silicon ratios (as measured by ESCA) ranging from 0.19 to 2.2. Glass with a C/Si ratio of 2 would be unnecessarily contaminated, while glass with a C/Si ratio of 0.2 would be as clean as can be readily achieved under reasonable working conditions. Clearly, the lower the ratio, the more desirable the glass surface becomes for studies of biological interactions with glass. The biomaterials scientist should be responsible for reducing unintentional contamination to the lowest possible levels and for insuring that all specimens or devices have reproducible (low) levels of contamination.

Also, consider that even though glass, platinum, and poly(ethylene terephthalate) might all be contaminated with an apparently similar layer of hydrocarbon-like material from the atmosphere, the essential properties characteristic of these three substances manifest themselves at their surfaces. "Clean" glass which has at least one monolayer of organic carbon material at its surface is "glass-like" and not "hydrocarbon-like" (e.g., polyethylene-like) in its interactions with proteins and cells.[2] The mechanisms by which the properties of a specific substance are propagated to the surface through contaminant films are not completely clear. Still, for moderately clean materials, the intrinsic properties of the material are apparent at the surface and can be exploited to study or influence biological systems.

3. Review of Biomaterials Contamination Studies

With these general surface contamination principles now set forth, it is worthwhile to proceed with a review of the scientific literature which attempts to deal with surface contamination on biomaterials and the biological implications of this contamination. There are, unfortunately, relatively few papers published on this topic.

3.1. Cleaning Agent Residues

In one of the earliest formal studies which considered contamination and its effects on bioreactivity, the processing of cross-linked poly(dimethyl siloxane) (silicone rubber) parts for heart-assist devices was closely scrutinized.[3] The manufacturer's procedure suggested soap solutions for cleaning parts formed from the molding resin (specifically, Ivory Soap was recommended). Therefore, experiments were performed to observe the efficiency of a pure soap, sodium stearate, in cleaning silicone rubber. The sodium stearate was found to be effective in removing impurities from the silicone rubber surface. However, gravimetric measurements indicated that the sodium stearate itself was strongly retained by the polymer. Even after exhaustive water washing, approximately 1 mg/cm^2 of sodium stearate still remained on the specimen. The nature of the adsorption kinetics suggested a mechanism whereby the soap first adsorbed to the surface and then diffused into the bulk.

The magnitude of the sodium stearate pickup was found to be strongly affected by the conditions under which the silicone rubber was cured. It was suggested that 2,4-dichlorobenzoic acid (2,4-DCBA), a decomposition product of the polymerization catalyst, was responsible for these differences. This substance may have been extracted from the silicone rubber simultaneously with the adsorption of stearate. Thus, the net weight increase attributed to stearate uptake appeared lower. The amount of 2,4-DCBA available for extraction was a function of the cure conditions, accounting for the pickup difference from polymer to polymer.

The significance for blood compatibility of the sodium stearate pickup was also addressed. The amount of stearate retained by a thoroughly rinsed specimen was considered to be sufficient to activate Factor XII, a blood coagulation protein believed to be responsible for contact-activated coagulation. However, no direct relationship between the presence of retained sodium stearate and blood compatibility was demonstrated. The authors did attempt to relate the presence of surface silica filler particles which were not completely coated by poly(dimethyl siloxane) to the tendency of the polymer to induce localized thrombosis.

More recently, another cleaning agent, sodium dodecyl sulfate (SDS), was also found to induce contamination during the "cleaning" of biomedical device surfaces.[4] Some manufacturers have been routinely "cleaning" poly(methyl methacrylate) (PMMA) intraocular lenses (IOLs) with SDS after polishing. Electron spectroscopy for chemical analysis (ESCA) showed the presence of a thin film, invisible by microscopic examination, that was apparently SDS (Figure 2). Using ESCA as a monitoring tool, a new wash procedure utilizing sodium hydroxide solution and sodium bicarbonate solution was developed that produced lens surfaces for IOLs indistinguishable from the surface of highly purified PMMA control specimens. ESCA also indicated that the NaOH solution would not induce any measurable degree of hydrolysis in the PMMA surface.

In the same study, a number of IOLs (various brands) rejected by an alert surgeon immediately after removal from their sterile packaging and just prior

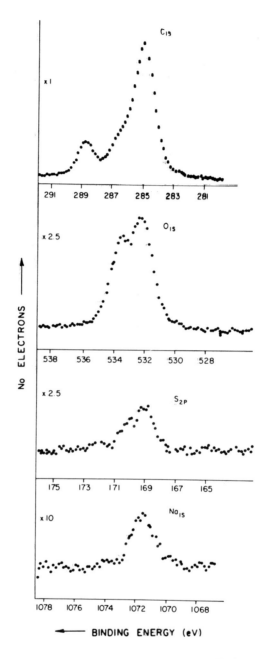

Figure 2. ESCA spectra of the surface of an intraocular lens. The increased hydrocarbon component of the C1s spectrum, the 169-eV position of the S2p peak (indicative of sulfate groups), and an Na/S ratio of 1 are all suggestive of sodium dodecyl sulfate contamination.

to implantation were examined by scanning electron microscopy (SEM) and energy-dispersive X-ray analysis (EDXA). Many particles, crystalline deposits, and smears were found on the lens surfaces. The compositions of these particles and surface debris were indicative of lint, SDS, polishing compounds, salt residues, and metal. Gross contamination of this sort, particularly on medical devices intended for humans, represents an inexcusable level of carelessness. Recommendations were offered for better quality control, filtered wash liquids, and a clean-room manufacturing and packaging environment.

In another IOL study,[5] gamma-ray sterilization of lenses hermetically sealed in vials of balanced salt solution was shown to possibly induce lens surface contamination. During irradiation, many organic compounds (polar and nonpolar) were formed in the storage solution. The investigators hypothesized that one of these new compounds was adsorbed onto the hydrophobic lens surface and later acted as an irritant after lens implantation.

Conventional steam sterilization[6] has been found to deposit hydrophobic organic and hygroscopic salt contaminants on biomaterial surfaces. Before and after autoclaving, detergent-washed metal plates, well-siliconized metal plates, and stearate-coated specimens were examined using SEM, internal reflection spectroscopy, and contact angle techniques. All results pointed to contamination deposition during the "sterilization" process.

Still another example can be cited of biomaterial surfaces contaminated by the agent intended to clean them. Triton X [poly(ethylene glycol) p-isooctylphenylether] was found on the surface of a Dacron fabric intended for use as a vascular replacement section.[7] The biological implications of this discovery were not discussed. Also, some filters used in biomedical studies contain Triton X and related compounds. The Triton X leached from these filters can contaminate solutions and biomedical devices immersed in those solutions.[8] Triton X exhibits marked tissue toxicity.

3.2. Environmental Contaminants

Two studies have considered the significance for biointeraction of the unavoidable hydrocarbon contamination present on all surfaces under atmospheric pressure conditions.[2, 9] The earlier of these studies addressed the resistance of glass surfaces to thrombus buildup before and after radio-frequency (RF) glow discharge cleaning in argon.[9] The vena cava ring, a short segment of tube implanted in a dog vena cava and monitored for thrombus buildup, was used to evaluate thromboresistance. After argon glow discharge treatment, all Pyrex glass vena cava rings remained unoccluded. Rings cleaned using conventional chromic acid methods all thrombosed in the same time period. Surface analysis in this experiment was performed using contact angle methods, infrared spectroscopy, and electron microscopy. From the results obtained from these analyses, it is impossible to attribute the improved thromboresistance of the glow-discharge-cleaned glass only to the removal of an organic layer. The authors point out that changes in the free radical abundance of the glass surface or changes in the ionic composition may have also affected the blood compatibility of the glass.

In a later study of the cleaning of glass surfaces for cell attachment studies, marked biological effects related to the cleaning process were not observed.[2] ESCA was used to monitor the effectiveness of a number of cleaning treatments on decreasing the surface ratio of carbon to silicon. The most effective cleaning method was UV/ozone exposure.[10] Acid cleaning solutions were also found to be reasonably effective in reducing the value of C/Si. However, different surface treatments showed no significant difference in altering the effectiveness of these surfaces as attachment substrates for Swiss 3T3 fibroblasts. The range of C/Si values studied in this experiment was 0.5–1.8. It is possible that this range was not wide enough to elicit a response (i.e., all surfaces looked like hydrocarbon contamination to the cells). Also, it is possible that the 3T3 cell line used in this experiment is relatively insensitive to the type of surface differences produced. Other researchers have reported differences in cell interaction depending upon how glass surfaces were cleaned and treated.[11,12] More detailed studies coupling thorough surface analysis with careful measurements of cell attachment are warranted to resolve this controversy.

In a study using a native whole blood assay,[13] it was shown that environmental contaminants could be solely responsible for the high blood reactivity of conventional glass bead columns. Columns that were cleaned by heating to 500 or 595°C for 18 h were benign toward flowing native whole blood for all variables measured. This heat treatment at or just above the melting point of glass was chosen to volatilize surface contaminants and to regenerate a fresh surface. ESCA spectra and zeta potential and wettability measurements of the untreated and heat-treated glass verified the effectiveness of this cleaning process.

3.3. Biocompatible Contaminants

Surface contamination can, in some cases, unexpectedly improve biocompatibility.[14–17] For example, Stellite 21, a cobalt–chromium alloy used for the support members in the Starr–Edwards heart valve, displayed high thromboresistance until thoroughly cleaned.[14,15] Contact angle measurements suggest that, as received from the manufacturer, the metallic components of the valve were coated with closely packed methyl groups with surface energies similar to those observed for surface-deposited, oriented, aliphatic fatty acid molecules. When this organic coating was removed by abrasion or flaming (decreased water contact angle), the thrombogenicity of the device increased. The manufacturer reported that the metallic components of the valve were polished with a tallow-based abrasive at the factory. The organic component of this abrasive was apparently lapped into the hard metallic surface, resulting in an abrasion-resistant organic coating responsible for the good performance of the valve in contact with blood.

Another illustration of fortuitous improvement in blood compatibility of synthetic materials due to surface contamination involved polyurethanes. This class of polymers holds great promise for use in long-term, totally implanted artificial hearts because of high resistance to flex-fatigue, excellent tensile strength, the ability to be synthesized with a wide range of properties, ease of fabrication, resistance to biodegradation,* and, finally, good blood compatibility reported in some cases. Extruded Pellethane and Renathane tubing were found, using the baboon arterio-venous (AV) shunt evaluation system, to produce low platelet damage.[16,18] By ESCA, the surfaces of these polymers were found to be predominantly hydrocarbon-like.[17,19] In the C1s ESCA spectra, a small peak in a position suggestive of amide-type groups was observed. This surface chemistry was unexpected for a polyurethane, which would be anticipated to exhibit a substantial polyether-type component (carbon singly bonded to oxygen) and a peak in the carbamate bond position. Upon thorough washing or solvent extraction, a surface of the type expected for a polyetherurethane material was obtained (Figure 3). However, based upon a correlation noted between the percentage of the surface in a hydrocarbon-like environment and platelet damage,[16,18] this surface would be expected to have substantially lower blood compatibility than the highly washed or unextracted polyurethane. Further analysis by gel permeation chromatography and infrared spectroscopy revealed that the material extracted from the polyurethane surface consisted of long-chain amide compounds. In subsequent discussions with the manufacturer, it was learned that such compounds are added to Pellethane to improve extrudability. Thus, the desirable blood compatibility of Pellethane and Renathane can be attributed to a surface contaminant, the extrusion lubricant.

Further analysis of the material extracted from polyurethanes demonstrated the presence of a low-molecular-weight polyether-rich fraction in the polyurethanes.[19,20] This type of material has been found to be surface-active in polyurethanes and can dominate the surface after the extrusion lubricant is removed.[20-23]

Organic silicone polymers are another common type of contaminant found in polyurethanes.[24,25] These polymers concentrate at the air interface of a cast polyurethane film. They are most likely introduced into the polyurethane through silicone-contaminated glassware and reaction vessels used during synthesis or fabrication. Contamination of reaction vessels and glassware readily occurs since

---

*This property of polyurethanes is presently the subject of some controversy. This controversy will not be dealt with in this work.

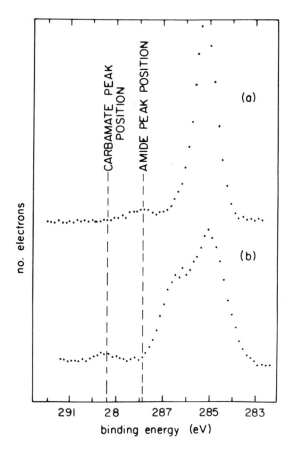

Figure 3. (a) C1s ESCA spectrum of Pellethane 2363-80A extruded tubing prior to cleaning or extraction. Note the predominantly hydrocarbon-like C1s peak and the absence of carbamate peak. (b) After extraction, the C1s peak is more characteristic of a polyurethane and the subpeak indicative of the carbamate linkage is clearly visible.

the silicone polymers are used as vacuum or stopcock greases, pump oil, and general lubricants. Once laboratories or apparatus are contaminated with silicone, it is extremely difficult to remove.

The implications of silicone contamination in polyurethanes are interesting since, in many cases, silicone polymers have been reported to improve the blood compatibility of polyurethane surfaces.[26-30] Other additives which migrate to the surface of polymers have also been intentionally added to polyurethanes to "contaminate" the surface with an overlayer of a desirable composition.[28,31]

4. Conclusions

This chapter has attempted to summarize the literature on surface contamination and its biological implications. Surface contamination has been found to be desirable for biomaterials in certain instances, undesirable in other situations, and, in some cases, without apparent effect on biointeractions. Often, contamination is not monitored in biomaterials studies and in industrial production. Thus, irreproducible results and unexpected consequences are common. The literature relating surface contamination to interactions with biological systems is small, and this is a fertile area for further investigation. The prospects of intentionally "contaminating" surfaces are promising and also worthy of further investigation.

Acknowledgments

The support of N.I.H. grants HL25951 and RR01296 during the preparation of this work is acknowledged. Helpful suggestions from J. D. Andrade have also contributed to this work.

References

1. K. L. Mittal, ed., *Surface Contamination: Genesis, Detection and Control*, Plenum Press, New York (1979).
2. B. D. Ratner, J. J. Rosen, A. S. Hoffman, and L. H. Scharpen, An ESCA study of surface contaminants on glass substrates for cell adhesion, in: *Surface Contamination: Genesis, Detection and Control* (K. L. Mittal, ed.), Vol. 2, pp. 669-686, Plenum Press, New York (1979).
3. E. Nyilas, E. L. Kupski, P. Burnett, and R. M. Haag, Surface microstructural factors and the blood compatibility of a silicone rubber, *J. Biomed. Mater. Res. 4*, 369-432 (1970).
4. B. D. Ratner, Analysis of surface contaminants on intraocular lenses, *Arch. Ophthalmol. 101*, 1434-1438 (1983).
5. D. W. Meltzer, Gamma ray sterilization and its effect on intraocular lenses, *Am. Intra-Ocular Implant Soc. J. 7*, 126-129 (1981).
6. R. E. Baier, A. E. Meyer, C. K. Akers, J. R. Natiella, M. Meenaghan, and J. M. Carter, Degradative effects of conventional steam sterilization on biomaterial surfaces, *Biomaterials 3*, 241-245 (1982).
7. P. Didisheim, M. K. Dewanjee, D. N. Fass, V. Fuster, K. E. Holley, M. P. Kaye, P. E. Zollman, and M. V. Tirrell, Blood Compatibility of Circulatory Assist Devices, Annual Report, June 30, 1981, National Heart, Lung and Blood Institute, National Institutes of Health, Bethesda, Md., 20014, PB 82 1295 37.
8. D. W. Meltzer, R. C. Drews, and A. S. Hajek, Millipore filters in ophthalmic surgery: A caution concerning their use, *Am. Intra-Ocular Implant Soc. J. 7*, 143-146 (1981).
9. R. E. Baier, V. A. Depalma, A. Furuse, V. L. Gott, G. W. Kammlott, T. Lucas, P. N.

Sawyer, S. Srinivasan, and B. Stanczewski, Thromboresistance of glass after glow discharge treatment in argon, *J. Biomed. Mater. Res. 9*, 547–560 (1975).
10. J. R. Vig, UV/ozone cleaning of surfaces: A review, in: *Surface Contamination: Genesis, Detection and Control* (K. L. Mittal, ed.), Vol. 1, pp. 235–253, Plenum Press, New York (1979).
11. L. Smith, D. Hill, J. Hibbs, S. W. Kim, J. Andrade, and D. Lyman, Glow discharge surface treatment for improved cellular adhesion, *Polym. Prepr., Am. Chem. Soc., Div. Polym. Chem. 16*, 186–190 (1975).
12. C. Rappaport, Some aspects of the growth of mammalian cells on glass surfaces, in: *The Chemistry of Biosurfaces* (M. L. Hair, ed.), Vol. 2, pp. 449–487, Marcel Dekker, New York (1972).
13. J. A. Bergeron, J. M. DiNovo, A. F. Razzano, and W. J. Dodds, Non-thrombogenicity of clean glass revealed by native whole blood assay in bead columns, *Thromb. Haemostas. 50*(4), 814–820 (1983).
14. R. E. Baier, R. C. Dutton, and V. L. Gott, Surface chemical features of blood vessel walls and of synthetic materials exhibiting thromboresistance, in: *Surface Chemistry of Biological Systems* (M. Blank, ed.), pp. 235–260, Plenum Press, New York (1970).
15. R. E. Baier, V. L. Gott, and R. C. Dutton, Thromboresistance of Stellite 21: The role of an adventitious waxy contaminant, *J. Biomed. Mater. Res. 6*, 465–470 (1972).
16. S. R. Hanson, L. A. Harker, B. D. Ratner, and A. S. Hoffman, Evaluation of artificial surfaces using baboon arterio-venous shunt model, in: *Biomaterials 1980, Advances in Biomaterials* (G. D. Winter, D. F. Gibbons, and H. Plenk, Jr., eds.), Vol. 3, pp. 519–530, John Wiley and Sons, Chichester, England (1982).
17. B. D. Ratner, ESCA studies of extracted polyurethanes and polyurethane extracts: Biomedical implications, in: *Physicochemical Aspects of Polymer Surfaces* (K. L. Mittal, ed.), Vol. 2, pp. 969–983, Plenum Press, New York (1983).
18. S. R. Hanson, L. A. Harker, B. D. Ratner, and A. S. Hoffman, *In vivo* evaluation of artificial surfaces with a nonhuman primate model of arterial thrombosis, *J. Lab. Clin. Med. 95*, 289–304(1980).
19. B. D. Ratner, ESCA and SEM studies on polyurethanes for biomedical applications, in: *Photon, Electron, and Ion Probes of Polymer Structure and Properties*, ACS Symposium Series (D. W. Dwight, T. J. Fabish, and H. R. Thomas, eds.), Vol. 162, pp. 371–382, American Chemical Society, Washington, D.C. (1981).
20. C. B. Hu and C. S. P. Sung, Surface chemical composition–depth profile of polyether polyurethaneureas as studied by FT-IR and ESCA, *Polym. Prepr., Am. Chem. Soc., Div. Polym. Chem. 21*, 156–158 (1980).
21. K. Knutson and D. J. Lyman, The effect of polyether segment molecular weight on the bulk and surface morphologies of copolyether-urethane-ureas, in: *Biomaterials: Interfacial Phenomena and Applications*, ACS Advances in Chemistry Series (S. L. Cooper and N. A. Peppas, eds.), Vol. 199, pp. 109–132, American Chemical Society, Washington, D.C. (1982).
22. R. W. Paynter, B. D. Ratner, and H. R. Thomas, Polyurethane surfaces—An XPS study, *Polym. Prepr., Am. Chem. Soc., Div. Polym. Chem. 24*, 13–14 (1983).
23. E. W. Merrill, E. W. Salzman, S. Wan, N. Mahmud, L. Kushner, J. N. Lindon, and J. Curme, Platelet-compatible hydrophilic segmented polyurethanes from polyethylene glycols and cyclohexane diisocyanate, *Trans. Am. Soc. Artif. Int. Organs 28*, 482–487 (1982).
24. S. W. Graham and D. M. Hercules, Surface spectroscopic studies of Biomer, *J. Biomed. Mater. Res. 15*, 465–477 (1981).
25. M. D. Lelah, L. K. Lambrecht, B. R. Young, and S. L. Cooper, Physiochemical characterization and *in vivo* blood tolerability of cast and extruded Biomer, *J. Biomed. Mater. Res. 17*, 1–22 (1983).

26. E. Nyilas, Development of blood-compatible elastomers. II. Performance of Avcothane blood contact surfaces in experimental animal implantations, *J. Biomed. Mater. Res. Symp. 3*, 97–127 (1972).
27. E. Nyilas and R. S. Ward, Jr., Development of blood-compatible elastomers. V. Surface structure and blood compatibility of Avcothane elastomers, *J. Biomed. Mater. Res. Symp. 8*, 69–84 (1977).
28. R. S. Ward, Jr., Development of thermoplastics for blood-contacting biomedical devices, *ACS Div. Org. Coat. Plast. Chem., Prepr. 42*, 227–231 (1980).
29. B. C. Arkles, IPN-modified silicone thermoplastics, *Med. Dev. Diagnostics Ind. 5*, 66–70 (1983).
30. B. Ashar, L. R. Turcotte, and L. A. Naturman, Development of a melt-processable copolymer for biomedical devices, *Trans. Soc. Biomat. 5*, 22 (1982).
31. R. S. Ward, P. Litwack, and R. Rodvien, Improved blood compatibility of modified thermoplastics, *Trans. Soc. Biomat. 5*, 46 (1982).
32. J. N. Mulvihill, J. P. Cazenave, A. Schmitt, P. Maisonneuve, and C. Pusineri, Biocompatibility and interfacial phenomena, *Colloids and Surfaces 14*, 317–324 (1985).
33. A. C. Beall, Biomaterials: Magnitude of the need, in: *Contemporary Biomaterials: Material and Host Response, Clinical Applications, New Technology and Legal Aspects* (J. W. Boretos and M. Eden, eds.), pp. 5–9, Noyes Publication, Park Ridge, New Jersey (1984).

12

Redispersion of Indoor Surface Contamination and Its Implications

Eric B. Sansone

1. Introduction

There are many industrial and laboratory processes that can produce contamination. This is true even when processes are partially or totally enclosed or when types of control other than enclosure are used. Worker exposure is possible when contamination is produced and escapes from primary containment. Once contamination has been produced, housekeeping can yield an environment in which exposure can take place by resuspension. Bad housekeeping, by imparting too much energy to the deposited contamination, can lead to resuspension immediately; it can also leave substantial residual contamination that may be reentrained subsequently. Even good housekeeping is likely to leave behind on surfaces residual contamination that may be reentrained. In any event, whatever contamination remains is available for further resuspension. It is this "secondary" exposure, rather than the exposure that occurs when the contamination is initially produced, that we are interested in here.

Surface contamination, whatever its origin, may become airborne and inhaled by workers or casual passersby. It may be transferred by contact with one's hands and from there to food or directly to the mouth and be ingested. It is also possible that surface contamination, however translocated, can penetrate intact skin or contact a wound and result in internal exposure. If portable materials are contaminated, they may be moved to another location, where the possibility of them being contaminated might not be recognized. Inappropriate handling of such materials could lead to exposure of people uninvolved with the contaminating events. Surface contamination can interfere with experiments

---

Eric B. Sansone • Environmental Control and Research Program, NCI-Frederick Cancer Research Facility, Program Resources, Inc., Frederick, Maryland 21701.

or production solely by its existence or by raising background contamination levels. With some types of radioactive contamination (sufficiently energetic beta emitters and gamma emitters) there may be an external radiation hazard from surface contamination. The possibility of exposure and its relationship to various forms of activity was recognized nearly a century ago.[1]

The scope of this review is confined to indoor contamination and its redispersion. A great deal of work has been done under the auspices of the U.S. Atomic Energy Commission (now the Nuclear Regulatory Commission). (A preponderance of the references related to redispersion are based on work with radioactive contaminants. This is due to the sponsorship of the Atomic Energy Commission and to the ease and sensitivity with which radioactive contamination can be detected.) Their interest has centered on radioactive contamination resulting from accidents and its redispersion, usually outdoors. Two representative reviews have been published by Mishima[2] and Sehmel.[3] While our interest is in redispersion of indoor contamination, we will refer to a few studies in which the relationship between indoor and outdoor contamination was investigated so that an estimate of the importance of resuspension indoors could be made. A review of the redispersion of indoor surface contamination has appeared before.[4] However, that work was of narrower scope than this and it is almost ten years old. Two symposia, principally concerned with indoor surface contamination and its redispersion, contain several contributions of interest. The symposia took place in Gatlinberg, Tennessee, in 1964[5] and in Bournemouth, England, in 1966.[6] Another, more recent symposium (1978), was primarily concerned with surface cleaning, characterization of surface cleanliness, and storage of clean surfaces or the kinetics of recontamination.[7]

This chapter does not consider aerosol dynamics, which is important for a full understanding of the factors governing deposition and resuspension of particulate matter. We refer the reader to several texts on this subject.[8-11]

An important point to bear in mind with regard to surface contamination, its redispersion, and its implications is that after a known contaminating incident, one will clean up. However, after an incident of which one is not aware or when there are continuing episodes of low-level contamination, decontamination or cleanup may not occur and surface contamination may accumulate.

2. Measurements of the Redispersion or Resuspension Factor (K)

The relationship between the airborne concentration of material redispersed or resuspended from a surface and the surface concentration of the material is called the redispersion or resuspension factor (K). Mathematically it is expressed as A/S, where A is the airborne concentration and S is the surface concentration of the material of interest. The units of this expression may appear in the literature as meters^{-1}, centimeters^{-1}, or as centimeters^2/meter^3.

Here, no matter how the original author has expressed the resuspension factor, we will present the data in meters^{-1}. The resuspension factor has been experimentally measured in operating facilities and in *ad hoc* experiments. A variety of radioisotopes, chemicals, and microbial materials have been used to measure the resuspension factor. Depending upon the tracer used, concentrations may be expressed in terms of disintegrations per minute, mass, or colony-forming units per unit volume if the material is airborne, and per unit area for surface concentrations. A resuspension factor of 10^{-6} m^{-1} means that of one million units of surface contamination per square meter, one unit per cubic meter becomes airborne.

The earliest work in which the concept of a redispersion or resuspension factor was used was that of Chamberlain and Stanbury.[12] They were interested in the potential hazard from inhaled fission products redispersed by rescue operations following an atomic bomb explosion. In an attempt to quantitate the potential hazard, they made brick and plaster dust radioactive with iodine-131. "Standard rescue operations" were studied, e.g., shoveling and cutting a hole of about two square meters in a work area. Resuspension factors were reported for active work in confined, unventilated spaces and active work in the open. These data were used a few years later by Dunster[13] to infer permissible levels of surface contamination from values of maximum permissible concentrations of radionuclides in air. (These data could also be used in another way: From maximal permissible concentrations of surface contamination and the K factors determined by Chamberlain and Stanbury, permissible levels of radioisotopes in air could be inferred.) Since these first reports, many studies of actual operations and *ad hoc* experiments have been made to yield data on the resuspension factor. The results from 25 reported experiments are summarized in Table 1. Some of the studies are discussed in more detail below.

Becher[17] concluded that surface contamination reflected operational changes or control inadequacies; that high surface contamination usually correlated with high airborne contamination; that surveys for surface contamination yielded a performance yardstick and an indication of control effectiveness; and that ". . . greatest personnel exposure should be related to locations with the most unconfined [contaminants]. . . ."

Jones *et al.*[20] found that "there was a progressive decrease in the resuspension factor with increase in particle size." They also found that the resuspension factor was higher by factor of 9 to 62 when cotton cloth overshoes were used instead of leather shoes. They supposed that scuffing with the cotton overshoes contributed to the increased air concentrations of contaminant.

Tagg[21] measured resuspension factors using both personal and area samplers to determine airborne concentrations under a variety of conditions. In all cases, the resuspension factors calculated from the personal samplers were greater than those calculated from the area samplers. The differences ranged from a factor of 2 to nearly 20.

Table 1. Resuspension Factors Reported in the Literature

Contaminant	Surface contamination measurement	Operating conditions	K (m^{-1})	Remarks	Reference
^{131}I-labeled brick and plaster dust	Total by ratemeter	Active work in open Active work in confined, unventilated space	1.8×10^{-6} 4.3×10^{-5}	Air samples at 4 and 6 ft Bulk of dust < 1 μm	12
UF$_4$ powder	Transferable by 100-cm^2 filter paper wipe	Normal With added ventilation and vibration	$2.5\text{-}19 \times 10^{-5}$ $1.3\text{-}2 \times 10^{-3}$	Air sampled up to 4 ft above floor	14
Ra	Transferable by wipe of about 100 cm^2	Normal	$7 \times 10^{-8}\text{-}1.5 \times 10^{-5}$	20-30-min air samples; data from 10 plants	15[a]
U	Total by ratemeter	Normal; floors vacuumed	$0.2\text{-}5.9 \times 10^{-5}$ $0.5\text{-}14 \times 10^{-4}$	Ore sampling plant Uranium reduction plant	16
U compounds	Transferable by wipe	Normal; 0.5-h air samples Normal; 8-h air samples	5×10^{-4} 3×10^{-5}	Air sampled from breathing zone	17[a]
Be and compounds	Transferable by 1-ft^2 wipe with 15-cm dia Whatman #41 filter paper	Loading and unloading Be blocks Cleaning Be blocks Preparation of Be cyclotron targets Be compound synthesis Warehouse inventory	1.5×10^{-2} $5\text{-}12 \times 10^{-3}$ 7.9×10^{-3} 9.3×10^{-3} 2×10^{-2}	Maximum air and surface contamination values; air sampled from breathing zone Estimated values Estimated values Estimated values	18[a]
U compounds	Transferable by wipe with 4-in square paper towel	Normal	$0.4\text{-}26 \times 10^{-5}$ $0.8\text{-}14 \times 10^{-4}$	8-h air samples 10-min air samples	19[a]
PuO$_2$	Total by ratemeter	No movement 14 steps/min 36 steps/min	$0.1\text{-}5 \times 10^{-8}$ $3.4\text{-}49 \times 10^{-6}$ $4\text{-}49 \times 10^{-5}$	10 air changes/h; data from painted and unpainted concrete, paper, and PVC	20

Material	Method	Activity	Value	Comments	Ref.
Fluorescent powder	Known amounts applied in aqueous propyl alcohol suspension	30 steps/min	$0.8–120 \times 10^{-6}$	9 air changes/h; data from bituminized paper	
α-Emitters	Total by ratemeter	Two walkers, 100 steps/min, contaminated floor	2.2×10^{-4} 5.9×10^{-5}	Personal air samples Area air samples	21
		Two walkers, contaminated clothing	4.9×10^{-4} 3.2×10^{-5}	Personal air samples Area air samples	
Microorganisms	Known amount of organisms applied	Walking	2.3×10^{-3}		22^a
ZnS powder	Not specified	Vigorous work with sweeping	1.9×10^{-4}	Asphalt tile floor; unventilated room	23
		Vigorous walking	3.9×10^{-5}		
		Light work	0.9×10^{-5}		
CuO powder	Not specified	Light work and light sweeping	7.1×10^{-4}	Circulating fans on	
PuO_2	Total by ratemeter	No movement	$0–1.5 \times 10^{-7}$	10 air changes/h; air samples at 3.5 ft; data from paper, PVC, and waxed and unwaxed linoleum floor	24
		14 steps/min	$0.6–20 \times 10^{-6}$		
		36 steps/min	$1.5–18 \times 10^{-5}$		
$Pu(NO_3)_4$	Total by ratemeter	No movement	$0–0.6 \times 10^{-7}$		
		14 steps/min	$0.3–1.3 \times 10^{-6}$		
		36 steps/min	$1–16 \times 10^{-6}$		
α-Emitters	Transferable by 900-cm² filter paper wipe	4–6 people moving; 9 air changes/h; concrete floor	1.2×10^{-4}	Air sampled at 4.5 or at 1.5 and 4.5 ft; K much smaller for waxed linoleum	25
	Total by ratemeter	2–4 people changing contaminated clothes; no ventilation; waxed paper floor	$0.4–8.8 \times 10^{-3}$ "small"–3.9×10^{-3}	Remaining contamination on clothes presumably fixed	
			$2.5–11 \times 10^{-4}$	Using results from personal rather than area air samplers	

(*Continued*)

Table 1. (*Continued*)

Contaminant	Surface contamination measurement	Operating conditions	K (m^{-1}) Pu	K (m^{-1}) U	Remarks	Reference
Pu and U	Transferable by filter paper wipe	Undisturbed	1.0	1.0×10^{-4}	Painted concrete, small room, no ventilation (Pu); concrete, large room, no ventilation (U)	26
		Fans on	13.5	1.3×10^{-4}		
		Fans on and continuous cart movement	97	1.45×10^{-4}		
Microorganisms	Transferable by impression plates	No details given		$1.0 – 5.6 \times 10^{-3}$	Class II or IV clean room; air samples by Andersen sampler	27[a]
Be	Transferable by filter paper wipe	2 men sweeping vigorously		4.2×10^{-4}	Unventilated storeroom with wooden floor; air samples at 3 ft; 90% of particles irrespirable	28
		2 men sweeping vigorously after vacuuming floor		1.0×10^{-2}		
UO$_2$ in ethanol	Total by ratemeter	Undisturbed		$\leq 1.7 \times 10^{-7}$	K decreased with increasing sampler height	29
		60 steps/min		$4.7 – 7.5 \times 10^{-6}$		
UO$_2$ powder		Undisturbed		$\leq 0.7 \times 10^{-7}$		
		60 steps/min		$1.0 – 1.7 \times 10^{-5}$		
Various α and β emitters	Not specified	No work		$\alpha: 2.5 \times 10^{-6}$	20 air changes/h; stainless steel floor	30[a]
				$\beta: 2.0 \times 10^{-6}$		
		Floors rubbed with dry cotton		$\alpha: 2.2 \times 10^{-4}$		
				$\beta: 6.8 \times 10^{-4}$		

Material	Description	Value	Conditions	Ref.
Chrysotile	Transferable by vacuuming	1.2–5.3×10^{-3}	Contaminated laboratory coat	31
$Ba^{35}SO_4$	Not specified	2.0–4.2×10^{-3}	Handling contaminated materials	32
		9×10^{-4}	Floor struck continuously; fan on	33
$^{89}Sr, ^{60}Co$	Known amounts applied in solution	Sr: 1.2×10^{-4} Co: 3.3×10^{-4}	Wind tunnel for 1 h	
			No significant differences between surfaces of steel, painted steel, stainless steel, vinyl plastic, and organic glass; Sr applied to glass as an aerosol gave K of 2.8×10^{-2}	
Be	Known amounts applied in aqueous suspension	0.2–13×10^{-6} 0.01–1.5×10^{-6}	Fan on Fan off	34^a
Ammonium fluoroberyllate	Known amounts applied in solution	0.3–10×10^{-6} 0.08–1.5×10^{-6}	Fan on Fan off	
			Reentrainment from a PVC surface; 1.8-m/sec air flow with fan; K values for surface concentrations from 0.67 to 67 mg/m^2	
Microorganisms ($S.\ aureus$)	Transferable by impression plates	1.2×10^{-3} 2×10^{-4} 3.5×10^{-3}	Air jet, 10 min Moist mop, 10 min Four walkers, 30 min	35
			Means from 7–10 experiments; no ventilation	
Pu	Not specified	4×10^{-5}	No details given	36^a
^{125}I	Transferable by 100-cm^2 wipe with swab soaked in 70% isopropyl alcohol	5.9×10^{-4}	During laboratory operations including protein iodinations	37
			Laboratory area	

$^a K$ factor calculated from reported data.

Fish et al.[23] used a copper oxide aerosol with a mass median diameter of 2 μm in a 90-minute experiment that included 20 minutes of light sweeping. The worker had sticky tape attached to various parts of his clothing. After the experiment, the tape was analyzed to see how much copper oxide had been resuspended and collected. The data are shown in Figure 1. Contamination is expressed as a fraction of the initial floor concentration, which was taken to be 1.0. Although deposition on the worker is not equivalent to what would be collected by an air sampler, the variation of concentration as a function of height can readily be seen.

Jones and Pond[24] found that the resuspension factor increased with activity (in their investigations K was based on total surface contamination rather than on removable surface contamination). Only 10 to 20% of the airborne activity was respirable. They found that the resuspension factor obtained for plutonium oxide, which had been applied in the form of an aqueous suspension, was greater than that obtained with plutonium nitrate, which was applied in solution in nitric acid.

Brunskill[25] observed no significant differences between resuspension factors measured at 1.5 and 4.5 ft above the ground but no particle size distribution data were presented for the material resuspended or the material collected by the air samplers. Brunskill also noted that the resuspension factors calculated using personal sampler results were usually lower than those obtained when static samplers were used. This may have been due to disproportionately high values obtained from the static samplers. He found that airborne activity was positively correlated with the activity of personnel.

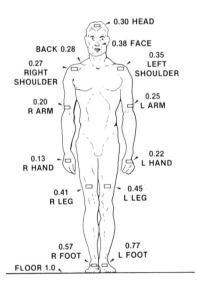

Figure 1. Deposition of resuspended CuO particles on clothing. Contamination is expressed as a fraction of the initial floor concentration, which was taken to be 1.0. (From Reference 23, with permission.)

Glauberman et al.[26] obtained data from a plutonium and a uranium plant. Both had concrete floors and similar ventilation conditions. They found K factors that differed by as much as a factor of 70. These differences were attributed to the size of the rooms. In one case, the room was about 2500 ft^3 in volume; in the other, where the K factor was smaller, the volume was approximately 1.4×10^6 ft^3. The authors felt that ". . . the severity of the disturbance . . . was far greater than would be expected in the normal plant situation." However, in spite of this, these disturbances did not yield maximum K values.

In some of the results reported by Mitchell and Eutsler,[28] two-stage air samplers were used. These indicated that only 6 to 10% of the airborne material was respirable. This finding indicates that not all the airborne material associated with resuspended surface contamination is available for inhalation or constitutes a potential health risk.

Cortissone et al.[29] concluded that since Brunskill's[25] and Tagg's[21] K values were higher than those reported by Jones and Pond,[24] the intensity of the disturbance and the characteristics of the floor covering had a greater influence on the resuspension factor than the concentration of surface activity. Their data also demonstrated that as the height of the air sampler above the floor increased, the K factor decreased whether the contaminant was dispersed as a powder or in the form of a solution. No particle size distribution data were reported, however. They found that the resuspension factor depended on the area contaminated and its relationship to the volume of the surroundings. After consideration of Jones and Pond's data,[24] they speculated that the resuspension factor did not continue to increase as activity increased because footsteps can tamp down, as well as resuspend, contamination. They noted that a contaminant applied in the form of a suspension was easier to redisperse than a contaminant applied as a solution.

Khvostov and Kostyakov[30] found that turning off the ventilation had little effect on the levels of air contamination. However, the elapsed time during which this observation was made was not given. They also found that air concentrations were increased 100-fold by rubbing the stainless steel floor with a dry wad of cotton.

Gorodinsky et al.[33] found that the resuspension factors obtained when aqueous solutions of strontium-89 and cobalt-60 salts were used were independent of the contaminated surface, the exposure time of the isotope on the surface (which ranged from 3 hours to 25 days), the velocity of air flow (which ranged from 0.3 to 1 meter per second), and humidity (which ranged from 65 to 95%). When a K factor based on strontium-89 aerosol contamination was determined, it was about two orders of magnitude greater than that obtained when an aqueous solution of strontium-89 was used.

Kovygin[34] calculated resuspension factors using surface contamination concentrations that varied over two orders of magnitude. He found that air concentrations did not increase in proportion to increasing surface contamination;

therefore, resuspension factors did not increase in proportion to increased surface contamination.

Wrixon et al.[36] stated that if there are no other sources of airborne contamination, the resuspension factor yields reasonable correlations between air and surface contamination for given conditions. They also noted that the resuspension factor was a function of surface type, the nature of the contamination, the extent of contamination, and the movement of air and personnel.

Dunn and Dunscombe[37] measured the mean air concentration and found that it was about ten times that estimated from the resuspension factor based on a centrally located sampler.

The data of Table 1 show that irrespective of the operating conditions or the technique used for surface contamination measurement, values obtained for the resuspension factor vary over a range of more than seven orders of magnitude: from 10^{-9} to more than 10^{-2} per meter. Examination of the data in Table 1 and the discussion above suggest a variety of causes for the very substantial range of resuspension factors observed. These include the vigor and frequency of the activity employed in the operating conditions; the height at which the air samples were obtained; whether the surface contamination measurement was of transferable or of total contamination; the nature of the contaminant (not only the particle size, density, and other physical characteristics but also whether the material was applied as a dry solid, a suspension, or a solution); the nature of the surface (i.e., porous or impervious and the effect this had on the degree to which the contamination on the surface could be reentrained); the ventilation, in both a macroscopic and microscopic sense; the size of the contaminated area in relation to the total volume of the experimental space; and the size distribution of the contaminant. Another factor that cannot be evaluated is what occurred when no data were specified.

The data in Table 1 show that there is considerable variation even when taking into account the fact that most of the data are averages for a greater or lesser number of runs. There are variations in resuspension factors even when similar procedures have been employed in parallel circumstances. According to Sehmel,[3] "even taking average resuspension factors is a risky concept because of the orders of magnitude of uncertainty within a single experiment in measured resuspension factors" and ". . . our ability to accurately predict airborne concentrations from either mechanically or wind-caused resuspension stresses is extremely poor."

3. Measurements of "Transferable" Surface Contamination

It has been frequently noted that it is not the total surface contamination but only the loose surface contamination that can be aerosolized and lead to inhalation exposure. Therefore, it is the transferable portion of the surface contamination that should be measured. However, the fraction of total surface contamination removed is often not known. This is the case regardless of the surface contamination sampling technique. The data available in the literature that indicate the removal efficiency of various surface contamination measurement techniques are shown in Table 2. The contaminant and the surface from which it is removed are also shown. Additional information for some of the studies cited appears below.

In observing the process of collecting wipe samples, Becher[17] noticed that the area wiped varied from 33 to 320 cm^2 and that the pressure applied varied from what he characterized as light to heavy. Despite these variations, the standard deviation from the mean for 80% of the wipe samples was about 40%.

Fish et al.[23] reported data that contrasted wipe samples, adhesive paper samples, and smair samples. The contaminant was settled thorium dioxide on stainless steel. As in other investigations,[29,47] the smair sampler recovered less material, at all particle sizes tested, than the other sampling techniques. The adhesive paper proved to be particularly efficient for particle sizes of 1.5 and 5 μm. The wipe sample was intermediate in performance.

Jones and Pond[24] found that the efficiency of removal for plutonium oxide applied in the form of an aqueous suspension was greater than for plutonium nitrate applied in solution in nitric acid.

Cortissone et al.[29] reported data obtained with the smair sampler. They found, as did Royster and Fish,[47] that the smair sampler always gave lower estimates of transferable surface contamination than constant-pressure or manual wipe samples. It was their view that the wipe test method using a constant-pressure sampler was fairly reliable. However, the variability among the samples taken at constant pressure was comparable to that among the wipe samples taken manually. The reproducibility of the samples obtained using the smair sampler was generally better than that for the other surface contamination measurement techniques.

Wrixon et al.[36] pointed out that surface contamination measurements were a good indication of the level of control but were not a substitute for air monitoring.

Barry and Solon[38] obtained wipe samples from uranium plates that had oxidized under normal circumstances. They found that each plate gave consistent results but plate-to-plate differences ". . . indicated that the amount of removable material is a fairly sensitive function of the past history of the plates, including manner and place of storage." Adhesive disks did not give results as reproducible as wipes.

Thomas et al.[39] assessed the efficiency of removal of microorganisms from visually clean and dirty dairy equipment. For clean equipment, they found that the first wipe removed from 30 to 60% of the total microorganisms collected. In successive wipes, 15 to 30%, 8 to 16%, 6 to 13%, and 6 to 12% were removed. For dirty equipment, from 20 to 40% was recovered in the first

Table 2. Measurements of Transferable Surface Contamination Reported in the Literature

Contaminant	Surface	Surface contamination measurement	Removal efficiency (%)	Remarks	Reference
UO_2	Uranium plates	Wipe[a]	No data	Values within 30% of mean	38
		Adhesive disks[b]	No data	Not as reproducible as wipes	
Microorganisms	Creamery equipment (clean)	Wiped twice with swab over 1 ft^2	30–60	Successive swabs yielded decreasing numbers of organisms	39
Micrococcus pyogenes	Nonporous china	Rinse—plate count	78–103	CV = 16%	40
		Rinse—membrane filter count	71–97	CV = 17%	
		APHA cotton swab	60–85	CV = 18%	
		Cotton swab	40–61	CV = 23%	
		Alginate swab	35–51	CV = 26%	
		Impression	43–50	CV = 7%	
^{204}TlCl	Resin tile	Wiped with 2.5-cm dia quantitative filter paper #5 using a mean pressure of 1 kg	4.0 ± 1.3	1 ml of an aqueous solution (pH = 5.4) was used; data are the mean and standard deviation for 5 samples	41
	Waxed resin tile		6.6 ± 1.5		
	Painted resin tile		9.9 ± 0.5		
	PVC		53.0 ± 9.9		
	Vinyl sheet		45.4 ± 4.9		
	Glass		42.1 ± 7.7		
Various radioisotopes	Waxed resin tile		1.7–37.3	^{137}Cs, ^{90}Sr–^{90}Y, ^{32}P, ^{60}Co, and U applied in HNO_3 solutions	
	Vinyl sheet		45.8–66.5		
Microorganisms	Aluminum	Wipe	95 ± 8	Average of 16–25 samples and standard deviation	42
		Trigger[c]	85 ± 9		
		Impression	46 ± 21		
		Rinse	95 ± 7		

	Oakwood	Wipe Trigger[c] Impression Rinse	37 ± 9 24 ± 12 3 ± 2 42 ± 18		
U	Smooth concrete or embossed metal plates	Wipe with Whatman #1 paper over 100 cm²	2-3		43
Pu	Plywood, Perspex, PVC, stainless steel, aluminum, linoleum, waxed protective paper	Wipe 100–1000 cm² using 10-cm square Whatman D.H.C. filter paper	11-20	Wipe pressure about 30 g/cm²	44
α-Emitters	Granolithic concrete	Ratemeter	8-53 mean = 39	Surfaces sampled before and after water wash	21
	Cotton	Wiped with dry filter paper over 100 cm²	2-17		
ThO₂	Stainless steel	Wipe[d] Adhesive paper[e] Smair[f]	96, 86, 49 96, 100, 68 58, 75, 10	Data for 1.5-, 5-, and 10-μm settled particles respectively; the 10-μm particles were agglomerates; constant area sampled	23
PuO₂	PVC Waxed linoleum Unwaxed linoleum	Wipe, no details	14 58 20	PuO₂ applied in aqueous suspension and dried	24
Pu(NO₃)₄	Paper PVC Waxed linoleum		0.1, 0.2 21, 29, 31 6	Pu(NO₃)₄ applied in HNO₃ solution and dried	
α-Emitters	Granolithic concrete	Dry filter paper wipes (6) over 900 cm²	1-3	Water wash removed 25%; subsequent detergent wash 43%	25
Ra	Not specified	Wipe	50-85	Calculated from reported data	45

(Continued)

Table 2. (Continued)

Contaminant	Surface	Surface contamination measurement	Removal efficiency (%) Dry	Removal efficiency (%) Wet	Remarks	Reference
[³H]Sodium acetate		Wipe (wet or dry) with Whatman #1 papers			Range for 3 replicates	46
	Shellstone		5–10	7–19		
	Fiberglass		20–30	26–32		
[³H]Paraffin	Shellstone		18–23	12–17		
	Fiberglass		15–30	5–6		
ThO₂	Various (see also Table 3)	Wipe[d]	24–75		Constant area sampled; particle size ~ 1 μm	47
		Adhesive paper[e]	44–86			
		Smair[f]	1–33			
Be	Wood	Wipe (back and forth) 1 ft² with 5" × 8" Whatman #41	3 of total, 20 of loose for each of three wipes over same area		Large portion of total remained in wood after washing with detergent	28
U	Preaflex	Smair[g]	0.06–2.6		U as UO₂(NO₃)₂ · 6 H₂O in HNO₃, UO₂ in C₂H₅OH, and UO₂ powder; standard deviation usually ≤ 15% of mean	29
		Wipe—constant pressure (8 g/cm²) over 100 cm²	20–67			
		Wipe over 100 cm² using 5.5-cm dia filter paper	42–70			
Various α and β emitters	Stainless steel	Wipe, no details	21			30
³H	Brass	Wiped with Whatman #3 filter paper soaked with glycerol	26.1 ± 4.4		Successive wipes removed 9.4 ± 1.1 and 7.2 ± 1.7%; ethylene glycol-soaked paper gave similar results, dry paper removed about half as much, and aluminum foil about a quarter	48

Contaminant	Surface	Method	Efficiency (%)	Notes	Ref.
[³H]Thymidine	PVC Stainless steel Glass Wood	Wiped with 2.5 cm dia filter paper 15 times across a 5-cm square	28.2 ± 4.8 86.1 ± 4.6 70.4 ± 5.4 4.2 ± 1.0	Ten replicates; three different filter papers used for wipes showed no marked differences	49
Aromatic amines	Stainless steel, concrete, painted wood	Wipe with 7-cm dia Whatman #42 filter paper moistened with 5 drops of methanol or ethanol over 500 cm²	No data	Lowest limits of detection were obtained with this technique	50
Methyl parathion	Smooth concrete	Wipe with alcohol-soaked gauze pads (4)	6	Contaminant recoverable after 5 months of periodic wiping	51
Various radioisotopes	23-mm dia unwoven fabric on adhesive tape	Remove tape from surface and count with appropriate detector	No data	Correlation coefficient = 0.82 for tape-smear comparison	52
Pb	Wood, painted or varnished Formica	Wiped "briskly" with a paper towel over 1 ft²	80–100 >95	Efficiencies calculated for surfaces sampled before and after scrubbing with water and a brush	53
Pb	Not specified	Wiped with a paper towel impregnated with 20% denatured alcohol and 1:750 benzalkonium chloride over 1 ft²	77 ± 2	Second wipe removed the remaining contamination	54
Microorganisms	Tile 1 Tile 2 Vitro-ceram PVC Inox	65-mm dia impression plates	39 30 42 24 40	Most reproducible results obtained when 200-g weight applied for 2 min	55

(Continued)

Table 2. (*Continued*)

Contaminant	Surface	Surface contamination measurement	Removal efficiency (%)		Remarks	Reference
			Po	Am		
^{210}Po and ^{241}Am		Wiped using 25-mm dia, 30-mg/cm^2 Toppan paper over 100 cm^2 using 0.2 kg/cm^2 pressure			1 ml of the nitrate dissolved in 0.1 N HNO$_3$ was spread over 100 cm^2 of the surface and dried; mean and standard deviation for 6 samples	56
	PVC		48.1 ± 1.4	42.3 ± 1.2		
	Aluminum		19.3 ± 0.5	20.4 ± 1.1		
	Glass		68.1 ± 2.0	69.8 ± 1.5		
^{125}I	Not specified	Wipe 100 cm^2 using swab soaked in 70% isopropyl alcohol	Assumed 10% of total removed			37
Microorganisms	Not specified	Polyester bonded cloth	97.2		Polyester recovery (90.4%) better than cellulose (72.0) or cotton swab (75.2) and large area sampled (~0.65 m^2)	57
		Cellulose cloth	97.5			
		Cotton swab	89.6			

Inorganic salts	Zinc	1.27-cm square Whatman 542 ashless, hardened paper moistened with distilled water placed on surface and removed after drying	100	Three successive samples removed all chloride, nitrate, sulfate, ammonium, potassium, magnesium, and sodium, and calcium cations from "normally contaminated" surfaces; from surfaces "heavily contaminated" with smoke, 3 samples removed all nitrate and ammonium and magnesium cations; pH variation from 4.0–5.8 had no effect	58
PbO	Formica	Wiped with moist Whatman #42 paper, 10 × 10 cm, over 100 cm²	86–91	Removal generally increased with increasing surface concentration from 64 to 730 μg/100 cm²	59
		Wiped with moist paper towel, 10 × 10 cm, over 100 cm²	74–84		

[a] Whatman #41 paper (1⅛ in dia) on #5½ rubber stopper; rubbed 10 times back and forth over width of plate (3¼–3½ m).
[b] Gummed paper (Simon Adhesive Co. #2) or gummed tape (3M Co. #216) 1⅛ in dia.
[c] Wipe using a standard area and pressure.
[d] Whatman #50 paper on #5 rubber stopper; rubbed over 5.8-cm dia sample location.
[e] 3.8-cm square paper pressed with #10 stopper against sample location.
[f] 5-cm² head held for 6 s on sample location (air flow = 30 m/s).
[g] 113-cm² head held for 5 min on sample location (air flow = 30 m³/h).

wipe. Successive wipes yielded 25 to 40%, 11 to 40%, 4 to 25%, and 1.5 to 8%.

Angelotti et al.[40] used a bacterial tracer. They compared six different sampling techniques and found that three techniques yielded high recoveries and average precision, two techniques yielded low recoveries and poor precision, and one technique yielded low recoveries but excellent precision. They found that wiping a number of similar sites or taking replicate samples yielded higher recoveries and better precision than sampling one site once. Wiping large areas averaged out variations in the distribution of contamination over the surface. They also found that the variance between samples was a function of the angle at which the sampler was used (swab techniques are frequently used for biological sampling), pressure of the wipe, variation in the area wiped, incomplete release of the contaminant collected, pipetting, and counting. They concluded that rinsing techniques were good because the entire surface was sampled, which yielded higher recoveries.

Yoshida et al.[41] assumed that the contamination removed by five wipes was the total amount of surface contamination and calculated how much of the total contamination the first wipe represented. This varied from nearly 40% for a synthetic resin tile to nearly 80% for poly(vinyl chloride). For permeable surface materials, marked differences were observed because of the salt concentration in the ammonium nitrite contaminating solutions. They also examined the effect of the nuclide on the amount of contaminant that could be removed from waxed synthetic resin tile and vinyl sheet. Strontium-90 and yttrium-90, cobalt-60, cesium-137, phosphorus-32, and natural uranium were applied in solution in nitric acid. The range of recoveries for the first wipe was from 1.7% for cobalt-60 to 37.3% for natural uranium on the waxed synthetic resin tile. On the vinyl sheet, the range was much narrower: from 45.8% for natural uranium to 66.5% for cobalt-60. The amount of cobalt-60 in nitric acid removed from a waxed synthetic resin tile surface was essentially independent of pressure over the range of 0.1 to 0.2 kg/cm^2.

Coretti,[42] in another study using biological materials as a surface contaminant, stated that the problem with all surface contamination measurement techniques is that the collected biological materials can become confluent when plated or that aggregates are obtained and yield only one colony. In both these cases, enumeration is difficult and uncertain.

Brunskill and Fletcher[44] studied the effect of wipe pressure on surface contamination measurement results. Five or six pressures up to 30 g/cm^2, typical of those pressures normally applied, were used. At higher pressures, the amount removed was relatively unaffected by pressure changes. The first wipe removed 11 to 20% of the total surface contamination. The second and third wipes removed significantly less at all pressures and for all surfaces tested (plywood, Perspex, PVC, stainless steel, aluminum, linoleum, and waxed protective paper).

Blatz and Eisenbud[45] investigated the radium watch and instrument dial painting industry in and around New York City. They found that monitoring surface contamination was often an effective control technique; it prevented airborne concentrations of contaminant from becoming a health hazard. Surface contamination was not directly related to airborne contamination; it was sometimes a result and sometimes a cause of the airborne contamination. It could even be both. Limits placed on surface contamination would probably have to depend upon factors related to airborne concentrations such as particle size, particle density, hygroscopicity, type of surface, ventilation, relative humidity, and traffic. They concluded ". . . there does not seem to have been established any reliable relationship between surface contamination and health hazard."

Royster and Fish[47] examined the amount of surface contamination removed as a function of the area sampled. They studied Formica and fiberboard surfaces using smair, adhesive, and wipe samples. Whereas the percent removed varied on both surfaces when adhesive and wipe samples were used, the smair sampler yielded results independent of the area sampled over the range of 20 to 200 cm^2. These investigators also examined for various surfaces the effectiveness of removal using these techniques (Table 3). Although there was no consistent agreement, the smair sampler results covered a wide range; the adhesive sampler results tended to be higher and were more consistent; and the wipe samples were related to surface roughness and, possibly, hardness. The

Table 3. Percentage of Total Radioactivity Removed from Surfaces by Different Sampling Techniques[a]

Composition of surface	% removed		
	Adhesive paper	Wipe	Smair
Polyethylene	70.3	56.6	10.9
Glass	75.0	64.6	27.2
Plexiglas	78.0	71.3	15.8
Fiberboard	73.4	23.5	6.6
Fiberboard, treated	75.9	34.4	20.0
Fiberboard, cleaned	56.9	23.5	9.0
Formica	73.4	70.6	26.5
Aluminum, painted	70.0	50.3	24.8
Aluminum, painted and treated	86.0	67.1	33.0
Asphalt tile	58.6	48.5	14.6
Asphalt tile, waxed	74.5	74.5	30.3
Concrete	55.5	39.5	22.0
Concrete, sealed and waxed	54.8	47.7	27.2
Concrete, greased	43.5	37.5	1.3
Stainless steel	67.7	50.5	10.5

[a]After Royster and Fish, Reference 47, with permission.

effect of personal variance was also examined. Each of 21 subjects took eight samples using each method. The variance among the wipe samples was greatest and that among the smair samples was least.

Eakins and Hutchinson[48] were also interested in the pressure applied when wipe sampling. Eleven health physicists used from 48 to 267 g/cm^2 pressure; the mean pressure was 160 g/cm^2.

Davis et al.[51] have reported the only contamination data for a pesticide. After an interval of one year, concrete blocks exposed to methyl parathion were fractured and the yellow degradation product of the pesticide, p-nitrophenol, could clearly be seen at depths of from 2.5 to 4.5 cm below the surface. The authors pointed out that physically sealing the surface may be effective (and it may be necessary) if one cannot decontaminate a surface completely.

Anzai and Kikuchi[52] adopted an unwoven fabric on an adhesive tape after trying paper tape, vinyl tape, and woven and unwoven fabric tape. A 23-mm diameter disk was taped to the surface to be monitored for subsequent collection and counting. The advantages of this approach were identified as the following: an integrated sample of the total contamination accumulated on the surface was acquired; the result did not depend on the composition or condition of the surface sampled; the technique was free from operator error or variability; complicated surfaces could be sampled; the tape could be applied to the sole of a worker's shoe to get a "walking smear"; and the technique was simple. They identified as the technique's disadvantage that it provided a fixed-point observation of unknown representativeness. They also noted that the technique measured total contamination, not loose contamination. They cited wipe sample drawbacks such as the physical condition of the surface, which affects collection efficiency; a complicated shape, causing the worker to change the pressure and angle of the sample leading to variability; person-to-person variations; and the inability to measure fixed contamination. These factors preclude comparisons among wipe sample data.

Vostal et al.[54] found that the results obtained from consecutive samples taken from adjacent sites using moistened paper towels did not differ by more than 20% in highly contaminated areas; larger differences on less contaminated surfaces were observed, however. Sample reproducibility among four investigators was fairly good. Standard deviations, expressed as percentage of the mean (coefficient of variation), were 20%, 24%, and 56% for samples obtained from floors, windowsills, and hands, respectively.

Hartemann et al.[55] used 65-mm diameter Rodac plates to sample microorganisms. They found that the most reproducible results were obtained when the Rodac plates were pressed onto a surface for two minutes under a 200-g weight. They also found that their data were reproducible for a given surface, but when surface irregularities were present, bacterial recovery became inconsistent.

Chavalitnitikul and Levin[59] used moist filter paper, moist paper towels, adhesive paper, adhesive cloth, and cloth tape to remove lead oxide particles less than 40 μm in diameter from Formica and plywood surfaces. They found that the efficiency of removal from plywood was less than that from Formica. They also found that adhesive tapes had a higher removal efficiency than filter paper or paper towels when used on rough surfaces. On smooth surfaces the removal efficiencies were similar when high wiping pressure was used; when minimum pressure was applied, the removal efficiencies for the adhesive tapes were lower than for the filter paper or paper towels.

Ayliffe et al.[60] concluded that since bacterial floor counts (in hospital wards) do not increase indefinitely, organisms eventually are removed or die at about the same rate they are deposited. They were unable to prove that under normal conditions mechanical or physical dislodgement of deposited bacteria occurred. Strong air currents or people exercising on contaminated floors did not resuspend much contamination. On the other hand, sweeping yielded large values for redispersion. They concluded that floor dust was not an important source of airborne infection. They also observed that about half the bacteria on the floor were deposited from airborne contamination; they believed the remainder came from shoes, wheels, and adjacent uncleaned areas. Redispersion, they felt, occurred by drafts, traffic, and procedures such as bed making, drawing curtains, sweeping, and vacuuming.

Vesley and Michaelsen[61] found that the numbers of microorganisms collected using Rodac plates depended on six factors: the age and renovation features of the institutions in which sampling was done, the condition of the floors sampled, traffic and dress control features (most of their work was done in hospitals), housekeeping procedures, the air-handling system, and work density. They noted several problems associated with the measurement of microbial surface contamination. These included surface variables such as porosity, absorption, and the adhesional characteristics of the contaminant; ecological influences such as nutrition, time, temperature, oxygen concentration, competition, and the effect of drying on the growth of microorganisms; and the number of cells from which the colony arose. Enumeration problems depended upon incubation sequence and the confluence of colonies. Finally, statistical difficulties included site selection (random is not the same as arbitrary), sample size, and variability within and among personnel.

In a study of the effect of floor coverings, Weber[62] dried a radioisotope solution on eight (unspecified) floor coverings after they were carefully degreased. The floor coverings were new or abraded to simulate wear. The isotopes used were cerium-144, europium-152 + europium-154, and cobalt-60 as chlorides; phosphorus-32 as $NaHPO_4$; and mixed fission products as nitrates. The contaminated floor coverings were flushed with deionized water, citric acid, sodium citrate buffer solution, or deionized water plus detergent for ten minutes. The amount of isotopes remaining was determined by ratemeter. From 1 to 99.9% was removed by these treatments. The results were strongly dependent on the nuclide applied, the cleaning solution used, and the smoothness of the surface.

As examples of the kinds of variability that can be seen in measurements of surface contamination, Cohen and Kusian[63] cited beryllium wipe sample practices at different facilities. At five different locations during the period 1956 through 1964, 1 ft^2 was wiped with dry filter paper, 12 in^2 was wiped with alcohol-wetted filter paper, 100 cm^2 was wiped with a dry paper (at two locations), and 1 m^2 was wiped with a dry paper.

Clearly, there are many variables affecting the determination of transferable surface contamination. One of them is the contaminant itself. Radioisotopes, chemicals, and biological agents have been used as contaminants. Variations in measurement may occur because of the specific chemical compound involved, its chemical state, the manner in which it is applied, the particle size of the material, and the degree to which the contaminant adheres to the surface. The nature of the surface itself may affect the amount of contamination that becomes airborne. Variables that might influence this are the composition of the surface, its roughness, and the local relative humidity. The measurement itself can have an effect. There is also variation within any individual technique. The area of surface wiped, the number of times it is wiped, the pressure used, the material used to take the sample, and whether the sampling material is wet or dry can affect the results.

Data reflecting the influence of many of these variables are included in Table 2. One variable for which data are conspicuous by their absence is relative humidity. Although Gorodinsky et al.[33] found that resuspension factors were unaffected by relative humidities in the range of 65 to 95%, Whitfield[64] reported very substantial effects of relative humidity on the adhesion of small particles (< 140 µm diameter) to polished metal foil surfaces (Figure 2). The

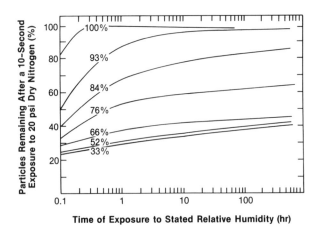

Figure 2. Effects of relative humidity on surface particle retention. (From Reference 64, with permission.)

effect of this variable on resuspension factors and surface contamination transfer deserves more attention.

Finally, Eisenbud et al.[15] concluded that surface contamination concentrations were not a good criterion for judging whether hazardous conditions exist, and Shapiro[32] stated that "although air entrainment tests are more realistic than transferable contamination tests in evaluating the surface contamination hazard, and correlate better with observed airborne levels, they simulate only partially the mechanisms by which particles are removed from surfaces and produce sustained airborne concentrations."

4. The Contribution of Resuspended Particulates to Exposure

The preceding sections have addressed the resuspension factor itself, some of the variables on which it depends, and the variables affecting surface contamination measurements in particular. There has been little work undertaken to learn just what contribution the resuspended material makes as a portion of total contamination. Those few data are discussed below and are supplemented by relevant inferences drawn by investigators who were primarily interested in resuspension factors.

Breslin et al.[43] simulated work conditions in a uranium extrusion plant to measure the contribution of resuspension to total dust exposure from all sources. Wipe samples, general (area) air samples, breathing zone (personal) air samples, breathing zone (area) air samples, and two-stage general (area) air samples were taken and the total amount of surface contamination was measured with a ratemeter. The amount of respirable dust collected during their simulation was about the same as that collected during actual operating conditions and so were the total suspended particulate matter data obtained by the several different sampling techniques. These findings suggest that the simulation was valid. The contribution of resuspension to total dust exposure ranged from 0.2 to 1%.

Ayliffe et al.[60] contaminated vinyl and terrazzo flooring and then disturbed the contaminated surface by having people exercise on it, by blowing on it with a jet of cold air (about 245 ft/min), or by sweeping it with a dry broom. The air samples collected during and after the disturbance led them to conclude that floor dust was not an important source of airborne infection and that resuspension was of minor consequence. No numerical estimate of resuspension's contribution was made, however.

Alzona et al.[65] cleaned the air in an experimental room with filters and collected airborne particulate matter indoors and outdoors simultaneously. The model developed to describe the data incorporated filtration, passage through the walls, and deposition on and resuspension from surfaces in the room. The indoor/outdoor ratio observed was consistent with zero resuspension. Indoor/outdoor concentration ratios were measured for calcium, iron, zinc, lead, and bromine. In three runs, when the room under normal conditions was compared with the same room with all surfaces but the windows covered with plastic sheet, these ratios were nearly identical. This suggested that all of the dust was from the outside and that the contribution of resuspension approached zero.

Schultz and Becher's[19] statistical analysis of the data obtained over an extended time in uranium hexafluoride feed manufacturing and uranium recovery facilities used data obtained from full-shift air samples, spot (ten-minute) air samples, surface contamination measurements, and urine analysis. They found positive, statistically significant correlations between surface contamination and urine analysis, spot air concentrations and surface contamination, and shift air samples and surface contamination. These results tend to confirm the usefulness of the resuspension factor, but they do not allow one to estimate the contribution of resuspension to total exposure.

Other data that will help us assess the importance and the implications of resuspension of indoor surface contamination are those flowing from accidents involving radiation. Data from perhaps two dozen incidents have been examined to see what fraction of the total activity resulted in exposure to personnel. The earliest data come from Franke and Hunzinger.[66] In a simulated accident, glove boxes containing known amounts of evenly spread plutonium oxide were shattered. Air samples (one near the glove box and the other at a distance from it) operated for an extended time at normal breathing rate. In two experiments, the amount collected by the air sampler near the box was 2×10^{-8} of the total amount of plutonium in the box. In other work, Franke et al.[67] took available information from 20 accidents and estimated the inhaled fraction of the total activity handled. In three of the accidents, the characteristics of the operation were uncertain. For the remaining 17, the inhaled fraction was apportioned in a matrix as shown in Table 4. The largest values observed for the inhaled fraction were 1×10^{-5}.

Brodsky[68] summarized ten events that yielded estimated fractional intakes of activity from less than 10^{-10} to 10^{-5}, except for one instance where the fractional intake was about 10^{-2}. However, this incident involved a spill of 0.078 curie and skin absorption was a possibility. In a later publication, Brodsky[69] examined additional information and suggested that, in practice, 10^{-6} seemed to be a reasonable upper limit for the fraction of material to which one may be exposed, barring skin absorption. He subtitled this paper "Is 10^{-6} a Magic Number in Health Physics?" One might ask whether 10^{-6} is a magic number in resuspension of indoor surface contamination.

In recent work, Sutter et al.[70] measured releases from freefall spills to estimate aerosols generated from accidents. Sodium fluorescein and uranium solutions and depleted uranium or titanium dioxide powder were dropped from heights of 1 and 3 m. The aerodynamic size of the particles corresponded to diameters of approximately 3.3 to 3.5 μm. The powder spills yielded approximately 1 part in 1000 by weight airborne when dropped from 3 m but only 1 part in 10,000 when dropped from 1 m. (High-volume samplers were located

Table 4. Inhaled Fraction of Total Activity Handled[a]

Nature of operation	Containment		
	Glove boxes	Ventilated areas	Nonventilated areas
Ambient temperature			
Solid compound	1×10^{-8}	2×10^{-7}	2×10^{-6}
	1×10^{-7}	4×10^{-7}	7×10^{-7}
		2×10^{-7}	
		1×10^{-6}	
		3×10^{-8}	
Aqueous solution		5×10^{-7}	
		4×10^{-7}	
		2×10^{-8}	
Elevated temperature		7×10^{-6}	
		1×10^{-5}	
		3×10^{-6}	
		2×10^{-6}	
		1×10^{-5}	

[a] After Franke et al., Reference 67, with permission.

at heights of 1 and 2 m; a high-volume impactor was located at 1.5 m. There was a significant sampler height effect, but no data were presented.) When solutions were dropped from 3 m the fraction airborne was about 10^{-5} to 4×10^{-5}; when dropped from 1 m the fraction ranged from about 10^{-6} to 3×10^{-5}. If one takes into account that these figures are for the total amount of material made airborne and not the quantity inhaled, these data tend to corroborate the previous data. This reinforces the idea that the inhaled fraction from an accident is likely to be of the order of 10^{-6} or smaller. This would appear to be a reasonable upper limit for material resuspended from indoor surface contamination.

Other investigators have made comments relevant to this issue. Blatz and Eisenbud[45] observed ". . . there does not seem to have been established any reliable relationship between surface contamination and health hazard." Utnage[16] said ". . . the dust contribution from floors is negligible compared to that from operational sources." Lannefors and Hansson[71] looked at the time variation patterns in indoor and outdoor bromine and lead concentrations and concluded that the particles entered from the outside mainly by infiltration. They said, "only a minor fraction seemed to be brought in and resuspended by the staff and children." Hambraeus et al.[35] calculated that the contribution redispersion made to the total bacteria airborne was less than or equal to 15%. Glauberman et al.[26] found that ". . . air contamination . . . was of minor significance with respect to occupational exposure." Dunn and Dunscombe[37] concluded that "... airborne radioactivity was higher during iodination than at other times, thus indicating that resuspension from contaminated surfaces was likely to be relatively unimportant."

5. Concluding Remarks

We have examined values of the resuspension factor reported in the literature for indoor experiments. We have seen that the resuspension factor is subject to many variables and that reported data range over several orders of magnitude even in the same experiment with relatively constant and reproducible conditions. One of the components of the resuspension factor, estimates of transferable surface contamination, is also highly variable and depends upon many factors. Estimates of the contribution of resuspended matter to the total aerosol burden are uniformly low. Related evidence suggests that the fraction of activity airborne and inhaled following an accident is a very small fraction of the total amount of starting material. Other environmental data[72-75] indicate that although contamination spreads from place to place, contact transmission associated with worker movement or movement of wheeled vehicles is more important than air dispersion of contamination. The implication is that resuspension of surface contamination in a health context is usually of minor, if not negligible, importance. It should be remembered, however, that health is not the only criterion. Surface contamination can have serious adverse effects in a number of environments and industrial applications, such as adhesion (soldering, welding), microelectronics (printed circuits), adsorption, lubrication, and clean rooms. In some of these situations resuspension cannot be ignored, even if it does play a minor role. Measurement of surface contamination is recommended as a means to learn whether primary control or cleanup is working and, if not, to discover that additional work needs to be done.

Acknowledgment

Research was sponsored by the National Cancer Institute under Contract No. NO1-CO-23910 with Program Resources, Inc.

References

1. T. M. Prudden, *Dust and Its Dangers*, G. P. Putnam's Sons, New York (1905).
2. J. Mishima, A Review of Research on Plutonium Releases during Overheating and Fires, Hanford Atomic Products Operation Report HW-83668, U.S. Atomic Energy Commission, Washington, D.C. (1964).

3. G. A. Sehmel, Particle resuspension: A review, *Environ. Int.* 4, 107–127 (1980).
4. E. B. Sansone and M. W. Slein, Redispersion of indoor surface contamination: A review, *J. Haz. Mater.* 2, 347–361 (1977/78).
5. B. R. Fish, ed., *Surface Contamination*, Pergamon Press, New York (1967).
6. Society for Radiological Protection, International Symposium on the Radiological Protection of the Worker by the Design and Control of His Environment, Society for Radiological Protection, Bournemouth, England (18–22 April 1966).
7. K. L. Mittal, ed., *Surface Contamination: Genesis, Detection, and Control*, Plenum Press, New York (1979).
8. N. A. Fuchs, *The Mechanics of Aerosols*, Pergamon Press, New York (1964).
9. H. L. Green and W. R. Lane, *Particulate Clouds: Dusts, Smokes and Mists*, 2nd Ed., Van Nostrand, New York (1964).
10. C. N. Davies, ed., *Aerosol Science*, Academic Press, New York (1966).
11. S. K. Friedlander, *Smoke, Dust and Haze: Fundamentals of Aerosol Behavior*, John Wiley and Sons, New York (1977).
12. A. C. Chamberlain and G. R. Stanbury, The Hazard from Inhaled Fission Products in Rescue Operations after an Atomic Bomb Explosion, Atomic Energy Research Establishment Report HP/R 737, Harwell, United Kingdom (June, 1951).
13. H. J. Dunster, Contamination of surfaces by radioactive materials: The derivation of maximum permissible levels, *Atomics* 6, 223–239 (1955).
14. J. C. Bailey and R. C. Rohr, Air-Borne Contamination Resulting from Transferable Contamination on Surfaces: Report K-1088, U.S. Atomic Energy Commission, Washington, D.C. (1953).
15. M. Eisenbud, H. Blatz, and E. V. Barry, How important is surface contamination? *Nucleonics* 12, 12–15 (1954).
16. W. L. Utnage, in: Proceedings of the Symposium on Occupational Health Experience and Practices in the Uranium Industry: Report HASL-58, U.S. Atomic Energy Commission, Washington, D.C. (1959), pp. 147–150.
17. A. F. Becher, in: Proceedings of the Symposium on Occupational Health Experience and Practices in the Uranium Industry: Report HASL-58, U.S. Atomic Energy Commission, Washington, D.C. (1959), pp. 151–156.
18. E. C. Hyatt, H. F. Schulte, R. N. Mitchell, and E. P. Tangman, Jr., Beryllium: Hazard evaluation and control in research and development operations, *A.M.A. Arch. Indust. Health* 19, 211–220 (1959).
19. N. B. Schultz and A. G. Becher, Correlation of uranium alpha surface contamination, airborne concentrations, and urinary excretion rates, *Health Phys.* 9, 901–909 (1963).
20. I. S. Jones, S. F. Pond, and D. C. Stevens, in: Society for Radiological Protection, International Symposium on the Radiological Protection of the Worker by the Design and Control of His Environment, Bournemouth, England (18–22 April 1966).
21. B. Tagg, in: Society for Radiological Protection, International Symposium on the Radiological Protection of the Worker by the Design and Control of His Environment, Bournemouth, England (18–22 April 1966).
22. O. M. Lidwell, in: *Proceedings of the 17th Symposium of the Society for General Microbiology* (P. H. Gregory and J. L. Monteith, eds.), pp. 116–137, Cambridge University Press, London (1967).
23. B. R. Fish, R. L. Walker, G. W. Royster, Jr., and J. L. Thompson, in: *Surface Contamination* (B. R. Fish, ed.), pp. 75–81, Pergamon Press, New York (1967).
24. I. S. Jones and S. F. Pond, in: *Surface Contamination* (B. R. Fish, ed.), pp. 83–92, Pergamon Press, New York (1967).
25. R. T. Brunskill, in: *Surface Contamination* (B. R. Fish, ed.), pp. 93–105, Pergamon Press, New York (1967).

26. H. Glauberman, W. R. Bootmann, and A. J. Breslin, in: *Surface Contamination* (B. R. Fish, ed.), pp. 169-178, Pergamon Press, New York (1967).
27. K. Kereluk, R. Meyer, and A. J. Pilgrim, in: *Surface Contamination* (B. R. Fish, ed.), pp. 333-344, Pergamon Press, New York (1967).
28. R. N. Mitchell and B. C. Eutsler, in: *Surface Contamination* (B. R. Fish, ed.), pp. 349-352, Pergamon Press, New York (1967).
29. C. Cortissone, O. Ilari, G. Lembo, and A. Moccaldi, La contaminazione dell'aria risultante da risospensione di contaminazione di superficie [Atmospheric contamination from resuspension of surface contamination], *Minerva Fisiconucleare, Giornal di Fisica Sanitaria Protezione Radiazione 12*, 63-79 (1968).
30. N. N. Khvostov and M. S. Kostyakov, Hygienic significance of radioactive contamination of working surfaces, *Hyg. Sanit.* (English ed.) *34*, 43-48 (1969).
31. R. F. Carter, The Measurement of Asbestos Dust Levels in a Workshop Environment, United Kingdom Atomic Energy Authority A.W.R.E. Report No. 028/70, Aldermaston, United Kingdom (1970).
32. J. Shapiro, Tests for the evaluation of airborne hazards from radioactive surface contamination, *Health Phys. 19*, 501-510 (1970).
33. S. M. Gorodinsky, D. S. Goldstein, U. Ya. Margulis, M. I. Rokhlin, V. A. Rikunov, Yu. A. Sevostiyanov, M. A. Sobolevsky, and V. A. Cherednichenko, Experimental determination of the coefficient of passage of radioactive substances from contaminated surfaces into the air of working premises (in Russian), *Gig. Sanit. 37*, 46-50 (1972).
34. G. F. Kovygin, Certain problems of substantiating the permissible densities of beryllium surface contamination (in Russian), *Gig. Sanit. 39*, 43-45 (1974).
35. A. Hambraeus, S. Bengtsson, and G. Laurell, Bacterial contamination in a modern operating suite. 3: Importance of floor contamination as a source of airborne bacteria, *J. Hyg., Camb. 80*, 169-174 (1978).
36. A. D. Wrixon, G. S. Linsley, K. C. Binns, and D. F. White, Derived Limits for Surface Contamination, NRPB-DL2, National Radiological Protection Board, Harwell, Didcot, Oxon, United Kingdom (1979).
37. M. J. Dunn and P. B. Dunscombe, Levels of airborne ^{125}I during protein labelling, *Radiation Protection Dosimetry 1*(2), 143-146 (1981).
38. E. V. Barry and L. R. Solon, Radioactive contamination sampling by smears and adhesive disks, *Nucleonics 11*, 60-61 (1953).
39. S. B. Thomas, E. Griffiths, K. Elson, and N. B. Bebbington, The suitability of swab tests for determining the bacterial content of dairy equipment, *Dairy Industries 20*, 41-43 (1955).
40. R. Angelotti, M. J. Foter, K. A. Busch, and K. H. Lewis, A comparative evaluation of methods for determining the bacterial contamination of surfaces, *Food Res. 23*, 175-185 (1958).
41. Y. Yoshida, Y. Sasaki, M. Murata, S. Izawa, and Y. Ikezawa, Experiments on measuring the density of surface contamination by the smear survey method (in Japanese), *Nihon Genshiryoku Gakkaishi 6*, 77-81 (1964).
42. K. Coretti, Über den Wert einiger bakteriologischer Methoden zur Ermittlung der Betriebshygiene in Fleischwarenbetrieben [The value of some bacteriological methods of determining the state of hygiene in meat processing plants], *Fleischwirtschaft 46*, 139-141, 144, 145 (1966).
43. A. J. Breslin, A. C. George, and P. C. LeClare, in: Society for Radiological Protection, International Symposium on the Radiological Protection of the Worker by the Design and Control of His Environment, Bournemouth, England (18-22 April 1966).
44. R. T. Brunskill and D. J. Fletcher, in: Society for Radiological Protection, International Symposium on the Radiological Protection of the Worker by the Design and Control of His Environment, Bournemouth, England (18-22 April 1966).
45. H. Blatz and M. Eisenbud, in: *Surface Contamination* (B. R. Fish, ed.), pp. 163-167, Pergamon Press, New York (1967).

46. J. R. Prince and C. H. Wang, in: *Surface Contamination* (B. R. Fish, ed.), pp. 179-183, Pergamon Press, New York (1967).
47. G. W. Royster, Jr., and B. R. Fish, in: *Surface Contamination* (B. R. Fish, ed.), pp. 201-207, Pergamon Press, New York (1967).
48. J. D. Eakins and W. P. Hutchinson, The Radiological Hazard from Tritium Sorbed on Metal Surfaces. Part 2: The Estimation of the Level of Tritium Contamination on Metal Surfaces by Smearing, Atomic Energy Research Establishment—R 5988, Health Physics and Medical Division, United Kingdom Atomic Energy Authority Research Group, Atomic Energy Research Establishment, Harwell, United Kingdom (1969).
49. A. Koizumi, Y. Bessho, T. Kikuchi, and Y. Yoshizawa, Measurement of tritium surface contamination by liquid scintillation counting of smear paper, *Radioisotopes 24*, 431-433 (1975).
50. R. W. Weeks, Jr., B. J. Dean, and S. K. Yasuda, Detection limits of chemical spot tests toward certain carcinogens on metal, painted, and concrete surfaces, *Anal. Chem. 48*, 2227-2233 (1976).
51. J. E. Davis, D. C. Staiff, L. C. Butler, and J. F. Armstrong, Persistence of methyl and ethyl parathion following spillage on concrete surfaces, *Bull. Environ. Contam. Toxicol. 18*, 18-25 (1977).
52. I. Anzai and T. Kikuchi, A new monitoring technique of surface contamination—The test surface method, *Health Phys. 34*, 271-273 (1978).
53. J. W. Sayre and M. D. Katzel, Household surface lead dust: Its accumulation in vacant homes, *Environ. Health Perspec. 29*, 179-182 (1979).
54. J. J. Vostal, E. Taves, J. W. Sayre, and E. Charney, Lead analysis of house dust: A method for the detection of another source of lead exposure in inner city children, *Environ. Health Perspec. 7*, 91-97 (1974).
55. P. Hartemann, C. Demange, M. F. Blech, and E. Thofern, in: *Proceedings of the 5th International Symposium on Contamination Control*, pp. 151-156, Verein Deutscher Ingenieure—Verlag GmbH, Düsseldorf, Germany (1980).
56. M. Takiue, Simple and rapid measurement of α-rays on smear samples using air luminescence, *Health Phys. 39*, 29-32 (1980).
57. J. R. Puleo and L. E. Kirschner, Speedy acquisition of surface-contamination samples, *Natl. Aeronaut. Space Admin. Tech. Briefs 6*(2), 174 (1981).
58. J. D. Sinclair, Paper extraction for sampling inorganic salts on surfaces, *Anal. Chem. 54*, 1529-1533 (1982).
59. C. Chavalitnitikul and L. Levin, A laboratory evaluation of wipe testing based on lead oxide surface contamination, *Am. Ind. Hyg. Assoc. J. 45*, 311-317 (1984).
60. G. A. J. Ayliffe, B. J. Collins, E. J. L. Lowbury, J. R. Babb, and H. A. Lilly, Ward floors and other surfaces as reservoirs of hospital infection, *J. Hyg., Camb. 65*, 515-536 (1967).
61. D. Vesley and G. S. Michaelsen, in: *Surface Contamination* (B. R. Fish, ed.), pp. 321-331, Pergamon Press, New York (1967).
62. J. Weber, in: Society for Radiological Protection, International Symposium on the Radiological Protection of the Worker by the Design and Control of His Environment, Bournemouth, England (18-22 April 1966).
63. J. J. Cohen and R. N. Kusian, in: *Surface Contamination* (B. R. Fish, ed.), pp. 345-348, Pergamon Press, New York (1967).
64. W. J. Whitfield, in: *Surface Contamination: Genesis, Detection, and Control* (K. L. Mittal, ed.), pp. 73-81, Plenum Press, New York (1979).
65. J. Alzona, B. L. Cohen, H. Rudolph, H. N. Jow, and J. O. Frohliger, Indoor-outdoor relationships for airborne particulate matter of outdoor origin, *Atmos. Environ. 13*, 55-60 (1979).
66. Th. Franke and W. Hunzinger, in: *Diagnosis and Treatment of Deposited Radionuclides* (H. A. Kornberg and W. D. Norwood, eds.), pp. 457-459, Excerpta Medica Foundation, Amsterdam (1968).

67. Th. Franke, G. Herrmann, and W. Hunzinger, in: *Proceedings of the First International Congress of Radiation Protection* (W. S. Snyder, H. H. Abee, L. K. Burton, R. Maushart, A. Benco, F. Duhamed, and B. M. Whearley, eds.) Vol. 2, pp. 1401–1406, Pergamon Press, London (1968).
68. A. Brodsky, Experience with intakes of tritium from various processes, *Health Phys. 33*, 94–98 (1977).
69. A. Brodsky, Resuspension factors and probabilities of intake of material in process (or "Is 10^{-6} a magic number in health physics?"), *Health Phys. 39*, 992–1000 (1980).
70. S. L. Sutter, J. W. Johnston, and J. Mishima, Investigation of accident-generated aerosols: Releases from free fall spills, *Am. Ind. Hyg. Assoc. J. 43*, 540–543 (1982).
71. H. Lannefors and H.-C. Hansson, Indoor/outdoor elemental concentration relationships at a nursery school, *Nucl. Instrum. Meth. 181*, 441–444 (1981).
72. E. B. Sansone, A. M. Losikoff, and R. A. Pendleton, Potential hazards from feeding test chemicals in carcinogen bioassay research, *Toxicol. Appl. Pharmacol. 39*, 435–450 (1977).
73. E. B. Sansone and J. M. Fox, Potential chemical contamination in animal feeding studies: Evaluation of wire and solid bottom caging systems and gelled feed, *Lab. Animal Sci. 27*, 457–465 (1977).
74. E. B. Sansone and A. M. Losikoff, Potential contamination from feeding test chemicals in carcinogen bioassay research: Evaluation of single- and double-corridor animal housing facilities, *Toxicol. Appl. Pharmacol. 50*, 115–121 (1979).
75. E. B. Sansone and A. M. Losikoff, Environmental contamination associated with administration of test chemicals in drinking water, *Lab. Animal Sci. 32*, 269–272 (1982).

13

Application of Pellicles in Clean Surface Technology

PEDRO LILIENFELD

> Shadow is the diminution of light by the intervention of an opaque body. Shadow is the counterpart of the luminous rays which are cut off by an opaque body.
>
> A shadow may be infinitely dark, and also of infinite degrees of absence of darkness.
>
> Leonardo da Vinci

1. Introduction

The protection of optically critical surfaces against the effects of particle contamination by means of transparent barriers has been a long-established approach applied to a broad range of cases, ranging from combustion monitoring[1] to space-borne[2] sensing. Such protective transparent barriers can take the form of cleanable windows, as in the case of the rather prosaic automotive windshield/wiper, or of flow screens, typified by the clean-air curtains incorporated in several types of gas[1] and aerosol[3] monitoring instruments.

Within this general category of methods, this chapter will concentrate on a more recent concept of optical protection for highly critical surfaces: the defocusing barrier. This approach consists of placing a transparent covering film at a position far removed from the focal plane of the optical system that contains the critical surface to be shielded from particle contamination. The volume bounded by the protective film and the protected surface is sealed (typically by a supporting frame), forming an integral assembly.

PEDRO LILIENFELD • MIE, Inc., 213 Burlington Road, Bedford, Massachusetts 01730.

One of the most important recent implementations of this technique is that of the pellicles* used in photolithographic fabrication of large-scale integrated semiconductor circuits. Although most of this chapter will be devoted to this technique, the discussion extends to another important and growing application of the optical defocusing barrier: the protection of optical data storage media such as computer, video, and audio disks. The photolithography pellicle constitutes a thin-film barrier, whereas the disk protection layer can be classified as a thick-film variation. The eye of most vertebrates can be considered an example of the natural evolutionary design of the pellicle concept. Here the protective film, certainly removed from the focal planes of the eyelens, is the cornea, which serves as the protective barrier against contamination by particles in the air whose gradual accumulation on the outer exposed surface is prevented by blinking.

The history of the two above-mentioned modes of application of the defocusing contamination barrier, i.e., the reticle/mask protection pellicle and the protective optical disk layer, can be traced back only a few years. The late 1970s saw the development of high-density optical data storage, as well as the rapid evolution of large-scale integrated circuit technology with its ever more stringent requirement of high fabrication yields in the face of decreasing feature dimensions.

This chapter will treat the application of pellicles in the manufacture of large-scale integrated semiconductor circuits, and will detail the optical configuration of pellicle-protected mask- and reticle-based photolithography, contaminating particle size considerations with respect to production yields, methods of attachment of pellicles to masks and reticles, inspection of pellicle-protected masks, and the tangible advantages derived from the use of pellicles. The chapter will conclude with a discussion of the principles involved in optical data storage and playback, and methods of protection of the recording media against contamination by small particles.

2. Pellicles in Integrated Circuit Fabrication

In order to introduce the pellicle method of contamination protection of masks and reticles, a brief description of photolithography in semiconductor component fabrication will be presented, as well as a discussion of the practical effects of particle contamination.

2.1. Semiconductor Fabrication by Optical Microlithography

At present the predominant method of fabrication of integrated semiconductor circuits is based on the optical lithographic technique, which is employed in one of several steps leading to the final operational "chip." Photolithography

---

*From the Latin diminutive of *pellis* (skin), i.e., *pellicula*. Pellicle-type optical beam splitters and polarizers will not be considered in this context.

applied to microcircuit manufacture was first developed in the 1960s, and as semiconductor devices continue to be reduced in size and/or the number of circuit elements per chip is increased, "optical microlithography not only stubbornly refuses to die" but "continues to comfortably dominate over alternative technologies."[4]

A semiconductor integrated circuit is, in general, an assembly of several planar layers. Each of these layers has a characteristic planar size of several micrometers and a thickness of one micrometer or less. These superimposed layers are constituted of semiconductor materials (generally silicon) with various doping impurities, insulating strata, and conductive films.

The fabrication of an integrated circuit chip consists of producing these stratified interleaving elements on the surface of a substrate (usually of silicon), layer by layer and subject to the required tolerances in three dimensions. Each stratum is created by a patterned transfer of matter through openings formed in a layer (typically of silicon oxide) superimposed on the substrate. The pattern of these openings is obtained by exposing an upper layer of photoresist material to visible or UV radiation. A mask or reticle, containing the master pattern, is imaged on the surface of the resist layer. These initial steps are shown, in very simplified form, in Figure 1.

The term *mask* is usually employed if the process involves actual size reproduction, i.e., if the mask pattern is imaged 1:1 on the resist-covered wafer. If the image on the wafer is a reduced version of that of the master pattern, the mask is referred to as a reticle.[5] Thus, the reticle contains an enlarged image of one or several circuits, and the complete pattern on the wafer is created by reduced projection of the reticle image accompanied by stepping the wafer in a very precisely preprogrammed sequence. Usual reticle-to-wafer reduction ratios are 5:1 and 10:1.

2.2. Methods of Projection

Initially, in the early 1960s, integrated circuit photolithography was based on contact printing. The photoresist-coated wafer was placed in direct contact with the imaged face of the mask. Although this method is still being used, it has been supplanted by more advanced configurations (projection printing and step/repeat) satisfying the increased stringency dictated by finer and finer geometries, as well as by the need to maintain or increase production yields. Problems encountered with contact printing resulted principally from the direct contact between the resist and mask surfaces, leaving residues and defects on both surfaces upon separation,[6] degrading yield, and severely limiting the useful life of the mask. Other problems idiosyncratic of this method of projection resulted from flatness imperfections affecting optical resolution.[5] Subsequent developments evolved soft-contact and proximity printing, i.e., gradual steps in the increasing separation of the mask from the wafer in order to avoid the problems of contact printing, albeit at the expense of feature resolution.

The next phase in the evolution of microlithography was projection print-

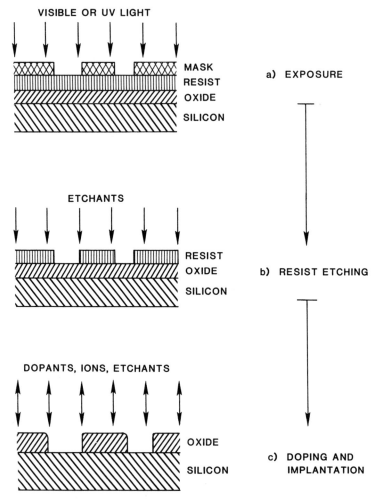

Figure 1. The initial steps in integrated circuit fabrication by microlithography. Contact printing shown as the exposure step.

ing, which eliminated the mask–wafer contact or proximity problems such as degradation by mutual contamination but posed a severe optical design challenge: image blurring. Two general projection-type optical configurations were developed to achieve the required imaging sharpness on the wafer: refractive and reflective optics. The complexities, cost, and other design disadvantages of the refractive approach resulted in the present preference for the projection version based on reflective optics. The essential components of a reflective projection printing system are shown in Figure 2. In practice, both the mask and the

Application of Pellicles

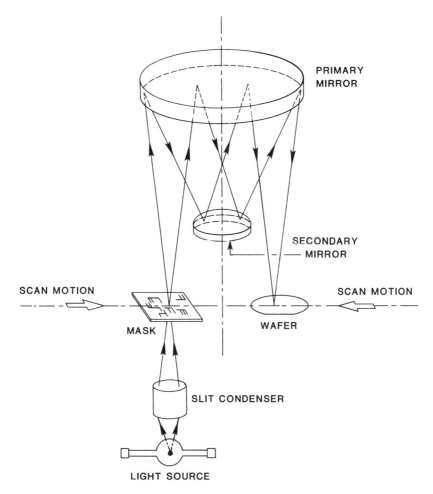

Figure 2. Essential optical components of a typical projection printing system.

wafer are scanned (i.e., advanced) concurrently in order to achieve complete coverage of the surface of the wafer. Replication is to a 1:1 scale, as mentioned before.

During the late 1970s, the direct-step-on-wafer configuration made its appearance. This system uses special illumination optics to gather a maximum amount of light from the source (i.e., a mercury arc lamp) and to direct the highly intense beam through the reticle, which is then imaged to a reduced dimension onto the wafer, which is stepped as mentioned previously.

Figure 3 depicts the optical configuration of a typical reticle-based wafer stepper system. The principal advantages of this approach are[6]:

- Improved correction for wafer distortion
- Improved immunity to contamination by dust on the mask (i.e., reticle) surface
- Improved resolution
- Easier exposure of large-surface wafers

Detailed discussions of the optical principles, performance characteristics, and other design details of the step-and-repeat microlithographic technique and apparatus are available in the extant literature.[7-9]

For both types of photolithographic systems, i.e., 1:1 projection and the stepper/reduction method, the illumination optics must be designed to achieve

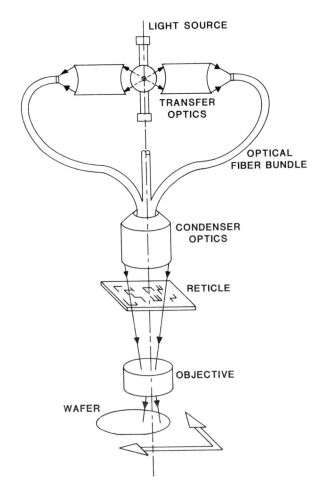

Figure 3. Essential optical components of a step-and-repeat printing system.

uniform irradiation at the mask or reticle. It is thus necessary to insert an optical element between the source of light and the condenser optics that prevents reimaging the source at the object plane. This is achieved by the use of optical fibers,[7,10] lenticular or channel type lenses,[11] or special slit optics.[5]

In the context of this chapter, one of the most important differences between the projection and the stepper methods of microlithography resides in the one-to-one replication of the mask master pattern on the wafer for the former method, as contrasted by the reduction from reticle to wafer (by the typical linear ratios of 5:1 or 10:1) for the latter of these two techniques. The consequences, in terms of mask/reticle surface contamination, of this difference will be discussed later in this chapter.

2.3. Feature Dimensions and Performance Limits of Optical Microlithography

Advances in optical microlithography during the last five or six years have kept pace with the concurrent demands for shrinking chip geometries and denser packing that have characterized this very-large-scale-integrated (VLSI) circuit period. It is generally recognized that the transition from VLSI to ULSI (ultra-large-scale-integrated) circuitry is presently (1985) underway. This terminology reflects, very approximately, the fact that feature dimensions or design rules are shrinking from the 1- to 2-μm range down to the submicrometer realm required to implement the transition to the 1-megabit RAM (random-access-memory) technology of the mid-to-late 1980s.[12]

The prevailing view on the ultimate dimensional limits of optical microlithography is that line widths as small as 0.75 μm are achievable and that eventually this technique will be extended down to about 0.5 μm by using shorter-wavelength (UV) light sources.[5]

Illumination wavelength is directly related to the optical resolution that can be achieved at the wafer surface, and thus the trend in microlithography has been towards shorter-wavelength systems. The most commonly used source, the mercury arc lamp, generates several emission lines in the overall range of about 300 to 600 nm. In general, and until recently, the mercury line at 436 nm (*G*-line) has been used prevalently in typical semiconductor microlithography systems. The trend towards decreasing feature size has been the driving force in the gradually increasing application of the so-called near-UV *I*-line of the mercury lamp emission spectrum at 365 nm,[9] and has stimulated investigations into the use of excimer lasers, in the deep-UV region (see Figure 4).

The minimum achievable line width/feature dimension is directly proportional to the wavelength of illumination and inversely proportional to the numerical aperture (N.A.)* of the projection objective.[9] Figure 4 is a plot of that

---

*Numerical aperture is defined as one-half the ratio of the clear aperture diameter and the focal length of a lens or a lens system.

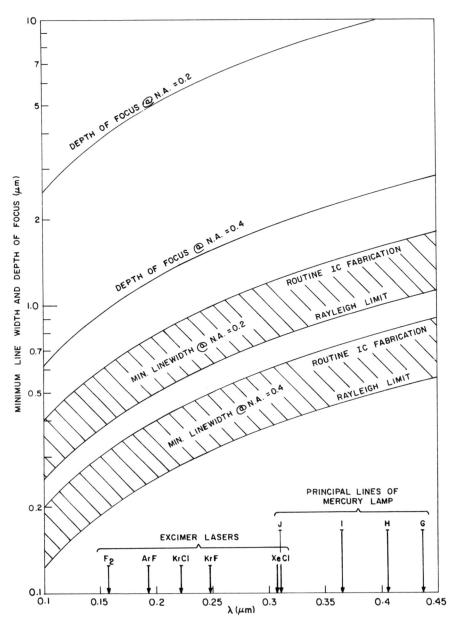

Figure 4. Minimum line width and depth of focus as functions of wavelength of illumination and numerical aperture (N.A.).

relationship showing, for two values of N.A., the ranges between the theoretical limits and the practically (i.e., routinely) achievable fabrication limits. Concurrently, however, as wavelength is decreased the depth of focus at the wafer surface decreases, demanding increasing stringency of the dimensional tolerances, stability, and uniformity of all elements of the system. Minimum line width/feature dimension as well as depth of focus will be related to particle contamination further on in this chapter. In addition, the role of illumination wavelength in the design of protective pellicles will be discussed in light of the trend towards use of UV and VUV (vacuum-ultraviolet) sources.[4,5,12]

2.4. Masks and Reticles in Semiconductor Microlithography

As discussed previously, the mask or reticle (depending on the type of projection system) constitutes the master whose pattern must be faithfully replicated on the wafer.

Masks and reticles usually consist of a square-shaped glass or quartz plate, with flatness tolerances (runout) of the order of 2 to 10 μm,[13] on which the mask coating is applied on one of its two faces. Typical dimensions of masks and reticles range from 2 × 2 inches (5 × 5 cm) to 6 × 6 inches (15 × 15 cm). The useful or "quality" area which is replicated on the wafer is usually circular and represents typically 50 to 60% of the total area of a mask.

The thickness of masks and reticles is related to their size. For example, 4-in (10-cm) units have a thickness that ranges from 0.06 in (1.5 mm) to 0.09 in (2.3 mm) whereas 6-in (15-cm) masks have a typical thickness of 0.09 in (2.3 mm) to 0.12 in (3.05 mm).[13] Most of the mask and reticle plates are beveled to prevent damage due to edge breakage.

Mask and reticle plates are coated with a thin material whose open areas allow the transmission of photons (or any other form of radiation used to expose the wafer) whereas the blocked areas prevent their transmission. Typically (but not exclusively), coatings are of chromium or of chromium oxide with thicknesses of the order of 30 to 100 nm.

The fabrication of masks and reticles is a relatively complex and costly process involving a multiplicity of steps,[14] as illustrated in Figure 5. As indicated in this process flow diagram, there are as many as six independent inspection and quality control operations during the task of producing a mask, and an equal number of rejection or repair "nodes." In addition, after shipping to the users' facility, incoming inspection is frequently performed to further ensure maximum yield in semiconductor chip fabrication.

2.5. Particle Contamination in Integrated Circuit Fabrication

As the dimensional demands in the fabrication of semiconductor integrated circuits have become increasingly stringent, the effects of particle contamina-

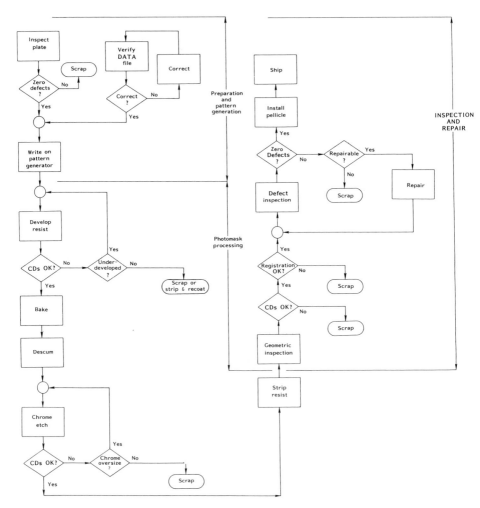

Figure 5. Mask/reticle fabrication process flow diagram. (CD stands for critical dimension.)

tion on the various fabrication processes have become more and more noticeable and destructive. Particle contamination is a very significant contributor to production yield degradation, and this contamination must be controlled, if not eliminated, at each of the steps involved, i.e., wafer preparation, deposition, microlithography, etching and stripping, implementation, diffusion/oxidation, etc. Thus, gases, processing chemicals, process fluids, and the air environment associated with these manufacturing steps must be as free as possible of extraneous particles. Within this chapter, the attention is concentrated on the protection of one of the crucial elements of the photolithographic process, i.e., the mask or reticle. Particles, both solid and liquid, that may deposit on the patterned surface of a mask may cause voids or bridges on the wafer (depending on whether positive or negative resists are used). The final effect of such imperfections may range from marginal increases in circuit resistance and a tendency to degraded performance to outright and catastrophic failure when the finished component is activated.

Contaminating particles can be of many differing origins. A typical list would include silicon, chromium and its compounds, human skin flakes, bacteria and other microorganisms, lint, facial makeup and powders, etc.

The presence of contaminating particles whose size is comparable to the line widths or line separations on the reticle pattern is unacceptable. The maximum dimensions of "acceptably" sized particles range from about 10% to 30% of the line widths[12, 16] and are thus comparable to the so-called critical dimensions of the mask patterns. Particles, as opposed to defects in the mask/reticle pattern itself, are frequently labeled as "soft" defects, presumably because they are not intrinsic to the mask structure and can thus be removed by a variety of cleaning methods. Although most of the investigations performed to assess the effect of particle contamination on semiconductor yield loss have been directed at deposits on the wafer itself, these models can be applied to corresponding effects of mask and reticle contamination. Figure 6 is a graph that relates airborne particle concentration to the total circuit defect density and to the production yield (for a 1-cm^2 chip area) as functions of the number of

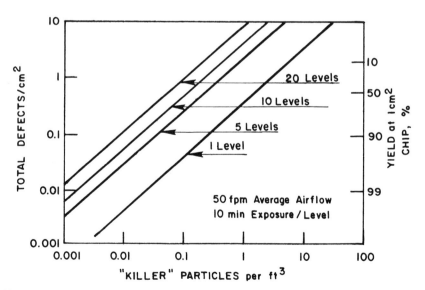

Figure 6. Semiconductor fabrication yield loss due to airborne particles, as a function of microlithographic masking levels on the wafer.[17]

levels of microlithography on the wafer.[17] The model depicted in Figure 6 is based on a 50-fpm (0.25-m/s) average air velocity over the surface and an exposure of 10 min/level. "Killer" particles are loosely defined as those whose size exceeds the maximum acceptable value referred to above.

The production yield can be calculated from the following equation[13]:

$$Y = 100 \left[1 + (D_0 A)^m \right]^{-1} \tag{1}$$

where Y is the yield in percent, D_0 is the area density of random defects on the wafer at each level, A is the die or chip area on the wafer, and m is the number of mask levels.

Approximate translation of the chip defect and yield parameters to corresponding contamination levels at the mask/reticle state can be performed by a direct scaling both in particle size and deposit density based on the mask-to-wafer projection reduction ratio. Thus, dimensions of particles on the mask are reduced to the corresponding defect size on the wafer, whereas particle area density on the mask is increased by the square of the projection reduction ratio. For example, if the defect density on the wafer is to be maintained at less than 1 cm^{-2}, the corresponding maximum "killer" particle area density on the mask of a 1:1 projection system would also be 1 cm^{-2}. In the case of a 5:1 reduction, the reticle contamination density must be maintained below $1/5^2$ cm^{-2}, i.e., less than one "killer" particle per 25 cm^2. For a 10:1 system this figure would be one particle per 100 cm^2. In summary, a 1:1 projection system imposes more stringent limitations on maximum permissible particle size on the mask but less stringent area density requirements, whereas the reduction stepper relaxes the demands on maximum particle size while requiring significantly decreased areal densities of contaminating particles.

As a consequence of the aforementioned criteria, a state-of-the-art requirement of better than 50% production yield on typically sized dies with wafer line widths of the order of one micrometer implies that a reticle used in a 10:1 stepper photolithographic system must be totally free of particles equal to or larger than 3 μm. The recent trend emphasizing the use of 5:1 systems[18] reduces the size of tolerable particles even further. These requirements impose extremely rigorous contamination control procedures, which, in practice and within a production environment, are very difficult to implement and maintain routinely.

2.6. Principles of Reticle/Mask Protection with Pellicles

The basic principle of the pellicle as applied to semiconductor circuit photolithographic fabrication was first introduced through two U.S. patents.[19,20] The first of these[19] described a thick-film approach, using a glass sandwich to encase the mask, whereas the second patent[20] dealt with the (now commonly adopted) thin-film pellicle configuration.

The theoretical background to the overall concept of contrast degradation in optical systems caused by defocusing has been treated in the literature,[21-23] providing the mathematical basis from which the optical parameters of the pellicle barrier can be determined. Further work on the theory is being pursued to refine some of the idiosyncrasies related to the diffractional interactions between the reticle pattern elements and the contaminating particles on the pellicle surface,[24] and design guidelines resulting from these analyses are treated further on in this section.

The basics of the application of pellicles to the protection of masks and reticles have been discussed in several papers[11, 25-30] during the last four years, and the material in this section is a distillation of that literature.

One of the earlier overviews by Hershel[25] describes the pellicle as a "thin, transparent membrane which seals off the mask or reticle surface from airborne particles and other forms of contamination." Figure 7 illustrates the basic configuration of the principal optical elements that include the photolithographic mask, the pellicle, and the wafer surface. Reference should be made to the overall system depicted in Figure 3.

Figure 7 shows, in simplified form, the optical geometry underlying the defocusing barrier principle as applied to a pellicle designed to protect the front or chrome side of a reticle. In some cases, depending on the optical parameters of the photolithographic system, pellicles may be mounted on both sides of a reticle. This applies especially to systems with large focal depths, i.e., with small numerical apertures.

One of the most important pellicle parameters influencing the degree of particle immunity afforded to the photolithographic process is the separation between the object plane containing the mask pattern and the pellicle surface on which the particles may deposit. This distance is commonly known as standoff (see Figure 7). This distance from the optical object plane, i.e., the mask pattern plane, determines the degree of defocusing imparted to the image of a particle lying on the pellicle. A useful measure of defocus is the effect of a pellicle-deposited particle on the light exposure at the image plane, i.e., at the wafer surface. The usual criterion is that the defocused "image" of a particle should reduce the light intensity at the wafer by less than 10%.

Hershel[25] provided the basic design relationship from which the standoff distance can be calculated, however, taking into account only the above-mentioned penumbral effect of particle image spreading at the wafer surface:

$$t_a = 3.57 \times 10^{-3} n f_\# d = 1.79 \times 10^{-3} nd/\text{N.A.} \qquad (2)$$

where t_a is the minimum standoff distance, in millimeters, required to produce a light attenuation of 10%, or less, at the wafer due to a particle on the pellicle; n is the index of refraction of the medium (usually air) between the object plane and the particle; d is the contaminating particle diameter, in micrometers; $f_\#$ is the f-number of the projection lens system; and N.A. is the numerical aperture at the reticle plane.

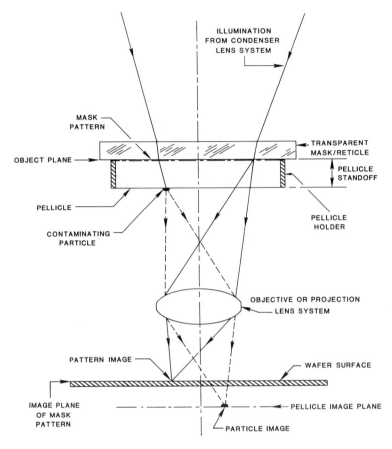

Figure 7. The pellicle within the essential photolithographic elements of a projection–reduction system.

In general, 1:1 projection systems have low f-numbers (typically 3 to 8) whereas step-and-repeat systems are designed with larger f-number values (typically 10 to 25). In the case of these reduction systems it is important, however, to distinguish between the numerical aperture at the reticle plane and that at the wafer plane (i.e., object and image planes, respectively). The optical parameters of equation (2) refer to the reticle plane, which is identical to the wafer plane only in the case of 1:1 projection systems.

Figure 8 is a graphical representation of equation (2) for a range of representative cases for both direct projection and step-and-repeat systems. Clearly, the smaller numerical apertures (at the object plane) of reduction systems impose significantly larger pellicle standoff distances. In addition, the shallower depth of field of lower f-number systems implies that the mask glass thickness

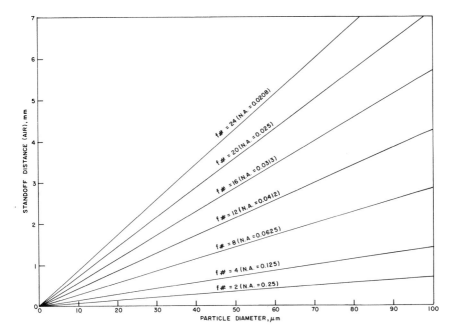

Figure 8. Minimum standoff distance in air as a function of particle size and f-number at mask for 10% light reduction at wafer (based only on shadow spreading).[25]

suffices to defocus any particles on the back side (nonpatterned) of the mask,[31] whereas pellicles may be required on both sides of large f-number configurations.

A more recent criterion for the standoff distance has been developed by Flamholz[24] based on diffraction theory. This model accounts for the image-blurring effect of a particle, i.e., the degradation in the sharpness of the feature edges as projected on the wafer surface (i.e., the image plane). The maximum acceptable degradation defined by Flamholz[24] is given by a blurring equivalent to a $\lambda/8$ defocusing of the image. This standoff analysis is based on a far-field assumption which applies whenever the following condition is satisfied:

$$t > \frac{2b^2}{\lambda} \qquad (3)$$

where t is the standoff, in micrometers; b is the pattern dimension on the mask, in micrometers; and λ is the wavelength, in micrometers.

The dimension b is exemplified by the line separation on the reticle which, for minimum feature dimensions of the order of 1 to 2 μm on the wafer and typical stepper reduction ratios of 5:1 and 10:1, corresponds to values of 5 to 20 μm. Thus, (for $\lambda = 0.4$ μm) t of equation (3) must exceed 0.125 to 2.0

mm,[24] a condition that is generally satisfied by typical commercially available pellicle standoffs.

Based on the diffraction theory mentioned above, the minimum standoff distance, t_d, is equal to:

$$t_d = \frac{4 \times 10^{-3} \, d}{\text{N.A.}} \qquad (4)$$

where N.A. again refers to the numerical aperture at the reticle plane.

Figure 9 is a representation of equation (4) for the same values of N.A. as shown on Figure 8, and applicable to 1:1, 5:1, and 10:1 photolithographic systems.

Because $t_d > t_a$, the diffraction criterion (i.e., Figure 9) should be applied whenever possible in the estimation of the minimum required pellicle standoff distance.

Typical pellicle standoffs range from 3 to 10 mm, depending principally on the numerical aperture of the projection optics. Although pellicles of that type should afford protection against particles as large as 50 to 100 μm in diameter (see Figure 9), image degradation exceeding the diffraction theory pre-

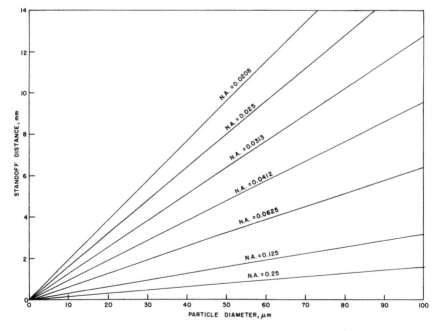

Figure 9. Minimum standoff distance as a function of particle size and numerical aperture at reticle for $< \lambda/8$ defocusing at wafer (based on far-field diffraction theory).[24]

diction has been observed for photolithographic underexposures greater than 10%.[24]

The above analysis and results refer to spherical particles on the pellicle. In the case of fibrous-shaped or elongated particles, it appears that the pellicle design criterion should be based on an equivalent optical size based on the area projected by the particle.

Most frequently, pellicles are attached to only one side (the pattern side) of masks used in 1:1 projections systems,[31] whereas two-sided pellicle protection is often required in the case of step-and repeat reduction systems.[11] The reason for this difference resides in the much smaller numerical aperture of the reduction systems which, in turn, results in a higher sensitivity to particle contamination on the nonpatterned face of the reticle.

2.7. Advantages Resulting from the Use of Pellicles

The principal beneficial result of and the primary justification for the use of pellicles on masks and reticles is the enhancement of the production yield in microelectronic circuit fabrication. Other related advantages associated with the use of pellicles, however, are very significant and are the result of the protection these membranes afford to the masks and reticles themselves. A pellicled mask does not have to be subjected to repeated inspection for contamination[28] and does not undergo the frequent cleaning processes required by unprotected masks. Mask/reticle insertion, transport, and inspection procedures frequently result in the addition of contaminant particles, and the required cleaning operations produce gradual thinning and degradation of the pattern on the reticle, reducing its useful life. Thus significant savings are obtained by the drastic reduction in the need for handling of masks and reticles when these are protected by pellicles.[32,27,13,33,25,28] The rupture of a pellicle membrane is then an unequivocal indicant of potential contamination and/or pattern damage.

In general, very significant semiconductor fabrication yield increases have been reported in the literature, although, as can be expected, the range of these improvements is wide. Rangappan and Kao[29] presented one of the earliest and most exhaustive reports on the positive effects of pellicle protection on fabrication yields; their particular yield improved from 18% to 25%, i.e., a relative enhancement of 28%. Overall, relative improvements have been reported to cover the staggering range of 5 to 400%.[27] Obviously the relative yield enhancement derived from the use of pellicles is very much dependent on the baseline cleanliness of a given fabrication process. Lent and Swayne[34] and Lent[35] report typical increases of four defects per square inch per insertion of unprotected masks, and zero increase of defects per mask insertion after attachment of a pellicle.

Until recently pellicle use was more prevalent in the U.S. than in Japan[36] but this trend seems to be changing in that Japanese IC manufacturers are increasingly relying on pellicle protection of masks and reticles.[30] This is understandably due to the increased stringency in contamination control dictated by ever decreasing feature dimensions. In essence, "pellicles are now commonplace in wafer fabrication."[37]

2.8. Optical Properties and Effects of Pellicles

2.8.1. Transmission Loss Mechanisms

Ideally the pellicle should be a lossless medium, i.e., with a transmission of 100%. Light attenuation implies an increase in exposure time and causes a concomitant decrease in production throughput.[25] In practice, however, several optical mechanisms preclude an idealized pellicle behavior: reflection, absorption, scattering, and interference losses. Each of these phenomena will be discussed briefly.

Reflection of light occurs at both pellicle surfaces, and the fractional loss r_0 can be calculated (without regard to interference effects) from the following relationship[25]:

$$r_0 = 2\left(\frac{n-1}{n+1}\right)^2 \tag{5}$$

The index of refraction, n, for most pellicle materials is in the range of 1.5 to 1.7 and thus the reflection losses will be of the order of 8 to 13%.

Interference or etalon effect[28] is caused by constructive and destructive wave interaction resulting from the reflections at the two surfaces of the pellicle. This interaction causes the typical oscillatory wavelength dependence of the transmission, an example of which is shown in Figure 10.[33] The approximate loss, r_e, associated with the etalon effect can be calculated (for normal ray incidence) from[25]:

$$r_e = r_0\left(1 - \cos\frac{4\pi nh}{\lambda}\right) \tag{6}$$

where h is the pellicle thickness in the same units as λ. Anti-reflection coatings applied to either or both pellicle surfaces tend to reduce the amplitude of the transmission oscillations, enhancing overall transmission.[28,33]

For high-quality nitrocellulose pellicles, the transmission losses caused by scattering are usually less than 2%; however, Mylar pellicles have been found to scatter as much as 10% at the shorter wavelengths. The scattering contribution to the overall attenuation by state-of-the-art materials, however, is less than 0.3% in the UV-4 band.

Light scattering effects are produced by minute surface and internal imperfections and by embedded particles.[25] Discrete particles internal to the pel-

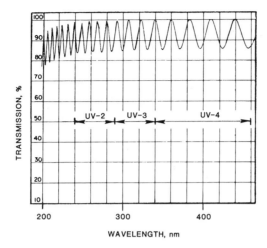

Figure 10. Spectral transmission of nominal 1.9-μm-thick pellicle (JTB/PE UV-2).[33] (Courtesy of the J. T. Baker Chemical Company.)

licle medium must be limited in size by the same criteria as those applied to surface-deposited particles discussed in Section 2.6 of this chapter.

Absorption losses in the pellicle are characteristic of the molecular structure of the film and usually increase with decreasing wavelengths, depending on the material. This wavelength-dependent behavior is exemplified in Figure 10, which shows the gradual roll-off of the average transmission for wavelengths below 260 nm for a pellicle material designed for operation at very short wavelengths.[33] Although absorption is, *per se,* a predictable phenomenon, its long-term effects can impair pellicle usefulness as discussed in the next subsection.

2.8.2. Optical Lifetime

The optical lifetime of these films (as distinguished from mechanical lifetime) is affected by the internal rupture of chemical bonds and by chemical activation of molecules. These absorption-associated effects become more pronounced at shorter and thus more energetic wavelengths.[28] As wavelength decreases, the tolerable absorption decreases for a given useful lifetime. For example, in the 230–300-nm region the peak transmission (i.e., the transmission in the absence of the etalon effect) of a pellicle must exceed 96% in order to ensure an acceptable lifetime.[33] Even at transmission peaks of 98%, eventual failure must be expected below approximately 290 nm (UV-2 band), whereas nearly indefinitely long lifetimes have been observed for exposure to wavelengths exceeding 340 nm (UV-4 band).

Nitrocellulose, a material used extensively in the UV-4 band (see Figure 10), exhibits rapid degradation due to actinic mechanisms when exposed to shorter wavelengths, and driven by the trend towards ever smaller wafer geometries, the search for pellicle materials capable of withstanding extended operation in the UV-3 and UV-2 bands has been escalated. A comprehensive review of the criteria to be fulfilled by candidate materials has been presented by Ward and Duly.[33] The principal conditions to be met are the following: homogeneity of refractive index; high tensile strength; high control of thickness uniformity; absence of pinholes, film gels, reticulation, and other defects; low transmission etalon amplitude; embedded particle size below 50 μm; minimal light scattering ($\leq 0.3\%$); high transmission down to 240 nm; and high degree of flexibility and rebound from stress elongation.

A desirable lifetime for UV-4 pellicles can be defined as the ability to withstand at least 10^5 exposures in the photolithography stage without significant degradation. Typical exposure times are of the order of ten seconds. For pellicles that must operate in the UV-2 band, more relaxed specifications are to be applied: lifetimes in excess of 10^4 exposures are usually considered acceptable.

Realistic lifetime testing of pellicle materials should be performed under representative duty ratio conditions since continuous (i.e., static) illumination can result in accelerated damage from excessive internal heating.[33]

2.8.3. Refraction

Refraction occurs whenever a light ray strikes the pellicle surface at an angle that differs from normal, i.e., if the angle of incidence is nonzero. Refraction within the pellicle then results in a shift of focus at the image plane. Systems with short depth of focus are thus more affected by the presence of a pellicle than those with a long depth of field. Referring to Figure 4, the depth of focus decreases with diminishing wavelengths and thus the defocusing effect of pellicles becomes more critical in the deeper UV.

The shift in focus Δy can be calculated from[28]:

$$\Delta y = \left\{ n - \cos \theta_i \left[1 - \left(\frac{\sin \theta_i}{n} \right)^2 \right]^{-1/2} \right\} \frac{h}{n} \qquad (7)$$

where θ_i is the angle of incidence. The maximum angle of incidence to which a pellicle is exposed can be obtained from the following relationship:

$$\theta_{i\max} = \tan^{-1} (2f_\#)^{-1} = \tan^{-1} \text{N.A.} \qquad (8)$$

As an example, assuming an $f_\#$ of 16 (N.A. = 0.0313) at the reticle, a pellicle thickness $h = 2$ μm, and an index of refraction $n = 1.5$, one obtains a maximum defocus (i.e., near the periphery of the useful pattern) of $\Delta y = 0.67$ μm, which is usually negligible compared with the typical depth of focus of several micrometers of most projection systems.

Other refraction-related problems may arise from pellicle tilt and from nonparallelism of the pellicle surfaces (i.e., thickness nonuniformities). In practice, however, neither of these geometric distortions becomes significant enough to cause appreciable defocusing effects through pellicle refraction.[28]

2.8.4. Effects of Thickness Nonuniformity

The etalon effect, discussed in Section 2.8.1 with respect to wavelength-dependent reflection losses, applies equally to pellicle thickness nonuniformities. Equation (6) indicates that increases in pellicle thickness (h) produce the same effect on light transmission as decreases in wavelength.

In general, thickness variations are more critical for monochromatic or narrow-band illumination, and their effects become more pronounced for decreasing ratios of pellicle thickness to wavelength.

Thus, the thickness uniformity required to maintain relative photolithographic exposure homogeneity is a function of the pellicle thickness itself, as well as the illumination wavelength. Exposure homogeneity requirements of 1–2% are typical, and thus thickness uniformities must be maintained, in some cases, within ± 0.01 μm.[28] By judicious selection of pellicle thickness for a specific illumination wavelength, this tolerance can be relaxed to as much as ± 0.3 μm (e.g., for a pellicle thickness of 2.85 μm operating in the UV-4 band).[33]

Figure 11 illustrates the effects of pellicle thickness and of illumination wavelength spectrum on the transmission of the type of pellicle material whose monochromatic spectral transmission curve is shown in Figure 10. Transmission sensitivity to pellicle thickness nonuniformities can be reduced significantly by means of anti-reflective coatings, as mentioned previously in the discussion on the etalon effect. Deposited MgF_2 is one of the most commonly used coatings, but problems arising from inadequate bonding to the pellicle and internal stresses in the coating have limited its application. The development of a proprietary anti-reflection coating devoid of these problems has been reported.[33]

2.9. Pellicle Materials and Mechanical Properties

The stringent requirements imposed on pellicles used in microlithography determine the type of materials of which they can be made. Initially, Mylar was the prevalent material but soon it was replaced by nitrocellulose which, until recently, was the most frequently used substance for this purpose. The requirement for shorter-wavelength microlithography has led to the recent (1985) appearance of several new materials, complementing and/or supplanting nitrocel-

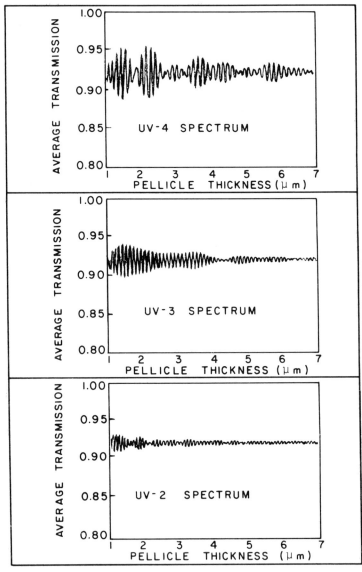

Figure 11. Computer simulation of the average integrated transmission vs. pellicle thickness for (a) UV-4 spectral region (340–460 nm); (b) UV-3 spectral region (290–340 nm); and (c) UV-2 spectral region (240–290 nm).[33] (Courtesy of Perkin-Elmer.)

lulose in this application. Ward and Duly[33] review in detail the criteria that must be fulfilled by such films and discuss the development of an amorphous organic polymer (JTB/PE) which exhibits broadband transmission characteristics and other favorable properties. Further details about the manufacturing methods used and the chemical composition of these pellicles are given in three recent patents.[38-40] Another polymer used for pellicles is parylene,[30] whose salient advantages are enhanced resistance to chemicals and better mechanical strength than nitrocellulose. Cellulose acetate is another widely used material for pellicles, especially for shorter-wavelength illumination (i.e., down to 240 nm).

Pellicles must withstand various types of cleaning procedures in order to remove very large particles (i.e., >50 μm, typically) deposited on the outer surface. Typical "strength" specifications given by pellicle manufacturers are: "film will withstand 40 psi from nitrogen gun at a distance of 6 inches"[41] or "will stand 30 psi from 3M 902 F gun at 4"."[42] In addition, the user will be informed that "film will withstand gentle washing with laboratory detergent and deionized water."[41]

2.10. Pellicle Mounting and Attachment

A pellicle is usually supported by a mounting ring or frame over which it is stretched and adhesive-bonded. This procedure typically requires the use of auxiliary locking rings, which are removed after bonding, in order to achieve uniform and repeatable radial tension of the thin pellicle membrane.[20] The height of the support ring or frame (which may be round or rectangular) determines the spacing between the pellicle and the mask/reticle surface. Other pellicle support frame fabrication methods have been developed such as those described by Duly et al.,[43] based on a thin metal membrane attached to a metal ring; the membrane is then coated with polymethylmethacrylate on one side and with a resist film on the other. After etching away a selected area of the metal substrate, the polymer film pellicle is left free.

In general, special care must be exercised in bonding pellicles to their supporting rings, in particular to prevent contamination of the inner surfaces of the pellicle[44] by the adhesive.

The use of transparent support rings has been proposed in order to permit grazing illumination for optical inspection of the mask surface after pellicle installation.[45] Most pellicle support rings/frames are metallic although other rigid and stable materials can be used.

Pellicles mounted on circular rings are used most frequently in projection-type photolithographic systems, whereas square-mounted pellicles are typical of step-and-repeat systems. The typical frame thicknesses (i.e., dimension perpendicular to the pellicle–mask gap) are in the range of 4 to 6 mm.

A most important procedure in the implementation of pellicle protection of masks and reticles is their installation or attachment. Reference will be made to the methodology detailed by Turnage and Winn[32] and to specific fixtures and mounting devices described in various relevant patents[20,46] and other literature.[47] It is essential that all steps in the installation and attachment of pellicles to masks should be performed under the most stringent clean-room conditions, with repeated visual inspection (under high-intensity collimated light illumination) of the inside pellicle surface (i.e., the side that will face the mask) and the mask surface itself. All work should be performed on a clean-air laminar flow bench, preferably with horizontally directed flow.[32]

Pellicles, which are usually stored in a sealed container,[48] should be held only by the edge of their frame or ring. The pellicle, after removal from its container (special handling tools are available for this purpose),[30] is placed in a special mounting fixture, various designs of which have been developed[46] and are commercially available.[41,42,49] The actual bonding between the pellicle ring or frame and the surface of the mask or reticle is performed by an adhesive on the pellicle ring which is protected by a backing layer prior to its attachment. The use of a nitrogen blow-off gun incorporating a high-efficiency particle filter and an antistatic ionizing blower (e.g., with a polonium-210 alpha radiation source) are recommended to clean off the critical surfaces of the pellicle and the mask.[32]

Special carriers for the insertion of the pellicled[32] (or pellicalised[29] or pellicized[13]) masks into the photolithographic systems are available with specific designs for each of the presently existing principal commercial types of projection and step-and-repeat units.[47] When not in use, pellicled masks and reticles must be stored in sealed contamination-free containers of which various types exist.[41,42,50] Tools for the removal of pellicles from masks are also available.[41,42,51]

2.11. Inspection of Reticles and Masks Protected by Pellicles

Although pellicles are highly effective in shielding the mask/reticle surfaces from contamination, these surfaces must be inspected for the presence of particles prior to and following pellicle attachment. Those inspection steps are mandatory because if any particles are enclosed by the pellicle, these contaminants are made quasi-permanent and may be cleaned off only after destructive removal of the pellicle (notwithstanding the mention by Fisher et al.[45] of a pulsed high-energy laser beam sub-pellicle particle elimination technique). Pre-pellicle installation inspection (and cleaning, if required, see Figure 5) ensures the proper initial conditions that must precede pellicle attachment. Post–pellicle installation inspection serves to confirm the results for the preceding inspection and determines whether the pellicle installation procedure itself resulted in any contamination, in which case the usefulness of and reliance on the pellicled mask becomes questionable. These inspection steps may be performed either at the mask/reticle manufacturer, if the pellicled mask is provided as a completed entity, or at the user's facility where pellicle installation may take place.

Mask and reticle inspection methods fall into two principal categories: pattern inspection and contamination detection. The former of these constitutes a detailed analysis of the pattern itself, usually performed by computerized comparison against another identical reference mask or with respect to stored data.[16,52] Any deviations from the reference pattern are thus detected. This is an exhaustive inspection that may require tens of minutes up to several hours.[53,54] It is customarily applied to the detection of "hard" defects, i.e., permanent pattern imperfections, internal bubbles, and inclusions as opposed to "soft" defects, i.e., removable surface-deposited particles. Inspection for these soft defects is preferentially (but not exclusively) performed by light scattering: either (until recently the most common method) by visual observation of the mask/reticle surface using grazing illumination by a collimated white light beam or, alternatively, by means of laser scanning.[55-57] The smallest particles that can be "seen" (with a reasonable degree of confidence) by the visual method are in the general range of 5 to 10 μm. The combined requirements for improved detection reliability and reduced particle size imposed by ever smaller dimensions and separations of the features to be printed on the wafer have spurred the development and use of automated laser scanning mask and reticle inspection systems capable of detecting particles as small as 2 μm in diameter (polystyrene reference spheres). Although particles down to 0.3 to 0.5 μm can be similarly detected on the unpatterned glass substrates used for photomasks, the repetitive pattern lines typical of many finished masks and reticles produce highly interfering optical signals due to constructive-wave diffraction effects. Although the general subject of detection of surface contamination[58] is treated elsewhere in this treatise, a summary description of the laser scanning detection methodology, in the specific context of mask/reticle inspection, will be included here.

As described in a recent review by Lilienfeld,[59] a typical laser scanning system, designed for mask and reticle inspection, incorporates a helium-neon continuous-wave laser with power outputs in the range of 2 to 10 mW, with focusing optics designed to generate a converging beam whose diameter at the point of illumination on the mask is of the order of 50 μm. Electromechanically driven scanner mirrors are used to produce the scanning motion of the beam at frequencies of the order of 100 to 300 Hz. Various different configurations have been developed to achieve illumination of the two faces of the mask/reticle: either using a single laser with alternate[55] or with sequential[57] illumination of the two surfaces or by means of two independent lasers.[56] Grazing illumination (i.e., large angle of incidence) is usually preferred in that it reduces general surface-scattered background intensity and interferences resulting from refraction/reflection of light reaching the opposite face.

Customarily, complete inspection coverage of the surfaces is obtained by combining the laser scan with uniform translation of the mask/reticle perpendicularly to the scan motion.

The optical configuration of a typical automated laser scanning inspection system with sequential illumination of the two surfaces of glass-plate substrates for masks and reticles is depicted in Figure 12.[59]

Several different detection geometries have been used in the present context: cylindrical objective lenses followed by fiber optics to transfer the collected light onto the photocathode of a multiplier tube[55,59-61] or multiple photomultiplier detector configurations "looking" along the scanning path of the laser beam impinging on the mask surface.[56,57] Both the angle of incidence and the angle of detection, with respect to the surface, are constrained if pel-

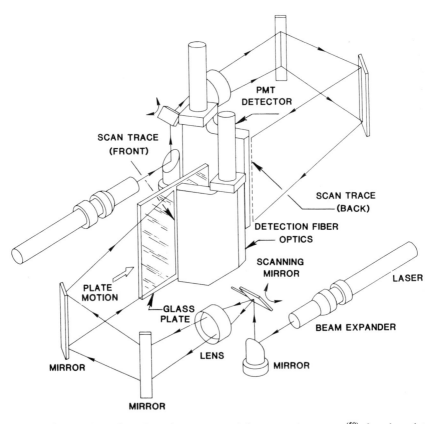

Figure 12. Optical configuration of an automated laser scanning system[59] for glass-plate substrates of masks and reticles (GCA model API-1000).

licled masks and reticles are to be inspected, because of the obstructing effect of the pellicle frame. Furthermore, inspection using very oblique angles with respect to the pellicle surface would result in excessive transmission losses and refractive distortions.

Several alternative optical techniques are used to discriminate against the aforementioned diffraction grating effect produced by the chrome pattern on the mask surface. The magnitude of the pattern-generated diffraction is a function of several parameters[57]: illumination light intensity, polarization and angle of incidence, size of the illuminated mask area, dimensions (separation, width, length) of the pattern lines, angle of these lines with respect to the plane of incidence, etc. The resulting noise at the detector(s) also depends on the scattering and acceptance angles, use of polarization analyzers, etc. Methods applied to reduce or eliminate pattern diffraction noise are (a) angular distribution discrimination[57] using multiple-detector sensing, (b) polarization discrimination,[56] and (c) dynamic spatial filtering.[61] In some cases combinations of these techniques are implemented.

The results of the scanning survey of the mask/reticle surfaces is typically displayed as a particle ''map''[55] on a video screen, indicating the location and size classification of each detected particle. Typical total inspection times, including the mask transport, scanning, and display functions are of the order of 10 to 100 seconds.

These laser scanning systems are either configured as independent inspection stations[56] or incorporated within microlithography units.[55,57]

The use of pattern inspection systems, as opposed to laser scanning devices, is generally limited to the detection of particles larger than 3 μm on pellicled masks because the pellicle standoff is incompatible with typical mask inspection optics,[62] although improvements in this technique are expected, extending its capabilities to smaller sizes.[16,53] An alternative method of pellicled mask inspection, compatible with mask analysis systems, has been reported by Asselt and Brooks.[62] It consists of generating photolithographically glass ''wafers'' starting with the pellicle-protected mask. These test ''wafers'' can then be subjected to a complete pattern analysis to uncover any printable defects.

Tests have been reported on comparisons between the quality of photolithographically processed silicon wafers generated with pellicled and unpellicled masks. Ausschnitt et al.[63] describe such a test, based on an electrical probe technique, that indicated that the presence of a 2.85-μm-thick nitrocellulose pellicle had no measurable effect on the distribution of line widths (with typical feature size of 1.6 μm) over the entire wafer surface, as obtained using a projection printer operating in the 350–450-nm wavelength regime; i.e., the presence of the pellicle did not degrade the optical resolution of the photolithographic process.

3. Protective Films on Optical Data Storage Media

Another important and growing application of the defocusing barrier method of protection against surface-deposited particle contamination is its use on optical write/read media such as digital data storage for computers, compact audio disks, video disks, optical cards, etc. The essential principles involved in the technology will be reviewed in this section.

3.1. Principles and Methods of Optical Storage

Practical versions of optical data storage devices were developed in the 1970s, principally in the form of video disks. These were originally designed as analog recording media where the video signal was frequency-modulation-coded on the disk.[64] This development was followed immediately by digital optical recording systems, initially for computer data storage[65,67] and then, more recently, in the form of audio disks (i.e., "compact disks" or CDs).[68] In addition, the experimental introduction of optical cards for the storage of personal information has been reported.[69]

Typical digital optical recordings are of "write once, read many times" (WORM) type, wherein the data storage is in the form of micrometer-sized pits or holes in a tellurium alloy metallic film deposited on a plastic substrate. On playback, the presence or absence of these pits determines the digital transitions between ones and zeros. Typical pit dimensions of compact digital disks are 0.5 μm wide, 0.11 μm deep, and 0.9 to 3.3 μm long. The pits are situated in a spiral track with a pitch or separation of 1.6 μm.[68]

There are two principal disk configurations to be considered in this context: the computer data storage disk and the compact audio disk. The former, which is depicted in Figure 13a, permits two-sided recording/playback, whereas the latter (Figure 13b) is designed for single-sided use.

Each of the two outer protective covers/substrates of the computer disk are made of (1.1 ± 0.1)-mm-thick transparent polymethylmethacrylate. The internal sealed air cavity (see Figure 13a) has a thickness of 0.5 mm. The tellurium alloy film, about 35 nm thick, is deposited on the inner surface of each of the transparent substrates. The recording laser beam passes through the substrate and vaporizes a minute amount of the metal film, forming a hole. The readout (i.e., playback) laser beam is strongly back-reflected by the unpitted tellurium, whereas at a hole, reflectivity is lowered significantly as the beam enters the sealed cavity, a miniature clean room.[70]

In the case of the compact disk, which is manufactured by injection molding using a master recording, a supporting plastic layer, containing the replicated steps, is coated with a reflective aluminum film which, in turn, is encap-

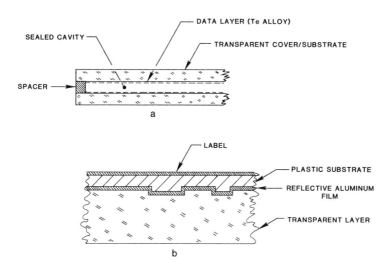

Figure 13. (a) Cross section of typical computer data optical storage disk. (b) Cross section of typical compact audio disk.

sulated in a transparent protective cover (1.2 mm thick) as shown in Figure 13b. Playback is performed by sensing the changes in beam reflection (by wave interference) that occur at the transition between adjacent steps.[68]

3.2. Optical Data Readout and Effect of Protective Layer

As mentioned before, signal pickup or playback in optical data recording systems is achieved by illuminating the micrometer-size pit track and sensing the resulting changes in reflectivity. If the surface containing the digitally coded information is left unprotected, contaminating particles settling on that surface are likely to be detected as reflectance discontinuities, causing error bits. Although some of these errors can be filtered out electronically, the overall reliability of such a recording could be impaired depending on the number, size, and location of such surface-deposited particles. Since the readout beam at the recording surface is of the order of 1 μm in diameter, particles on that surface that are larger than about 0.4 to 0.5 μm could affect the system performance. Although experimental optical data storage systems have been operated in clean rooms, the generalized use of such recording media (especially the compact digital disk) would have been entirely impractical if such a particle-free environment were to be required.

The transparent protective layers of the optical data storage disks (Figures

13a and 13b) thus serve to provide functional immunity from contaminating particles before, during, and after recording in the case of the WORM computer version, and after compact audio disk fabrication. This immunity from particle-related effects is particularly important in the latter case because compact disks are handled in a completely uncontrolled environment (the home) on a potentially very frequent and long-term basis. The role of the protective layer includes, in addition, the prevention of any other "hard" damages to the information-bearing surface, i.e., scratches, indentations, fingerprints, etc. The transparent layer thus constitutes a thick-film version of the thin-film pellicle protection configuration discussed in detail in Section 2 of this chapter. The basic design criterion is very similar to the pellicle case: the creation of a barrier that constrains any potentially interfering contaminant to an optical plane re-

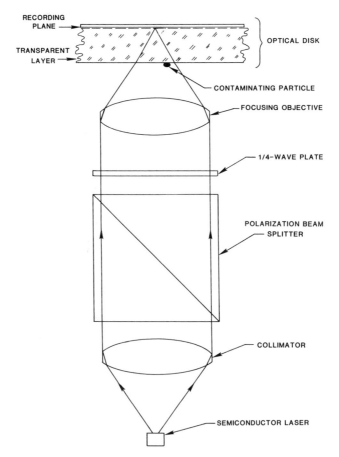

Figure 14. Optical configuration of compact digital disk pickup (simplified, not to scale, and showing only illumination components).

moved far enough from the critical surface that, through their out-of focus location, deposited particles will have a negligible effect on the information to be retrieved from the digital recording medium.

The basic optical configuration of a typical compact disk player is shown in Figure 14. For clarity, only the illumination components are depicted and the automatic focusing components are omitted. The light source is an AlGaAs semiconductor laser emitting at a wavelength of 800 nm. The beam splitter and quarter-wave retarder are required to separate the illumination beam from the reflected signal beam.[68] The focal length of the collimator is typically 17 mm whereas the objective lens system has a focal length of only 4 mm. The effective focal depth of the optical pickup is only 2 μm. The spot size of the illumination beam on the outer surface of the transparent layer is 0.8 mm in diameter, and consequently particles and other surface defects smaller than approximately 0.5 mm have no effect on the digital signal readout (the effects of larger obstructions are reduced or eliminated by electronic error correction codes); this is illustrated qualitatively in Figure 14.

It can be foreseen that future, more advanced, digital recording media such as those required for erasable magneto-optical data storage[71] will incorporate thick-film transparent defocusing barriers of the same type as in the case of the nonerasable media described in this section. Magneto-optical recording based on the Kerr effect can be expected to place more stringent requirements on immunity from particle contamination because of the signal-to-noise idiosyncrasies of that technique.

4. Summary and Future Prospects

This chapter presents a review and description of the practical implementation of the defocusing barrier technique designed to impart operational immunity from surface particle contamination to two classes of optically based processes: microlithographic fabrication of large-scale integrated semiconductor circuits and optical digital data storage/playback. The general concept is based on creating a transparent physical barrier between the critical surface and the contaminating particles such that these are constrained to lie in a plane that is sufficiently removed from the focal plane of the optical system. Consequently, the presence of these particles becomes optically harmless with respect to the process to be protected.

The theory, critical parameters, practical implementation, and beneficial effects of pellicle shielding of masks and reticles used in microlithography have been presented, as well as a brief discussion on the transparent protective layer incorporated in computer data storage media and compact audio disks based on optical digital recording and playback.

This method of protection from particle contamination represents a general design philosophy wherein the presence and surface deposition of such contaminants is accepted but their deleterious effect is neutralized through an appropriate optical configuration.

The future developmental refinements of pellicles applied to semiconductor photolithography will most probably be directed at new materials capable of transmitting light in the deep-UV region with minimum losses in order to maximize useful lifetime. Although new semiconductor fabrication technologies may be developed (e.g., electron beam lithography) that do not require the use of pellicles, the presently installed base of optical systems and their replacement cost should ensure the continuing application and further evolution of these protective devices.

The defocusing barrier will, as mentioned before, become an integral and inseparable design element of optical and optically related high-density data recording/storage media.

In general, it can be expected that the pellicle concept of optical protection against particle contamination will be applied with increasing frequency to other critical optical imaging systems, especially in those cases involving large numerical apertures, i.e., short focal lengths combined with large lens/mirror diameters. It can also be foreseen that the design of certain optical systems will be influenced by the need to incorporate pellicle-type barriers, rather than adding such devices as *a posteriori* solutions to particle contamination problems.

References

1. P. Tomic, New Concepts of Optical Windows in Process Streams, GCA Final Report on U.S. Environmental Protection Agency Contract No. 68-02-3168 Work Assignment No. 109 (April, 1984), 25 pp. Available from GCA/Technology Div., Bedford, Mass.
2. S. A. Hoenig, Electrostatic dust protection for optical elements, *Appl. Optics 21*(3), 565–569 (1982).
3. P. Lilienfeld, High concentration dust mass monitor, *Particulate Sci. Technol. 1*, 91–100 (1983).
4. K. Jain, Laser applications in semiconductor microlithography, *Lasers and Applications 2*(9), 49–56 (1983).
5. G. Pircher, Submicron lithography, in: Proceedings of 9th International Vacuum Conference and 5th International Conference on Solid Surfaces, Madrid (Sept. 26–Oct. 1, 1983), pp. 427–444.
6. D. J. Elliott, *Integrated Circuit Fabrication Technology*, McGraw-Hill Book Company, New York (1982).
7. J. Roussel, Step and repeat wafer imaging, *SPIE Proceedings on Developments in Semiconductor Microlithography III 135*, 30–35 (1978).
8. G. L. Resor and A. C. Tobey, The role of direct step-on-the-wafer in microlithography strategy for the 80's, *Solid State Technol. 22*(9), 101–108 (1979).
9. V. Miller and H. L. Stove, Submicron optical lithography: *I*-Line wafer stepper and photoresist technology, *Solid State Technol. 28*(1), 127–136 (1985).
10. H. B. Lovering, Optics in Microelectronics, Kodak Publication G-45 (1975), pp. 66–71.

11. T. L. Hershey, Pellicles on wafer steppers with lenticular optics, *Solid State Technol.* 26(7), 89–94 (1983).
12. A. C. Tobey, Semiconductor microlithography through the eighties, *Microelectronic Manufact. Testing* 8(4), 19–20 (1985).
13. D. J. Elliott, *Integrated Circuit Mask Technology*, McGraw-Hill Book Company, New York (1985).
14. R. M. Shoho, Fabrication of microelectronics reticles, *Solid State Technol.* 22(2), 75–79 (1979).
15. J. J. Greed, Photomask and reticle making for VLSI, *Microelectronic Manufact. Testing* 6(7), 22–24 (1983).
16. P. H. Singer, Photomask and reticle defect detection, *Semiconductor International* 8(4), 66–73 (1985).
17. C. M. Osburn, Aerosol control in semiconductor manufacturing, paper presented at the First International Aerosol Conference, Minneapolis, Minn. (September 17–21, 1984). Extended abstract in: *Aerosols—Science, Technology and Industrial Applications of Airborne Particles* (B. Y. H. Liu, D. Y. H. Pui, and H. J. Fissan, eds.), p. 673, Elsevier, New York (1984).
18. P. S. Burggraaf, Reduction reticle trends: Emphasizing 5×, *Semiconductor International* 7(8), 58–63 (1984).
19. G. Abraham and G. Bergasse, Projection Printing System with an Improved Mask Configuration, U.S. Patent 04063812 (December 20, 1977).
20. V. Shea and W. J. Wojcik, Pellicle Cover for Projection Printing System, U.S. Patent 04131363 (December 26, 1978).
21. W. H. Steel, Etude des effets combinés des aberrations et d'une obturation centrale de la pupille sur le contraste des images optiques, *Revue Optique* 32(1), 4–26 (1953).
22. W. H. Steel, The defocused image of sinusoidal gratings, *Optica Acta* 3(2), 65–74 (1956).
23. H. H. Hopkins, The frequency response of a defocused system, *Proc. Royal Soc. A 231*, 91–103 (1955).
24. A. Flamholz, An analysis of pellicle parameters for step-and-repeat projection, *Proceedings of SPIE on Optical Microlithography III 470*, 138–146 (1984).
25. R. Hershel, Pellicle protection of IC masks, *Semiconductor International* 4(8), 97–106 (1981).
26. R. Winn and R. Turnager, Pellicles—an industry overview, *Solid State Technol.* 25(6), 41–43 (1982).
27. R. Iscoff, Pellicles—a means to increase die yield, *Semiconductor International* 5(9), 95–108 (1982).
28. T. A. Brunner, C. P. Ausschnitt, and D. L. Duly, Pellicle mask protection for 1:1 projection lithography, *Solid State Technol.* 26(5), 135–143 (1983).
29. A. Rangappan and C. Kao, Yield improvement with pellicalised masks in projection printing technology, *Proceedings of SPIE on Optical Microlithography—Technology for the Mid-1980s 334*, 52–57 (1982).
30. R. Iscoff, Pellicles 1985: An update, *Semiconductor International* 8(4), 110–115 (1985).
31. P. S. Burggraaf, Wafer steppers: Considering the issues, *Semiconductor International* 5(4), 57–78 (1982).
32. R. Turnage and R. Winn, Attaching pellicles to photomasks in a production environment, *Microelectronic Manufact. Testing* 6(7), 31–32 (1983).
33. I. E. Ward and D. L. Duly, A broadband, deep UV pellicle for 1:1 scanning projection and step and repeat lithography, *Proceedings of SPIE on Optical Microlithography III 470*, 147–156 (1984).
34. J. Lent and S. Swayne, The Implementation of a Pellicle Mask Protection System into an Established Production Area, Kodak Publication G-136 (1982), pp. 93–99.
35. J. Lent, Pellicle mask protection for 1:1 projection aligners, *Motorola Technical Developments 2*, 22–23 (1982).

36. K. W. Edmark and G. Quackenbos, An American assessment of Japanese contamination-control technology, *Microcontamination* 2(5), 47-53, 125 (1984).
37. R. L. Ruddell, Resist and mask trends, *Semiconductor International* 7(7), 104-108 (1984).
38. I. E. Ward and P. M. Papoojian, Pellicle Compositions and Pellicles Thereof for Projection Printing, U.S. Patent 04499231 (February 12, 1985).
39. I. E. Ward, Polyvinyl Butyrate Pellicle Compositions and Pellicles Thereof for Projection Printing, U.S. Patent 04482591 (November 13, 1984).
40. I. E. Ward, Pellicle Compositions and Pellicles Thereof for Projection Printing, U.S. Patent 04476172 (October 9, 1984).
41. Micropel Products Bulletin, EKC Technology, Inc., Hayward, Calif.
42. Advanced Semiconductor Products Data Sheet No. 109, Santa Cruz, Calif. (June 9, 1982).
43. D. L. Duly, H. Windischmann, and W. D. Buckley, Method of Fabricating a Pellicle Cover for Projection Printing System, U.S. Patent 4465759 (August 14, 1984).
44. P. R. Carufe and J. R. Kraycir, Photomask pellicle support ring design, *IBM Technical Disclosure Bulletin* 27(1B), 769 (1984).
45. D. W. Fisher, V. Shea, P. Trongo, and W. Wojcik, Transparent ring for low angle pellicle inspection, *IBM Technical Disclosure Bulletin* 23(2), 526 (July, 1980).
46. C. M. Walwyn and D. E. Bohonos, Pellicle Mounting Fixture, U.S. Patent 04443098 (April 17, 1984).
47. A. B. Patel and E. Wojciekfsky, Mounting of mask with pellicle, *IBM Technical Disclosure Bulletin* 26(8), 4036-4037 (1984).
48. Y. Yen, Dustfree Packaging Container and Method, U.S. Patent 04470508 (September 11, 1984).
49. Tau Laboratories, Inc., Products Bulletin (1982), Riddings, Derby, England.
50. A. K. M. Miller and R. Mason, Container for Masks and Pellicles, U.S. Patent 4511038 (April 16, 1985).
51. J. W. Conant, Pellicle Ring Removal Method and Tool, U.S. Patent 04255216 (March 10, 1981).
52. P. Chipman, Qualifying reduction reticles, *Semiconductor International* 7(8), 68-73 (1984).
53. P. S. Burggraaf, 1× Mask and reticle technology, *Semiconductor International* 6(3), 40-45 (1983).
54. R. A. Simpson and D. E. Davis, Detecting submicron pattern defects on optical photomasks using an enhanced EL-3 electron-beam lithography tool, *Proceedings of SPIE on Optical Microlithography—Technology for the Mid-1980s 334*, 230-237 (1982).
55. G. Quackenbos, S. Broude, and E. Chase, Automatic detection and quantification of contaminants on reticles for semiconductor microlithography, *Proceedings of SPIE on Integrated Circuit Metrology 342*, 35-43 (1982).
56. M. Shiba, M. Koizumi, and T. Katsuta, Automatic inspection of contaminants on reticles, *Proceedings of SPIE on Optical Microlithography III 470*, 233-239 (1984).
57. A. Tanimoto and K. Imamura, Reticle contamination monitor for a wafer stepper, *Proceedings of SPIE on Optical Microlithography III 470*, 242-249 (1984).
58. K. L. Mittal, ed., *Surface Contamination: Genesis, Detection and Control*, Vols. 1 and 2, Plenum Press, New York (1979).
59. P. Lilienfeld, Optical detection of particle contamination on surfaces—a Review, *Aerosol Sci. Technol.* 5(2), 145-165 (1986).
60. L. McVay and P. Lilienfeld, Automatic Detector for Microscopic Dust on Large-Area Optically Unpolished Surfaces, U.S. Patent No. 4402607 (September 6, 1983).
61. E. T. Chase, S. V. Broude, and G. S. Quackenbos, Surface Inspection Apparatus, *U.S. Serial No. 682794 patent pending* (filed December 18, 1984).
62. R. V. Asselt and G. Brooks, Technique for Inspecting Photomasks with Pellicles Attached, Kodak Publication G-136 (1982), pp. 158-162.

63. C. P. Ausschnitt, T. A. Brunner, and S. C. Yang, Application of wafer probe techniques to the evaluation of projection printers, *Proceedings of SPIE on Optical Microlithography—Technology for the Mid-1980s 334*, 17–25 (1982).
64. G. Bouwhuis and J. J. M. Braat, Video disk player optics, *Appl. Optics 17*(13), 1993–2000 (1978).
65. G. C. Kenney, D. Y. K. Low, R. McFarlane, A. Y. Chan, J. S. Nadan, T. R. Kohler, J. G. Wagner, and F. Zernike, An optical disk replaces 25 mag tapes, *IEEE Spectrum 16*(2), 33–38 (1979).
66. D. C. Kowalski, D. J. Curry, L. T. Klinger, and G. Knight, Multichannel digital optical disk memory system, *Optical Eng. 22*(4), 464–472 (1983).
67. R. McFarlane, G. Blom, A. Chan, S. Chandra, E. Frankfort, G. Kenney, D. Low, and J. Nadan, Digital optical recorders at Mbit/s data rate, *Optical Eng. 21*(5), 913–922 (1982).
68. S. Miyaoka, Digital audio is compact and rugged, *IEEE Spectrum 21*(3), 35–39 (1984).
69. J. Hecht, Optical memory for personal computers, *Lasers and Applications 4*(8), 71–76 (1985).
70. H. Brody, Materials for optical storage: A state-of-the-art survey, *Laser Focus 17*(8), 47–52 (1981).
71. M. Hartmann, J. Braat, and B. Jacobs, Erasable magneto-optical recording media, *IEEE Trans. Magn. 20*(5), 1013–1018 (1984).

Index

Adhesion (metal), effect of organic film thickness on, 216
Adhesion of iron
  effect of oxygen on, 207
  effect of various hydrocarbons on, 213
Adsorbed water and contact electrification, 240-241
Air cleanliness classes, 125
Air-conditioned telephone office, particulate contaminants on contacts exposed, 188
Aluminum, oxide thickness on, effect of UV/ozone exposure, 20
Aluminum film, Auger spectra of, before and after UV/ozone cleaning, 6
Atmospheric (lab air) contamination, effect on thermocompression bonding of gold, 217

Biocompatible contaminants, 254-256
Biomaterials and surface contamination, 247-259
Biomedical applications of synthetic polymers, 248
Bubble cleaning to clean liquid surfaces, 40-42

Carcinogenic contamination detection, 115
Carcinogenic materials, absorption of, into mouse skin, 113-114
Clean liquid experiments, materials for, 47-49
Clean rooms, classification of, particle size distributions for, 126
Clean surface technology, application of pellicles in, 291-325
Cleaning of liquid surfaces, techniques for, 27-52

Cleaning of solid surfaces
  by Hydroson system, 53-69
  by UV/ozone, 1-26
Cold welding, threshold deformations for, 206
Compact audio disk, cross section of, 319
Computer data optical storage disk, cross section of, 319
Contact (electric) performance, effect of surface contamination on, 179-203
Contact electrification, effect of surface contamination on, 235-245
Contact interface, microscopic view of, 181
Contact resistance
  determination of, 198-200
  —load characteristic curves, 182, 195
  probe for determining contaminants on printed circuit boards, 202
  probes for the detection and characterization of contamination, 197-202
  of various materials after fretting in air, 192
Contact resistance-load characteristic curves, 182-195
Contact resistance probes for the detection and characterization of contamination, 197-202
Contact surface, schematic illustration of, 181
Contacts exposed indoors, particulate contaminants on, 188
Contaminants, biocompatible, 254-256
Contaminants (surface), *see also* "Particle surface contaminants" and "Surface contaminants"
  and biomaterials, 247-259
  characterization of, by luminescence using UV excitation, 103-122
  classification of, 209-210

328　　**Index**

Contaminants (*cont.*)
　and contact electrification, 235–245
　detection and characterization of, by contact resistance probes, 197–202
　and device failures, 123–148
　and electric contact performance, 179–203
　ionic, methods of measurement of, 71–101
　and performance of HVDC insulators, 149–178
　removal of
　　by Hydroson system, 53–69
　　by UV/ozone, 1–26
　and solid state welding of metals, 205–234
Contaminants (surface), removed by UV/ozone, 8–9
Contamination, see also "Contaminants"
　increase of, by personnel, 129
　indoor, redispersion of, 261–290
　of skin, detection and characterization of, 103–122
　sources of, 182–194
Contaminometer for ionic surface contamination, 87–90

Data (optical) storage media, protective films on, 318–321
Deformation welding, 222–229
　effect of surface preparation on, 229
Device failures and particulate surface contamination, 123–148
Diffusion welding, 220–222
Discharge growth, 160–167
Distillation for cleaning liquids, 37–40
Dynamic liquid surface phenomena, origin of, 28–31

Egan's method for ionic residue measurement, 76–80
Electric contact performance, effect of surface contamination on, 179–203
Epitaxial growth, effect of particulate contamination on, 135–136

Fatty acids on steel surfaces, friction transition temperature of, 215
Fiberoptic luminoscope for monitoring surface contamination, 109–111
Flashover process
　general overview of, 152–154
　prevention of, 168–170

Fluorescent tracer detection technique, 112–113
Fluorosensors (laser-based), remote sensing with, 111–112
Four-wire technique for measuring contact resistance, 200
Fretting, 190–192
Friction, effect of organic films on
　kinetic, 214
　sliding, 215
Friction transition temperature, 215

Gold, thermocompression bonding of
　effect of atmospheric (lab air) contamination on, 217
　effect of photoresist contamination on, 217
Gold plated contacts, contact resistance of, effect of manufacturing process contaminants on, 193
Gold-to-Gold thermocompression bonding, effect of UV/ozone cleaning on, 19

Hobson and DeNoon method for extraction of ionic residues, 81–83
HVDC insulators, performance of, effect of surface contamination on, 149–178
Hydrocarbons, effect of, on adhesion of iron, 213
Hydroson cleaning, 53–69
　commercial applications, 57–59
　comparison with ultrasonic and megasonic cleaning, 65–67
　recent developments, 60–64
　safety and economy, 59–60
　size and cost of equipment, 65

Impurity contamination in silicon, 144–145
Indoor surface contamination, redispersion of, 261–290
Insulator performance
　and effect of surface contamination, 149–178
　test methods to evaluate, 168
Integrated circuit fabrication
　particle contamination in, 299–302
　pellicles in, 292–317
Intraocular lens, ESCA spectra of the surface of, 252
Ion Chaser for ionic surface contamination, 86–87

Index

Ionic contamination, effect on contact electrification, 239-240
Ionic measurement, 72-76
Ionic surface contamination, methods of measurement of, 71-101
  Contaminometer, 87-90
  Egan's method, 76-81
  Hobson and DeNoon method, 81-83
  Ion Chaser, 86-87
  Ionograph®, 90-97
  Omega Meter® 83-87
Ionograph® system for ionic surface contamination, 90-97
Iron, adhesion of, effect of various hydrocarbons on, 213

Kinetic friction, effect of organic films, 214

Laboratory air contamination, effect on thermocompression bonding of gold, 217
Laser (automated) scanning system, optical configuration of, 316
Laser-based fluorosensors, remote sensing with, 111-112
Laser burning to clean liquid surfaces, 44-45
Liquid surface phenomena (dynamic), origin of, 28-31
Liquid surface tension, physical origin of, 29
Liquid surfaces
  history of clean, 32-33
  nature of surface contamination of, 33-36
  notion of surface cleanliness of, 32
  techniques for cleaning, 27-43
    bubble cleaning, 40-43
    further distillation, 39-43
    laser burning, 44-45
    primary distillation, 37-39
    solid adsorption, 44
    surface skimming and talc cleaning, 43-44
Luminescence technique for surface detection, 104-107
Luminescence using UV excitation for characterization of surface contaminants, 103-122
Luminoscope for monitoring skin contamination, 109-111

Manufacturing process contaminants, effect on contact resistance of gold plated contacts, 193

Masks in semiconductor microlithography, 299
Megasonic cleaning, comparison with Hydroson system, 65-67
Mercury discharge lamps (low pressure), principal wavelengths of, 3
Metal adhesion, effect of organic film thickness on, 216
Metals, welding of, role of surface contaminants in, 205-234
MOS devices, reliability data for, 144
MOS gate oxides, failure mechanism in, 137-144
Mouse skin, absorption of carcinogenic materials into, 113-115

Omega Meter® for surface ionic contamination, 83-86
Optical data storage media, protective films on, 318-321
Optical microlithography
  feature dimensions and performance limits of, 297-299
  semiconductor fabrication by, 292-293
Optical properties of pellicles, 308-311
Optical storage, principles and methods of, 318-319
Organic films, effect of,
  on coefficient of friction, 214
  on kinetic friction, 214
  on metal adhesion, 216
  on sliding friction, 215
Oxygen, absorption spectrum of, 4
Ozone, absorption spectrum of, 4

Particle or particulate contamination, sources of, 124-131
Particle or Particulate surface contaminants or contamination 123-148, 135-136, 154-158, 186-189, 210, 282, 299-302
  on contacts exposed indoors, 188
  and device failures, 123-148
  in integrated circuit fabrication, 299-302
Particle retention (on surfaces), effect of relative humidity, 282
Particles, deposition of, on insulators, 154-158
Pellicle (s)
  applications of, in clean surface technology, 291-325
  in integrated circuit fabrication, 292-317

Pellicle (s) (cont.)
  materials, 311–313
  mounting and attachment, 313–314
  optical properties and effects of, 308–311
Personnel, increase of contamination by, 129
Phosphorimetry (room temperature), surface detection by, 117–120
Photomask defects, probable sources of, 135
Photomasks, particulate contamination on, 131–135
Photoresist contamination, effect on thermocompression bonding of gold, 217
Photoresist stripping by UV/ozone, 12–13
Polycyclic aromatic hydrocarbons (PAH) detection, 103–122
Polymers (synthetic), biomedical applications of, 248
Printed circuit boards, contact resistance probe for determining contaminants on, 202

Radioactive contaminants, redispersion of, 261–290
Redispersion of indoor surface contamination, 261–290
Redispersion or resuspension factor, measurements of, 262–270
Relative humidity, effect on retention of particles, 282
Reliability data for MOS devices, 144
Retention of particles, effect of relative humidity, 282
Reticles in semiconductor microlithography, 299

Semiconductor microlithography, masks and reticles in, 299
Silicon, impurity contamination in, 144–145
Skin contamination detection and characterization, 103–122
"Skin-wash" method to study workers' skin contamination, 108–109
Sliding friction, effect of organic films, 215
Sodium dodecyl sulfate contamination, 252
Solid state welding of metals
  mechanisms for elimination of surface barriers during, 219–228
  role of surface contaminants in, 205–234
Solution conductance, measurement of, 74
Spill spotter for detection of surface contamination, 111

Surface barriers during solid state welding, mechanisms for elimination of, 219–228
Surface cleaning
  by Hydroson system, 53–69
  of liquids, techniques for, 27–52
  by UV/ozone, 1–26
Surface contaminants or contamination (see also Contaminants or Contamination, and Particle surface contaminants)
  and biomaterials, 247–259
  characterization, by luminescence using UV excitation, 103–122
  classification of, 209–210
  and contact electrification, 235–245
  detection by spill spotter, 111
  and device failures, 123–148
  electric contact, detection and characterization by contact resistance probes, 197–202
  and electric contact performance, 179–203
  indoor, redispersion of, 261–290
  ionic, measurement of, 71–101
  and performance of HVDC insulators, 149–178
  radioactive, 261–290
  removal
    by Hydroson system, 53–69
    by UV/ozone, 1–26
  and solid state welding of metals, 205–234
  transferable, measurements of, 270–283
Surface particle retention, effect of relative humidity, 282
Surface skimming to clean liquid surfaces, 43–44

Talc cleaning to clean liquid surfaces, 43–44
Thermal diffusion, 189–190
Thermocompression bonding of gold
  effect of atmospheric contamination on, 217
  effect of photoresist contamination on, 217
  effect of UV/ozone cleaning, 19
Tin, deformation welding of, effect of contaminants on, 219
Transferable surface contamination, measurements of, 270–283

Ultrasonic cleaning, comparison with Hydroson system, 65–67

Index

UV "black light" for surface detection, 107–108
UV excitation, luminescence using, for characterization of surface contaminants, 103–122
UV/ozone cleaner, schematic drawing of, 12
UV/ozone cleaning of surfaces, 1–26
  apparatus for, 2
  applications of, 17–19
  effect of, on gold-to-gold thermocompression bonding, 19
  effects other than cleaning, 20–21
  facility construction for, 16–17
  mechanism of, 13–14
  safety considerations for, 14–16
  schematic representation of process of, 13
  in vacuum systems, 14
  variables of, 2–13

Water (adsorbed) and contact electrification, 240–241
Water surface, cleaning techniques for, 27–52 (*see also* Liquid surfaces)
Welding
  deformation, 222–228
  diffusion, 220–222
Welding of metals
  mechanism for elimination of surface barriers during, 219–228
  role of surface contaminants in, 205–234
Workers' skin contamination study by "skin-wash" method, 108–109